# Tripping On Utopia

# Tripping On Utopia

## Margaret Mead, the Cold War, and the Troubled Birth of Psychedelic Science

BENJAMIN BREEN

GCP

New York Boston

Grand Central Publishing
Hachette Book Group
1290 Avenue of the Americas, New York, NY 10104
grandcentralpublishing.com
@grandcentralpub

First Edition: January 2024

Grand Central Publishing is a division of Hachette Book Group, Inc. The Grand Central Publishing name and logo is a trademark of Hachette Book Group, Inc.

The publisher is not responsible for websites (or their content) that are not owned by the publisher.

The Hachette Speakers Bureau provides a wide range of authors for speaking events. To find out more, go to hachettespeakersbureau.com or email HachetteSpeakers@hbgusa.com.

Grand Central Publishing books may be purchased in bulk for business, educational, or promotional use. For information, please contact your local bookseller or the Hachette Book Group Special Markets Department at special.markets@hbgusa.com.

The maps on pages xi and xii utilize imagery found in vintage maps from the 1920s through 1940s that have entered the public domain. These maps were digitized by the David Rumsey Map Collection, David Rumsey Map Center, Stanford Libraries.

Library of Congress Cataloging-in-Publication Data
Names: Breen, Benjamin, 1985- author.
Title: Tripping on utopia : Margaret Mead, the Cold War, and the troubled birth of psychedelic science / Benjamin Breen.
Description: First edition. | New York : Grand Central Publishing, 2024. | Includes bibliographical references and index.
Identifiers: LCCN 2023036568 | ISBN 9781538722374 (hardcover) | ISBN 9781538722398 (ebook)
Subjects: LCSH: Hallucinogenic drugs—Research—United States—History—20th century. | Anthropology—Research—United States—History—20th century. | Mead, Margaret, 1901-1978. | Bateson, Gregory, 1904-1980. | United States—Civilization—20th century. | Cold War.
Classification: LCC BF209.H34 B744 2024 | DDC 154.4—dc23/eng/20230908
LC record available at https://lccn.loc.gov/2023036568

ISBNs: 9781538722374 (hardcover), 9781538722398 (ebook)

Printed in the United States of America

LSC-C

Printing 1, 2023

*To Roya and Yara*

We are faced with the problem of building a new world.

*—Margaret Mead and Gregory Bateson, 1942*

Thoughts are like Stars in the Firmament, where some are fix'd, others like the wandering Planets; others again are only like Meteors, which when their Substance is wasted, their Light goeth out... Memory is like the Moon.

*—Margaret Cavendish,* The World's Olio *(1655)*

# Contents

## Part III

## THE NOISE (1959–80)

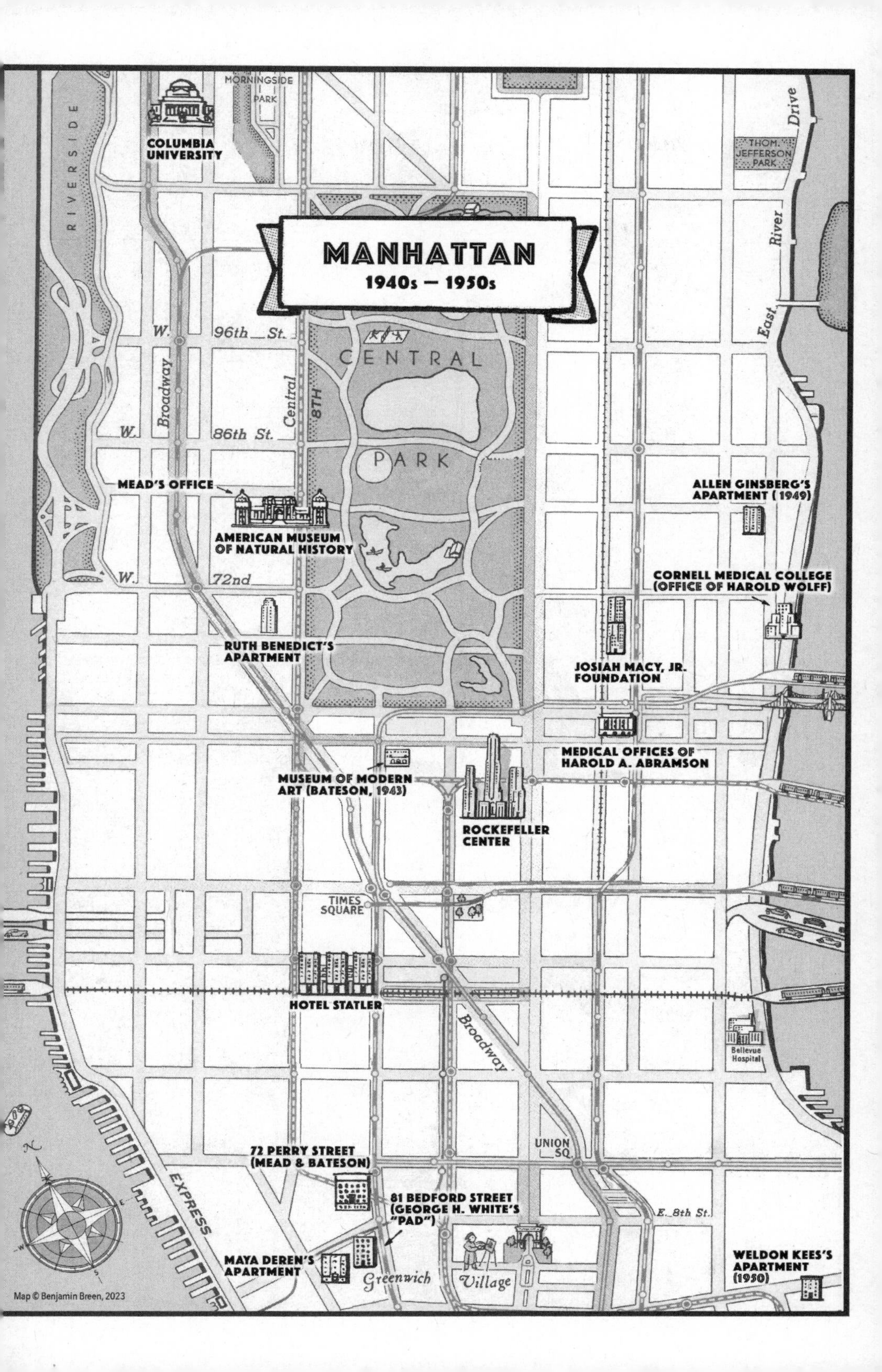
MANHATTAN
1940s — 1950s
MORNINGSIDE PARK
COLUMBIA UNIVERSITY
RIVERSIDE
THOM. JEFFERSON PARK
Drive
River
East
W. 96th St.
W. 86th St.
W. 72nd
Broadway
Central
8TH
CENTRAL PARK
MEAD'S OFFICE
AMERICAN MUSEUM OF NATURAL HISTORY
ALLEN GINSBERG'S APARTMENT (1949)
CORNELL MEDICAL COLLEGE (OFFICE OF HAROLD WOLFF)
RUTH BENEDICT'S APARTMENT
JOSIAH MACY, JR. FOUNDATION
MEDICAL OFFICES OF HAROLD A. ABRAMSON
MUSEUM OF MODERN ART (BATESON, 1943)
ROCKEFELLER CENTER
TIMES SQUARE
HOTEL STATLER
Broadway
Bellevue Hospital
UNION SQ
72 PERRY STREET (MEAD & BATESON)
81 BEDFORD STREET (GEORGE H. WHITE'S "PAD")
E. 8th St.
EXPRESS
MAYA DEREN'S APARTMENT
Greenwich Village
WELDON KEES'S APARTMENT (1950)
Map © Benjamin Breen, 2023

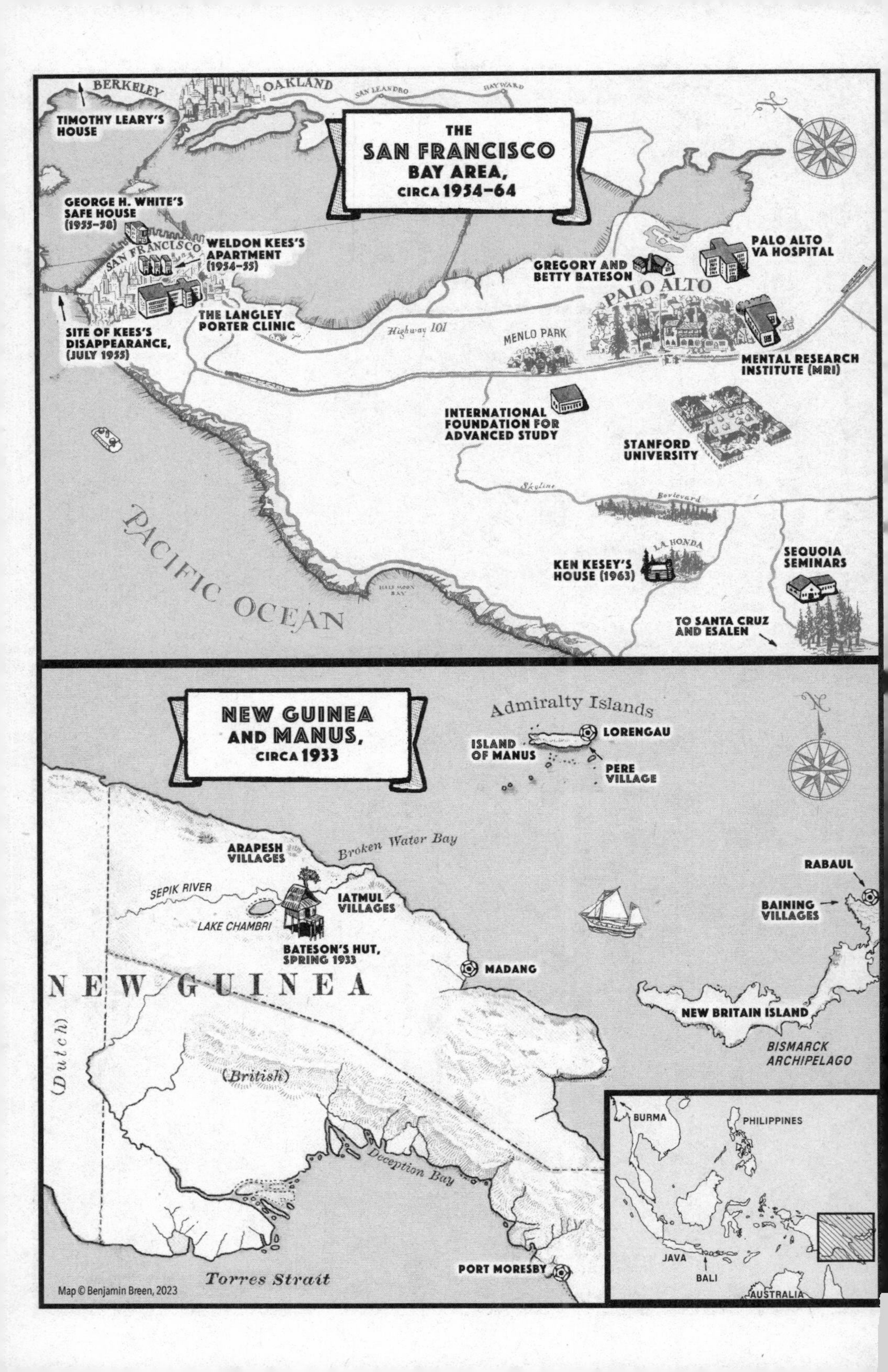
THE SAN FRANCISCO BAY AREA, CIRCA 1954–64
BERKELEY
OAKLAND
SAN LEANDRO
HAYWARD
TIMOTHY LEARY'S HOUSE
GEORGE H. WHITE'S SAFE HOUSE (1955–58)
SAN FRANCISCO
WELDON KEES'S APARTMENT (1954–55)
THE LANGLEY PORTER CLINIC
SITE OF KEES'S DISAPPEARANCE, (JULY 1955)
Highway 101
MENLO PARK
GREGORY AND BETTY BATESON
PALO ALTO
PALO ALTO VA HOSPITAL
MENTAL RESEARCH INSTITUTE (MRI)
INTERNATIONAL FOUNDATION FOR ADVANCED STUDY
STANFORD UNIVERSITY
Skyline
Boulevard
PACIFIC OCEAN
HALF MOON BAY
LA HONDA
KEN KESEY'S HOUSE (1963)
SEQUOIA SEMINARS
TO SANTA CRUZ AND ESALEN
NEW GUINEA AND MANUS, CIRCA 1933
Admiralty Islands
ISLAND OF MANUS
LORENGAU
PERE VILLAGE
ARAPESH VILLAGES
Broken Water Bay
SEPIK RIVER
IATMUL VILLAGES
LAKE CHAMBRI
BATESON'S HUT, SPRING 1933
MADANG
RABAUL
BAINING VILLAGES
NEW GUINEA
NEW BRITAIN ISLAND
BISMARCK ARCHIPELAGO
(Dutch)
(British)
Deception Bay
Torres Strait
PORT MORESBY
BURMA
PHILIPPINES
JAVA
BALI
AUSTRALIA
Map © Benjamin Breen, 2023

# Tripping On Utopia

PROLOGUE

# The Future That Never Arrived

> Our Father, We thank Thee for the torch of learning which has lighted our paths in time past... from Stonehenge, to the Parthenon, to Rockefeller Center, to the Gregory Bateson Building, artistic permanence has been added to mankind.
>
> —*prayer offered "at the time of the completion of the Gregory Bateson Building" (*Journal of the Senate, Legislature of the State of California, *May 4, 1981)*

AT THE LOWEST POINT of the Great Depression, a work of genius came to life in a half-built Manhattan skyscraper. It was late March 1933. Outside the building, a cold spring rain fell on the unemployed, whose numbers nationwide had grown to include nearly one quarter of the United States' working-age population. Inside, Diego Rivera was creating a masterpiece. It started with simple lines drawn onto a wall in the entrance hall of 30 Rockefeller Center. The lines became figures. The figures filled with brilliant color. And as they did, it became clear that the fresco Rivera called *Man at the Crossroads* had the potential to be remembered for the next hundred years. More than remembered. It was an artwork that seemed to *predict* the coming century.

The composition centered on two crossed ovals that resembled the beams cast by searchlights, or perhaps the arms of an X chromosome. Rivera filled one of the ovals with a crimson tangle of cells and nerves. The other became a window into deep space: a maelstrom of glowing stars. Cutting through both was a tower of futuristic gears, like something from Fritz Lang's *Metropolis*. Human figures crowded around the machine. On one side, a group of workers stared through a huge lens into what Rivera called "a night-club scene of the debauched rich." Above, ranks of soldiers marched in gas masks. At bottom left, Charles Darwin peered out enigmatically beside an X-ray machine.

Detail from Diego Rivera's *Man, Controller of the Universe*, 1934, Palacio de Bellas Artes, Mexico City (a re-creation of his *Man at the Crossroads*). *Courtesy Wikimedia Commons.*

Most prominent of all was a massive figure at the center who was dressed like a technician. Two huge glass lenses flanked him. It was this figure, this cosmic engineer, who seemed to control the painting's machinery. Our technology would make us gods, Rivera seemed to be saying. And as humanity entered the crossroads of a new world, the cruelties of past centuries would be magnified along with our power.

Rivera's most important viewer was the enormously wealthy man who owned the building in which he worked: John D. Rockefeller Jr. At first, Rockefeller was thrilled by the mural's progress. This changed when he realized that an idealized portrait of Vladimir Lenin would dominate one side of the composition. There was outcry from the building manager (who worried about scaring off tenants) and then from the press. Work halted. Then, on the night of February 10, 1934, a team of men entered the lobby. The day before had been the coldest ever recorded in New York City, with at least six fatalities in the city's heatless tenements. Stripping off their winter coats to reveal coveralls not unlike those of the godlike central figure, the men pulled Rivera's painting off the wall.

For three years, the walls of 30 Rockefeller Center remained blank. And then in 1937, a new design began to fill them. Gone were Rivera's gemlike hues—the radioactive greens, the cosmic blues, the rainbow plumes of the bird of paradise perched at Darwin's side. In came a collection of 1930s clichés: steel beams rising on cranes, smokestacks puffing skyward, all rendered in a drab palette of beige and gray.

The new painting was called *American Progress.*

--------

This book uncovers the secret history of the first wave of psychedelic science, a radical field that sought to harness the power of psychedelic substances to explore human consciousness, pioneer groundbreaking therapies, and even transform global society. It is a story of human ingenuity and ambition, of triumphs and tragedies, and of a future that was always just out of reach. Like *Man at the Crossroads*, psychedelic science challenged the norms of its time, pushing the boundaries in ways that were both beautiful and unsettling. And like Rivera's mural, the legacy of psychedelic science has been painted over.

There were good reasons for this mass forgetting: abuses of medical ethics, devil's bargains with militarism, and disillusionment with utopian promises. But before it became a failed utopia, it was a beautiful dream. Restoring it to view can help guide us in the present, as we confront yet another crossroads between techno-utopian ambition and messy reality.

The people and the events chronicled here formed the opening chapter of a mass improvisation with our own brain chemistry that continues to this day. Taken together, they make it clear that the first era of global experimentation with consciousness-expanding substances took place much earlier than commonly thought, in the 1920s through the 1950s, rather than the 1960s and 1970s.

Put another way: Timothy Leary and the Baby Boomers did not usher in the first psychedelic era. They *ended* it.

More than anyone else, it was the group of interdisciplinary scientists connected by the anthropologists Margaret Mead and Gregory Bateson in the years surrounding World War II who shaped the development of psychedelic research from the 1930s onward. This group included anthropologists, drug researchers, psychologists, neuroscientists, and pioneers of early computer science. Brought together over a period of two decades in conferences funded by the Josiah Macy, Jr. Foundation, this "Macy circle" shared a commitment to applied, interdisciplinary science that not only challenged conventional norms but explicitly sought to create a new global culture. The Macy circle was the seedbed for the flowering of psychedelic science in the 1950s, when substances like psilocybin, peyote, and LSD emerged as potential tools not just for calming anxiety or for aiding psychotherapy, but as a "shortcut" to transformative cultural change and expanded collective consciousness.

This era of utopian drug research was not just located in the laboratories of scientists. It was a mass movement, involving thousands of people in ways both hugely positive and enormously destructive: A Los Angeles housewife who took LSD on television in 1957 and turned to look into the eyes of an older male scientist, saying simply, "I feel sorry for you." A young man from Harlem who became an unwitting guinea pig in the CIA's illicit testing of psychedelics at a government facility in Kentucky. The NASA-funded attempt to teach dolphins how to speak that fell apart after the lead scientist began injecting both himself *and* the dolphins with large doses of LSD. Other psychedelic users in the 1950s included twelve men and women who volunteered to sample LSD in Baghdad, Iraq; thirty medical students in Budapest who spent the summer before the 1956 Hungarian Revolution injecting one another with a newly discovered psychedelic known as DMT; the Hollywood icon Cary Grant and the writer Anaïs Nin; and a young Juilliard-trained opera singer and refugee from Nazi Germany who spent part of her LSD trip being interviewed by Margaret Mead regarding the possibility that psychedelics could unlock psychic abilities.

Taken together, these stories cast a vibrant, surreal, and at times troubling new light not just on the history of psychedelic science, but on the twentieth century itself.

Running through this extraordinary history was the intellectual partnership and tumultuous love affair that Margaret Mead—the most famous, and most polarizing, scientist of her generation—shared with her third husband, the British anthropologist Gregory Bateson. Their influence extended in so many directions it's sometimes difficult to track. "She was truly interested in *everything*," one of Mead's friends remembered. "It was a radically original mind." In the half-light of Cold War espionage and atomic age paranoia, Mead and Bateson nurtured friendships with everyone from an ex–Soviet spy to a religious leader on a small island off the coast of New Guinea, from the scientist at the heart of the CIA's drug experimentation program to a dolphin that appeared on the television show *Flipper*. They studied apocalyptic cults and schizophrenia, sex hormones and space travel, world peace and mescaline.

Mead and Bateson and their circle matter to the history of psychedelics for one reason above all: their shared vision of science as a tool for expanding human consciousness. Mead's primary goal in life, she once said, was "to keep the future safe." Achieving this, she believed, meant enlarging the collective "awareness" of the human species so that we could "learn consciously to create civilizations within which an increasing proportion of human beings will realize more of what they have it in them to be." The 1920s had been a decade of scientific marvels: new discoveries like X-ray machines revealed hidden realities beneath the surface; miracle drugs like penicillin cured deadly diseases; radio waves carried voices across oceans; airplanes soared above clouds; skyscrapers rose above cities; and quantum mechanics defied all common sense. Science, in short, had *already* transformed the world. And it seemed poised to go even further—promising changes not just in the technologies of everyday life, but in the very experience of being human. When the Great Depression came, it did nothing to dampen Mead's utopian hopes. It catalyzed them. The troubled years of the 1930s convinced her that a new kind of science must be developed that would intervene directly in the world—a science that would *save* the world.

One of Margaret Mead's key mentors was a Macy Foundation executive named Lawrence Frank. And it was Frank who defined the problem in a way that Mead and Bateson would both return to for decades afterward: the world, Frank declared, was "a sick society in need of treatment." The economic collapse of 1929 and its aftermath had shattered millions of lives and livelihoods, creating widespread unemployment, hunger, homelessness, and despair. Mead saw this as a crisis of culture, not just of capitalism. She believed that science offered a path—the *only* path—for preserving cultural diversity in a world threatened by the terrifying uniformity of totalitarianism, the rapacious demands of still-powerful empires, and the potential devastation of a second world war. And as Lawrence Frank had advised, Mead and Bateson saw their science not just as a tool for diagnosis, but as a *treatment in itself.* They imagined the creation of a new, global culture built on vast diversity rather than uniformity. Mead spoke of a "cultural evolution" that would push humanity past the old boundaries of race, nation, and gender—and even of states of consciousness. Mead and Bateson saw themselves as intervening in a unique moment of global crisis, one that would forever afterward "define the paths along which future generations will be able to advance."

It was a vision of the future that did not go according to plan.

--------

This is a history that resonates with our present moment of rapid cultural change and renewed interest in psychedelic research. But it is also strikingly unfamiliar—above all, perhaps, because of its enormous optimism. Carl Sagan was once asked why we should expect that extraterrestrial life might be sending signals into deep space. "All they need is one Margaret Mead," he replied. Mead saw herself as "a listening post," collecting and synthesizing information from every domain, like a central computer. Science had "introduced another level... of awareness" to the story of humanity, she believed. It had shown us our cultural diversity and our unconscious drives, but more than this, it had allowed us to *reprogram* our own minds and societies. Margaret Mead had come of age in the era of Freud, when the unconscious was thought to define human behavior. But she believed that scientists like herself were in the process of a revolutionary transformation in collective consciousness, making those invisible drives and motivations *visible* in ways that could end ancient divisions, heal trauma, and unlock human potential. *We* were the machines becoming conscious.

We *who*, though? Who was excluded by this vision, and who was positioned to exploit it?

The history of psychedelics in the twentieth century has almost always been told as a story dominated by white American men, and above all by Timothy Leary and Richard Alpert (Ram Dass). These figures appear in the pages of this book, to be sure. But throughout, I have tried to set them in a larger context. (DMT, to take one example, was closely studied by Brazilian ethnobotanists and Czech psychiatrists before it reached the counterculture of the United States.) As anthropologists, Mead and Bateson exemplified this global perspective. Mead spent most of 1953—a pivotal year for psychedelics—on an island off the coast of New Guinea, where she was studying the development of an apocalyptic religion that she called "the Noise." When she returned to New York in 1954 and became involved in LSD research, her perspective was colored by this experience as much as it was shaped by domestic concerns of Americans in the 1950s.

Margaret Mead's importance to this story—and a hint of the dark side of the world she helped bring together—is made visible in two photographs from the 1954 Macy conference on altered states of consciousness. Both depict the same scene, with a curious difference. The first photograph is a standard group portrait, showing the various attendees of the conference. Their eyes are open; they are fully alert. The second, which was unpublished and distributed privately to attendees, shows the same group with eyes closed, miming a state of trance. In the center of both sits Margaret Mead. Around her is a who's who of psychedelic science: At her right side is Roy Grinker, one of the inventors of narcosynthesis, an experimental drug treatment for trauma. Sitting at bottom left, bald and wearing an amiable smile, is Harold Abramson, the man at the center of the CIA's still-mysterious drug testing program.

Group portrait of the participants in the fifth Macy conference on "Problems of Consciousness," March 22–24, 1954, Princeton, NJ. *Courtesy Josiah Macy, Jr. Foundation.*

And the "unconscious" version distributed privately to participants. *Courtesy David Kornetsky.*

The photo, in other words, testifies not just to Mead's centrality, but to her secrets. Of the eighteen people in the photo from that spring day in 1954, over half performed psychedelic research, much of it classified. One of these, the neuroscientist Mary Brazier, once said, "What we, as a group, have got from these conferences does not, I think, emerge for an outside reader. It is all between the lines."

Bateson and Mead would likely have been surprised to find themselves as the protagonists in a history of psychedelics. Why center this narrative on them rather than the chemists and pharmacologists who discovered the drugs, or the patients who took them? The story told "between the lines" of psychedelic science—the story told in private photographs, in cryptic diary entries, in forgotten manuscripts and scribbled letters and classified memos—is a story that to a surprising degree centers Mead and Bateson and their circle. Mead's archive includes countless references to her decades-long interest in peyote. And then there's the 1954 document entitled "Preliminary LSD Memo," which describes her participation in one of the earliest American trials of lysergic acid diethylamide. Meanwhile, Gregory Bateson was directly responsible for Allen Ginsberg's first LSD trip and played a key role in the birth of psychedelic psychiatry in 1950s Silicon Valley.

As important as their direct role was their intellectual impact on psychedelic researchers and writers. Aldous Huxley read Mead carefully as he wrote *The Doors of Perception* following his mescaline experiments in the 1950s. Timothy Leary's earliest published work as a scientist was, his collaborator recalled, directly inspired by Bateson. And in one of his first speeches about psychedelics, Leary quoted Mead, while behind the scenes he tried to convince her to take psilocybin with him.

Today, in a new era of global crisis and amid the threat of a new cold war, the same drugs are back. In so many ways, we seem to be returning to where we started with these mysterious, fascinating, and deeply misunderstood substances.

We have much to learn—and to remember.

## Part I

# THE PATTERN

*(1923–49)*

Let the taro turn the mouth towards it,
Towards plentifulness,
Towards greatness…
Let him eat my taro.
He must do no evil.

*—magical incantation recorded by*
*Margaret Mead in the village of Pere,*
*island of Manus, Admiralty Islands, 1929*

Margaret Mead at the American Museum of Natural History, 1930. *Courtesy New York Historical Society/ Getty Images. Photographer: Irving Browning.*

CHAPTER 1

# Prophet of the Long Future (1923–30)

It is not frustrated sex, it is frustrated gentleness that is so hard to bear when one is working for long months alone in the field.

—*Gregory Bateson, circa 1933, as quoted by Margaret Mead*

MARCH 14, 1933, FOUND Diego Rivera sketching out the earliest designs for *Man at the Crossroads.* Half a world away, in a small village in the interior of New Guinea, a thirty-one-year-old Margaret Mead was confronting a crossroads of her own. She did so while staring at a blank piece of paper.

The paper, like everything else in the village on the shores of Lake Chambri, had grown slightly damp in the tropical humidity. Outside the walls of her hut, the sounds of life could be heard—birdsong, the crying of babies, perhaps a distant sound of splashing water from the canoes on the lake. But Mead's mind was far away. She was remembering Manhattan in 1923, the Grand Canyon in 1925, Paris in 1928. She was remembering scenes from her life with a woman she had fallen passionately in love with ten years earlier—a woman who was also her most important mentor, the person who had convinced her to become an anthropologist in the first place.

But what she had to say to Ruth Benedict was not just another profession of love. It was a confession. On that day in March, Margaret Mead was falling in love with Gregory Bateson, pulling away from her second husband, Reo Fortune (the man with whom she shared her hut in New Guinea), and laying the foundation for the beliefs that would drive the rest of her life. She was beginning to see herself as engaged in an epic scientific project of expanding human consciousness.

Writing to Benedict, Mead did her best to make sense of what was happening. She felt for Bateson, she said, "a 16th year old delight with a large amount of childlike play in it," but also "a strong maternal feeling" mixed with an emotion she had experienced only once before—when she had felt it for Benedict herself. She felt her excitement about Bateson becoming not just love or passion, but a new sense of "consciousness of myself—and a sense of the Heaven to be seen through the stained glass window." She had seen in him a kind of utopia. And it scared her. The feeling was "strictly inappropriate," she wrote now to Benedict, while Bateson was spending two weeks away in a neighboring village.

It was an agonized and agonizing letter. In the months to come, even as Margaret Mead remained married to Reo Fortune, the references to Bateson and to their mysterious joint intellectual project grew and grew. She knew that she had to continue to tell Ruth Benedict about Gregory and the new life they were beginning to envision, a life as vivid and strange as the mural that had begun to fill the walls of Rockefeller Center in those same months. She had to tell her because she was the only person in the world who could truly understand.

## New York City, 1923

They had first met almost exactly ten years earlier, at Columbia University in 1923. At that time, Ruth Benedict was obsessed with visionary states. Before long, the young Margaret Mead was obsessed with Ruth Benedict.

A petite woman with luminous eyes and sardonic black eyebrows, Benedict had something of the look of a spiritual medium. She had many of the interests of one, too. Her fascination was with cultural patterns that deviated from the norm of American society in the 1920s—and, above all, with forms of sexual identity and altered states of consciousness that her own society rejected as madness. This was not just a set of academic interests. Benedict saw herself as helping to create a new kind of science that would allow humanity to navigate the challenges of the modern world. In the spring of 1923, she was particularly focused on the role of visions in Indigenous societies of North America: Hallucinatory vistas induced by vision quests on the Great Plains. Torture visions caused by self-inflicted pain, a characteristic practice of some North American Indigenous societies—barbs lacerating the tongue or penis, thorns piercing the soft flesh of the arm, all performed not as punishment but as a means of transcending the limits of the body. Visions induced by alcohol or the bitter seeds of jimsonweed, the devil's trumpet. Visions from the divine cactus, peyote.

For a sheltered young woman from the suburbs of Philadelphia, Mead's first encounter with Benedict and her work was both thrilling and frightening. Frightening, because it threw Mead's carefully ordered life into doubt. (As precocious in her relationships as in everything else, Mead had become engaged at age seventeen to Luther Cressman, the earnest son of her high school science teacher. And Cressman, with his hopes of becoming a country pastor, was not exactly the right person to speak to about such things as Benedict's account of the "wide-spread Plains torture pattern known as 'Feeding-the-sun-with-one's-body.' ") Thrilling, because Benedict's work resonated with Mead's deep-seated sense of otherness. On the surface, Mead's life goals at the time seemed borrowed from the pages of *Little Women*. She hoped to raise five children with Cressman and then to establish a small progressive school. But not long after she matriculated at Barnard College in 1922, with the experimental culture of New York City in the Jazz Age approaching its peak, Mead's vision of her future shifted. She began to imagine a life as a politician, or life as a poet—or as an anthropologist, a roving, eternal outsider whose work derived from their ability to see the world askance.

This last possibility had been kindled in one of Mead's first-year classes, a survey course taught by Professor Franz Boas of Columbia University. An expert in the languages and cultures of the Arctic Circle, Professor Boas cut a striking figure even before he opened his mouth. Born into a secular Jewish family in Prussia, a young Boas had attempted to signal his social belonging by joining one of the fencing clubs that proliferated in the elite universities of Imperial Germany. He fought more than his fair share of duels: even after four decades, the large mustache that bisected his face did little to hide his rapier scars. But it was the mind behind this forbidding exterior that was the true marvel—encyclopedic in scope, humorous in attitude, and deeply iconoclastic in its fixations. "My whole outlook upon social life," Boas once said, "is determined by the question: How can we recognize the shackles that tradition has laid upon us? For when we recognize them, we are also able to break them."

Ruth Benedict served as Boas's teaching assistant in the lecture class Mead attended. But it was a tragic event in February 1923 that first brought Mead and Benedict into close contact. Mead's circle of friends at Barnard was a group of sexually adventurous, hard-drinking young women—if not flappers, then certainly flapper adjacent—who called themselves the Ash Can Cats. When one of them, Marie Bloomfield, fell into a deep depression, Mead appointed herself Bloomfield's caretaker. One day, Mead found the door to Bloomfield's dorm room locked. There was no answer when she knocked. After getting help to break down the door, Mead discovered that Bloomfield had killed herself with cyanide.

Ruth Benedict (*left*), Margaret Mead (*center*) in the mid-1920s, and Franz Boas (*right*) in an undated photograph, sitting in Benedict's apartment. *Courtesy Barnard College Special Collections and Archives; Library of Congress.*

The incident became a major news story, drawing scrutiny to the Ash Can Cats and to Mead. Bloomfield's "face stared up at us from trampled newspapers on the subway floor," Mead remembered. Barnard's administrators wanted Mead to testify that her friend had been insane, a strategy to absolve the school of responsibility. Mead resisted. And she shared her conflicts with Benedict. Fifteen years older, Mead's former teaching assistant now became a confidante. "If you can get away, come yourself," Benedict wrote to her. "I've nothing all day that can't be put off. I shall be thinking of you today." Mead visited her that evening.

By April, Benedict had convinced Mead to become a cultural anthropologist like her. "I need a companion in harness," Benedict said. "I have nothing to offer but an opportunity to do work that matters."

The priceless knowledge of thousands of generations, thousands of distinct cultures, was being lost, Benedict said. Entire cultures were being mowed down by the violence of colonial empires and flattened further by the homogeneity of modern life. Every day that passed was a day in which an ancient language or unique artistic tradition might disappear. Cultural anthropology was not about collecting dead relics to gather dust in museums. It was about salvaging the distilled knowledge of millions of lives—hard-won lessons that might one day help shape humanity's collective future.

The pitch worked. After Mead married Cressman in September 1923, she spent the night of their honeymoon writing a seminar paper for Boas's class. "Anthropology had to be done now," Mead decided. "Other things could wait." Mead's father, a professor at Penn, was predicting a second global war by 1939. From the moment she accepted Ruth Benedict's offer, Mead wrote, "I was to be single minded, trying to do as much field work as possible before the next war." The urgency of what she saw as this precious window, this period of fifteen or so years in which to perform what later scholars have called "salvage anthropology," yielded an obsessive focus. "Even the political upheaval of the Depression and later of the Spanish War went in a sense unnoticed," she said, "for I was living in a time perspective of a hundred years, and preparing the materials with which we would, hopefully, be better prepared for that long future."

In other words, when Mead earned her master's degree in anthropology in 1924, she had more than one audience in mind. This work was not just for the people of her own time. It was also for the people who would inhabit that "long future."

--------

Benedict saw the study of culture as a way of moving beyond racism, beyond the dangerous myths of blood. Though this view would become commonplace after World War II, it was radical in its time. If human nature was a book, then nearly everyone up to that point had been seeing only a single page: the culture into which they were born. Benedict and Mead wanted to read the whole thing. "The purpose of anthropology," Benedict once said, "is to make the world safe for human differences." By studying radically different cultures, the anthropologist could access a vast "laboratory of social forms." But crucially, this was not a laboratory in which the scientist was an impartial observer, staring down into sterile test tubes. Scientists, after all, were parts of the cultures they studied, moving *in* the world rather than outside it. Benedict and Mead were not just trying to understand such things as "deviance" or altered states of consciousness in other societies, other lives. They were also trying to understand themselves.

In 1925, Mead and Benedict planned a trip to the Grand Canyon that they kept secret from their colleagues. Mead had settled on a research project for her dissertation—the study of adolescence in Polynesia—and was journeying west to board a steamship in San Francisco that would carry her to nine months of solitary fieldwork on the island of Samoa. Benedict, meanwhile, was traveling to study the Zuni people of the southwestern pueblos, whom she had come to see as a sober counterpoint to the visionary cultures of the Great Plains.

"Ruth and I got different things out of the Grand Canyon," Mead wrote. "She was most impressed by the effort of the river to hide, a torturing need for secrecy which had made it dig its way, century by century, deeper into the face of the earth."

But Margaret Mead didn't see anything tortured about the Grand Canyon. She saw *potential*—potential as boundless as her own future, as potent with promise as humanity itself.

"The part I loved the best," she recalled, "was the endless possibilities of those miles of pinnacled clay."

--------

Like Ruth Benedict, Margaret Mead saw anthropology as a hunt for the hidden patterns that shaped our lives. But unlike her mentor and lover, she would come to see this work as part of something that extended *beyond* science—a project that was part of an epic story of humanity's transformation into a species capable of reshaping its destiny. Boasian anthropologists had decentered the Western sense of self. This "new kind of consciousness" of humanity's complexity, as Mead once put it, allowed humans to guide the evolution of our societies in a self-aware fashion, aiming intentionally for the creation of more peaceful, more tolerant social orders. This was nothing less than a science of expanded consciousness, an explicitly utopian project of species-wide uplift.

Margaret Mead became a scientist in 1923. Within three years, she was already well on her way to being a *famous* scientist. It seemed to happen almost effortlessly. She just happened to be good at talking, and the press happened to enjoy listening to her. In the fall of 1926, for instance, before she had even published her findings from her fieldwork in Samoa, a *New York Times* article announced her work to the world under the eye-catching headline "Scientist Goes on Jungle Flapper Hunt."

Mead's chosen topic of female adolescence in a traditional society turned out to be perfectly tuned to the interests of a 1920s America that was quickly becoming dominated by young people. The word "teenager" had been coined just two years before Mead reached Samoa. Bestselling novels like Sinclair Lewis's *Babbitt* chronicled a modern form of family in which alienated parents presided over children who hid their anxiety behind the gloss of consumer culture. Against this backdrop, Mead's Samoan youths stood out. Carefree, mentally healthy, sexually active, and unburdened by adolescent angst, they offered not just a contrast to the social structure of the West but a self-evidently superior alternative. Or so, at least, Mead claimed in her book *Coming of Age in Samoa* (1928), which was an immediate bestseller. Franz Boas summed up her argument in the book's foreword: "What constitutes courtesy, modesty, very good manners, and definite ethical standards is not universal. It is instructive to know that standards differ in the most unexpected ways." The "modesty" part of Boas's formulation was the key: Mead's work implied, with unusual frankness, that the sexual norms of 1920s America were merely one manifestation of sexuality among many possible—and preferable—alternatives.

In the aftermath of her Samoan success, Mead sailed easily into a career as a public intellectual. She wrote brilliantly, in enormous quantities, seemingly on any topic that came to mind. And she cultivated a persona that would with time become iconic. With her close-cropped hair, youthful demeanor, and unconventional clothes, the young Margaret Mead had something of the air of Amelia Earhart.

She also had a personal life that was nearly as complicated as those of the silent film stars that the newspapers of the 1920s fawned over. In 1926, while traveling via steamship from Australia to France, where Luther Cressman awaited her, Mead fell in love with a young anthropologist from New Zealand named Reo Fortune. Lanky, overtly masculine, and almost cartoonishly brooding, Fortune was the gentle Cressman's opposite. Their shared interests ran deeper than anthropology. Since her youth, Mead had cultivated an interest in psychic phenomena. She was proud of her ability to go into a trance, fascinated by the unconscious, and a believer in extrasensory perception (ESP). Reo Fortune, she learned, was not just a cultural anthropologist of New Guinea but a trained psychologist who had written a book called *The Mind in Sleep*. He shared her fascination with altered states of consciousness. At the end of their voyage, Mead walked down a gangplank in Marseilles to find Luther Cressman holding a bouquet of flowers. Reo Fortune was at her side.

Boas had sent her to Polynesia. Soon after divorcing Cressman and marrying Fortune, Mead again followed a research path determined by someone else, traveling to Fortune's favored fieldwork site on the island of Manus, off the coast of New Guinea. But by 1930, having cemented her reputation with *Coming of Age*'s success, "Mrs. Reo Fortune" finally had the chance to choose her own research site.

After consulting with Ruth Benedict, she chose to follow her footsteps into the peyote visions of the Great Plains.

# Girl Ethnologist to Live With New Guinea Tribes

## Margaret Mead of American Museum to Study Ways Of Primitive Folk by Becoming One Of Them.

WASHINGTON, Sept. 6.—(AP)—Youth in primitive places is happier, more free from stress, than in civilized communities, according to the observations of Margaret Mead, 26, assistant curator of ethnology of the American Museum of Natural History.

Miss Mead will live among the natives of a little island off the coast of New Guinea this winter. In 1925 she spent nine months living the native life on the isolated island of Tau, Samoa. Her object was to study the life of adolescent girls in the islands.

She found that in a society where the small intimate family was replaced by a large relationship group with much looser emotional ties adolescence was painless, unaccompanied by crises or strain.

Children in the Samoan islands, she observed, are admitted at an early age to a full knowledge of life, death and sex.

MARGARET MEAD.

A newspaper profile of Mead that appeared in the *Pittsburgh Post-Gazette* in September 1928—one of many articles of the era that described Mead as a "girl ethnologist" or "girl anthropologist." *Public domain.*

CHAPTER 2

# Society Is the Patient (1930–33)

## Nebraska, 1930

MEAD AND FORTUNE FIRST saw the town of Macy, Nebraska, through the windshield of a Ford Model A in June 1930. Though it was Mead who had organized and funded the trip, she kept a low profile. That evening, Fortune told the people of the settlement that he was an anthropologist who had come from New York City to study their customs. As they drove out to the peripheral villages of the Omaha Reservation—Rosalie, Pender, Walthill—Mead was understood to be the anthropologist's unemployed wife.

That summer was a uniquely gloomy time to begin new fieldwork. The financial panic of 1929 was starting to look less like a stock market crash and more like an economic collapse. Newspapers recorded drought conditions across the Great Plains. And in Europe, Hitler's National Socialists had become Germany's fastest-growing political party. Among the Omaha, however, this newest wave of bad news simply added to a ledger that was already overflowing with decades of exploitation and state-sanctioned violence.

Mead arrived in Nebraska hoping to find a culture filled with survivals from a rich precolonial past. What she found instead, she thought, was a cautionary tale. Omaha society, it seemed to her, had been fractured by colonization. The shards of what had been were not easy to piece together.

The hallucinogenic cactus known as peyote, which has been used by Indigenous cultures of North America for over five thousand years, was a case in point. Despite fierce opposition from Christian authorities following Spain's conquest of the Aztec Empire in the 1520s, peyote had a habit of reappearing following periods of suppression. And each time it did so, it seemed to take on a new social role. In an epic poem from sixteenth-century Mexico, for instance, peyote figures as a dangerous tool of divination:

> The peyote, Lord, watch out for it!
> This is a thing not found in Spain.
> When you drink it, you can know—in an evil way—
> however much you wish to know about all the world.

Calling it a "widely introduced vice" that induced "mental images, fantasies and hallucinations," the Mexican Inquisition banned peyote in 1632. Yet throughout the colonial period, a wide range of people—not just Indigenous healers but also African slaves, soldiers, and even-would be rebels—continued to use the cactus.

By the nineteenth century, after great effort on the part of Mexican and American authorities, peyote use was forced underground. But it never disappeared, as Franz Boas's first student, Alfred Kroeber, learned firsthand. Kroeber had studied the peyote cactus (and sampled it himself) during his fieldwork among the Arapaho people in 1900, noting "a marked effect on the general feeling of the person, giving the impression of stimulating especially the intellectual faculties." Around this time, peyote use was spreading rapidly from northern Mexico into the Great Plains region. Indigenous communities in these states began taking the drug in a newly formed religion called the Native American Church, which combined a monotheistic belief system with the ritual use of peyote.

Ruth Benedict's 1922 article "The Vision in Plains Culture" (one of the works that inspired Mead to become an anthropologist) had offered an especially influential new perspective on peyote. Western scientists, Benedict noted, assumed that altered states of consciousness *reduced* the capacity for thought—that they were a form of intoxication. But what if the pursuit of visionary states was, in some cultures, a constructive goal? In these "Dionysian" cultures, as she dubbed them, visionary or trance states offered a source of new knowledge, mutual understanding, and social cohesion, not an escape into fantasy. Few outside the circle around Franz Boas agreed, however. Even as Mead first began to study the peyote in 1922, a bill to ban it was moving through Congress. This was the product of a decades-long campaign led by an alliance of evangelical missionaries and Prohibition-era antidrug crusaders. An article in the *New York Times* that appeared in January 1923 gives a taste of the rhetoric used by these groups. The eminent newspaper described peyote as a "false god" used in a "cult of death."

Among the Omaha whom Mead spoke with, there was an evident pride in the homegrown religion of the Native American Church and its cactus sacrament. But there was also fear that outsiders would portray them as immoral drug users. "Peyote is just dope," one of Mead's informants, a woman named Fay, told her. "It just makes you see things and prophecy [*sic*]. Often they're sad." Fay used peyote herself and believed the cactus had cured the typhoid fever of a boy she knew, but she also claimed that "peyote has ruined the tribe" due to the expense of holding the sacramental rites.

Mead's most important source on peyote was George Phillips, an Omaha man in his forties. Phillips told Mead that he had taken twelve buttons the night before. He explained:

> It works in your brain and makes you see things you never think of, makes you see when you have done wrong. Sometimes it makes you see twenty five or fifty years in the future. There are some people who can prophecy [*sic*] what the weather is going to be. Both men and women can take it.

Sometimes, though—and here the frustrating ambiguity around peyote came to the fore—newcomers "don't know how to take it and they eat it and eat it and it goes to their brain," Phillips admitted. "I was like that."

The exact mechanism by which mescaline (the active chemical in the peyote cactus) influences the brain is still poorly understood. But we know that it has something to do with the remarkable similarity of the mescaline molecule to naturally occurring neurotransmitters such as serotonin. By interfering with a specific serotonin receptor associated with the frontal cortex, mescaline powerfully alters the subjective experience of consciousness. Perception of time is altered, long-buried emotions rise to the surface, beautiful or terrifying visions play out in the mind's eye. Through millions of years of evolution, in other words, a tiny, rather unobtrusive cactus growing in the Chihuahuan Desert began producing a substance that can profoundly transform how humans think... for about ten hours, at least.

"Doesn't somebody warn the newcomer about how to take it?" Mead asked.

"No," said Phillips. "That's something you got to find out for yourself... Only those with a good heart can understand how to take it."

Then he told her a story about a U.S. congressman who visited the Omaha on a fact-finding trip about peyote. Local evangelicals thought the drug was demonic, and the congressman apparently did, too. But, Phillips added, the visitor "got pretty sleepy waiting for someone to go crazy."

"Congressman wasn't shown any of this," Mead hastily scribbled beneath the typewritten transcript of her interview. "All quiet decorum. Fake on Congressman."

The book that Mead wrote about her time with the Omaha, *The Changing Culture of an Indian Tribe* (1932), reflected the ambivalence of the Omaha men and women she met. Peyote, Phillips said, allowed some to see up to fifty years into the future. But Mead quoted another Omaha who said peyote "took the Indian way back, hundreds of years." Mead herself argued that "the peyote cult" was an innovative response to modernity rather than a remnant from a vanishing world. It was, indeed, a tool for the creation of a new, hybrid culture. By linking together "Christian and aboriginal Indian religious elements," peyote had successfully "reassembled the antagonistic elements of the tribe."

This idea—that drugs could be a tool for the creation of a new culture—was genuinely original. Peyote and related tools for the alteration of consciousness, like psilocybin mushrooms, were usually imagined as survivals from a premodern past. But what if they were, instead, pathways to a glorious future?

## Sepik River, New Guinea, 1932–33

Mead never returned to the study of peyote. But the core insight of that encounter stayed with her. The peyote research of 1930 was a step toward what she came to see as her life's goal: creating a new culture that expands humanity's collective consciousness. That goal came into focus in a place that could not have been more different from the sunbaked brown plains of Nebraska, a place of torrential rains and breathtaking biodiversity: northeastern New Guinea.

It was in New Guinea that Mead first came to see herself as a scientist who would not just *study*, but would push forward the limits of human consciousness. What were the potential ranges of personality types? Of sexuality? Of mental states? At stake was the creation of an entirely new vision of the human that moved beyond Western frames of reference. Also at stake: her marriage to Reo Fortune. For it was in New Guinea, at the end of 1932, that she met the intellectual and romantic partner who, more than any other, would define the rest of her life. In New Guinea she met Gregory Bateson. Mead and Bateson, as they later wrote, came to believe they were engaged in "building a new world" that would synthesize "the old values of many contrasting and contradictory cultural systems into a new form which will use but transcend them all."

But that was apparent only with the benefit of hindsight. At the time she first arrived in New Guinea, in December 1931, she was mostly just trying to get from one day to the next. Fieldwork before the age of air transportation was not unlike an Arctic expedition. Mead and Fortune traveled via a motorized canoe laden with room-sized mosquito nets, crates of ammunition for hunting birds, precious packets of antimalarial pills, thousands of cigarettes and matches (these being by far the most important currency for bartering), and many months' supply of notepaper, typewriter ribbons, Leica film, India ink, and colored pencils. They sought out the most remote and unstudied societies they could find—the so-called unpacified peoples at the farthest periphery of the European colonial empires.

Though they were steadfastly opposed to imperialism, Mead and Fortune carried some mental baggage with them, too. They imagined their work in wholly benevolent terms: a race to record cultural riches before European colonizers reduced unfathomably complex social orders into a monoculture of plantation laborers, gold prospectors, and mining company representatives. It was in some ways a naïve view, one that cast the anthropologist as a quasi-heroic savior figure. Yet in the context of her time, Mead's anthropology was genuinely revolutionary in its embrace of cultural differences. This helped her build rapport with the peoples she lived with, even if it didn't fully cancel out her sometimes condescending assumptions about "primitive societies"—or Fortune's explosive temper.

By late 1932, as they began their second year in the field, Mead knew what to expect. There were stretches of happiness, making new discoveries and writing productively, living off crocodile eggs and coconut milk bread. But their temperaments were fundamentally different. It was humanity's potential that fascinated Mead—and, even more, her own potential role in fulfilling it. Fortune was drawn no less powerfully to humanity's latent evil. The New Zealander tolerated Mead's careful study of how girls braided their hair or cooked medicinal herbs. But this work was not for him. Fortune's books, which bore titles like *Sorcerers of Dobu*, avoided the peaceful plains of life. Instead, they rappelled down into a culture's subterranean depths.

For a time, this clash between their characters had been a source of electric tension. But it was also corrosive, eating away at the bond between them like an acid etching into metal.

In the form of Gregory Bateson, that acid met its catalyst.

--------

In early 1932, word reached Mead and Fortune that a young English anthropologist had settled some miles upriver. Gregory Bateson, age twenty-seven, had the same University of Cambridge mentor as Fortune. Though Fortune said that he barely knew the younger man, Mead noted evidence of a rivalry.

Odd scraps of news about Bateson began to reach their camp. He was said to be living among the Iatmul people. Here, apparently, he was collecting phallic flutes used by an Iatmul secret society and studying a ceremony known as *naven*, a gender-bending rite of passage in which a teenage boy is symbolically birthed by his mother's brother, who wears female clothing. In May, a woman in the city of Rabaul (capital of the Australian-administered Territory of New Guinea) wrote to Margaret Mead that Bateson had been spotted returning from the field alongside the wife of a colonial official. The tryst was, she declared, "a scandal."

By this time, Mead was entering the first and (so she later claimed) only depressive period of her life. She had rushed into marriage with Reo Fortune following a doctor's pronouncement that she would never be able to bear children. The problem was a "tipped uterus," Mead wrote. "If I attempted to have a child, I was told, I would always miscarry." Cressman had the makings of a good father and a boring husband; Fortune was just the opposite. Despite this, Mead still nurtured secret hopes for the pregnancy that science told her was impossible. But hope was not enough, and she had already suffered a miscarriage while in the field.

At the height of Mead's depression, the newest issue of the anthropological journal *Oceania* arrived, and an article by Gregory Bateson was listed in the table of contents. The Iatmul people with whom Bateson was now living struck Mead as fascinating and deeply impressive. Bateson's skills as an anthropologist did not. Mead wrote to Ruth Benedict that he was "incredibly bad" at fieldwork. The year before, Bateson had published the first results of his research on New Britain, an island off the coast of New Guinea. After over a year living among the Baining people, Bateson had nothing publishable besides a short article called "Further Notes on a Snake Dance of the Baining" that forlornly described "the enormous number of snakes used."

As Mead dug further into his background, she uncovered the hallmarks of a dilettante. Bateson had been born four years after the death of Queen Victoria, at the apex of the British Empire's power, into a family that stood to benefit from that power more than most. His father, William Bateson, was not just a famed biologist (the first to coin the term "genetics") but also a child of privilege who sat on the board of the British Museum. The Darwins, the Huxleys, the Batesons: this was the snobbish pinnacle of Victorian England's scientific elite. Like Jane Austen characters in lab coats, the three families had known one another for decades and were continually plotting advantageous marriages. Bateson once told his mother that Mead had an "intelligent... almost female Darwin face," and he meant it as a compliment.

What Mead didn't yet know was that Bateson shared her sense that science was somehow tied up with the destiny of the human species. He also shared Mead's sense of otherness. William Bateson had taught his three sons that there was something weighty and mysterious about a life devoted to science.

It was a collective labor through the generations, an intellectual cathedral that rose through history toward some glorious end that was unknowable to the humble individuals building it.

The future William Bateson imagined for his children fell to pieces before Gregory reached adulthood. On October 14, 1918, mere weeks before the end of World War I, the oldest son, John Bateson, died in a pointless infantry charge of the British Army. Four years later, on the anniversary of John's death, the middle son, Martin, held a gun to his temple in central London and pulled the trigger. He was standing directly beneath the famous statue of Anteros in Piccadilly Circus: the Greek god charged with punishing those who scorn love.

In the aftermath, the weight of William Bateson's expectations fell on his sole surviving child, eighteen-year-old Gregory. His brothers had broken the pattern of scientific greatness. Now he, alone, was left to fulfill it—and he was failing.

--------

That fall, Mead wrote to Professor Boas to announce that she and Fortune were planning to track down their new rival. Afterward, Mead, who was struggling with a sprained ankle, malaria, and tropical dysentery, wrote a sarcastic note about the plan to Ruth Benedict. She pictured "how surprised Bateson will be—for no one expects a woman anthropologist to be beautiful at all, to say nothing of being so surpassingly lovely."

Gregory Bateson in New Guinea, 1929, as photographed by Sarah Chinnery. *Courtesy National Library of Australia.*

But when they met, it was Mead who was surprised. She had pictured an arrogant amateur. Instead, Bateson offered "the first cherishing words" that she'd heard in months. Noticing her malarial fever when she and Fortune arrived at his riverside hut, Bateson offered her a chair to rest in. Then, handing her a drink, he said he liked her most recent book. It was a simple kindness, but that was enough.

From her seat in the shade, Mead observed this awkwardly tall Englishman. Almost immediately she concluded that he was unlike anyone she had ever met. At the edge of the Iatmul village of Kankanamun, Bateson inhabited a "crazy screen room that had a tree growing through the roof so that his cat—and of course the mosquitoes—could come and go at will." His boots were often untied, his sandy brown hair uncombed. "Gregory Bateson is an awfully nice chap—6 foot 4 inches and very handsome," wrote the anthropologist Sarah Chinnery around this time, "but he gets terribly careless about his dress." In contrast to Fortune's unerring directional sense, Bateson was notable for his tendency to get lost while tracking the paths of small animals. And he also shared Mead's sexual fluidity. As an undergraduate at Cambridge, his closest mentor had been an older man named Noel Porter, a self-taught sexologist who was known for his prominent limp, his velvet trousers, and his collection of "phallic jewelry," and for serving as what Bateson called a "lay psychiatrist" to those around him. Bateson strongly hinted that he and Porter had once been lovers; in one letter, he compared their relationship to the one between Mead and Benedict. Porter once said he was drawn to Bateson's "appealing tender wistful touchingness."

When Fortune proposed they spend the upcoming Christmas holiday together with a nearby group of Australians, Bateson gratefully accepted. For the next two days, the three anthropologists steered their motor canoe downriver. It was a hypnotic journey. The river water was serene and dark, studded here and there with pink lotuses and, beneath them, the iridescent glimmer of Sepik rainbowfish. Mead began to realize that something was already shifting in her life. She was "intoxicated" by Bateson's jumps between physics, biology, and geology. Bateson was no less fascinated by Mead, who had earned a PhD and published three books during the five years that Bateson spent struggling to complete his master's degree.

On Christmas Day, Reo Fortune got uncharacteristically drunk on champagne and went to bed early. Bateson and Mead stayed up all night talking. The next morning, Fortune realized that the trio's dynamic had suddenly shifted. "By then," Mead recalled, "Gregory and I had already established a kind of communication in which Reo did not share."

--------

There was something "impish" about Bateson, a friend once observed. Something that, if pushed, could become "sort of devilish." He would play mental games with people, spiraling into ever more complicated ideas in a way that could be intellectually generous but also infuriating. Eclecticism was his defining trait. After he died, an archivist tasked with listing the subjects in one of Bateson's letters recorded over one hundred topics, including "planetary ecology," "drugs," "homosexuality," "colonial systems," "God, lack of faith in," "Shiva, dance of," "contradiction of systemic rules & values," "General Motors," "prostitute," "survival," "T'ai Chi," and "Navaho child being sung into coma." The letter was two pages long.

With Mead, however, this trait met its match. She could keep up with anyone. After that first night it was all one continuous dialogue, a process of falling in love that also, from Mead's perspective, seemed to be culminating in a momentous new discovery about the science of human nature.

But what exactly *was* that discovery? Mead had an intuition that she and Bateson were grasping at something important but hidden, something just beyond knowability. The spark came, once again, from Ruth Benedict, who was mailing drafts of her newest work. Indigenous peyote users and female shamans in Siberia, Benedict wrote in her essay "Anthropology and the Abnormal," were labeled as deviants by Westerners due to their "liability to trance." Yet they were also figures "of great power and importance" within their own societies. Benedict also pointed to homosexuals and those who crossed gender boundaries (the term "transgender" did not yet exist, so Benedict spoke of the "man-woman") as individuals who could occupy positions of honor in other cultures but were rejected by Western society. Mead knew that Benedict was not just writing about Siberian shamans; she was writing about herself.

Benedict privately self-identified as *androgyne*, combining the Greek words for both man and woman.

The essay ended by arguing that these people were not "sick," despite Western society's labeling of them as such. Perhaps the society that excluded them was *itself* "a sick civilization." The trouble was that "no one civilization can possibility utilize in its mores the whole potential range of human behavior." In other words, *some* outgroup would always be excluded. That those exclusions—of gender, sexuality, race, class, personality—were made by culture, not by biology, did not remove the fact that they persisted in every society ever recorded.

But what if Benedict was wrong? What if science could develop a new culture to accommodate *all* human potentials?

Working in an eight-by-eight-foot mosquito room in the village camp near Lake Chambri, Bateson and Mead began to develop "a new formulation of the relationship between sex, temperament, and culturally expected behavior." Social and sexual identities comprised a vast spectrum of forms, far beyond the binary of male and female, they theorized. These forms could then be mapped in a kind of personality chart they called "the Squares." On the *y* axis of the chart they plotted "Northern" (masculine-presenting) and "Southern" (feminine-presenting) personality types. A perpendicular line, the *x* axis, tracked two additional categories of personality: "Fey" and "Turk." The former mapped onto narcissism, individuality, and introversion; the latter to extroversion and "care for others." Recording these traits on a chart was just a means to an end. For, they believed, by scientifically identifying how personality cut across racial and sexual categories, they would be able to overcome the barriers these categories imposed on the world. A man and a woman from utterly different cultures might, in their system, be revealed as far more alike than two women or two men from the *same* culture.

Mead's first attempt at writing about the Squares is a strange document. It begins with a description of the "assumptions" on which "our present culture" was based. These included the idea that there were "two types of human beings, one male and the other female," and that these were "essential physical type[s]" in which biology determined destiny. Men were thought to be systematic, women intuitive; men active, women passive; and so forth. Those who deviated from this standard, Mead wrote, were told "that they were 'homosexuals' or 'alcoholics.'" This, at least, was the prevailing wisdom. "It was wrong," Mead typed. "Masculine" or "feminine" personality types need not correspond to what Mead called the "mechanical" traits of having a penis or a uterus. The "perfect female temperament," she wrote, could be "found almost equally in each gender."

It was a genuinely radical theory. Although historians of sexuality have tended to see the distinction between the concepts of *gender* and *sex* as one developed in the 1950s, Margaret Mead was clearly heading in that direction by 1933. Yet this first collaboration between Bateson and Mead never saw the light of day. It was, Mead ultimately decided, not just an overly simplistic idea, but an expression of her own worst trait: a dangerous tendency to make sweeping claims about world-changing breakthroughs based on scanty evidence. "It was the closest I've ever come to madness," she concluded years later. Bateson agreed. "All three of us together"—he, Mead, and Fortune—were, he said, "pretty well psychotic."

In rejecting gender as a determinant factor in human personality, the Squares did not achieve their larger goal of unifying humanity. On the contrary, it threatened to splinter humanity into four new pieces—Northern and Southern, Fey and Turk—to join the divisions that already existed. When Mead sent an early report of her supposedly revolutionary discovery to Franz Boas, he urged her never to publish it. Its sweeping claims were not just a potential embarrassment for Mead or for himself, Boas thought. They might even contribute to a new kind of bigotry.

--------

The earliest manuscript describing the theory of the Squares bears the names not of Mead and Bateson, but of "M. Mead and R. Fortune." Perhaps Mead, guessing what people back in New York City might think, had thought it wiser to include the name of her husband rather than that of a man she had met only a few weeks earlier. Whatever the reason, Reo Fortune angrily rejected any association with the document. He crossed out his name no less than seven times, his pen nearly tearing the paper. "I have nothing to do with this—RFF," he scrawled alongside them. "It was no good."

Reo Franklin Fortune was so dismayed by the Squares that he wrote to Bronislaw Malinowski, one of the world's leading anthropologists, to complain about the "dangerous madness" being cooked up by his wife and Bateson in their mosquito hut. Fortune blamed his new rival for what he considered Mead's "terrifying" mental breakdown, adding that he feared "for her mind's permanent condition."

It is true that Mead's writing about the Squares was close to messianic in tone. She had apparently been reading Aldous Huxley's *Brave New World* (1932), comparing her idea to the revolutions of "Jesus Christ and Henry Ford." But Fortune, too, had become unhinged. "Bateson egged it on to make me sicker," he wrote to Malinowski.

His anger culminated in a shocking act of physical violence. One day in their camp near Lake Chambri, Fortune tried to pull the Squares document out of Mead's hand. He lunged violently for the paper, pushing her to the ground. And his wife, as it turned out, was pregnant. Afterward, Mead suffered another miscarriage.

"Threats of assault and battery," Mead wrote years later, in a passage in her memoirs that she removed before publication, were "part of Reo's approach to the world." But this time the New Zealander had crossed a line. She remained with Fortune a few months longer, living for a time with him and Bateson in a shared flat in Sydney as they attempted to write up their field notes and make sense of what had happened on the shores of Lake Chambri. But at the end of the summer of 1933, Margaret Mead set sail, alone, for New York City. Reo Fortune remained behind in Australia; Bateson returned to England.

"I feel singularly free and lonelier than I ever have in my life," Mead wrote to Bateson that October, "sleeping in my big flat all alone, although I have hardly spent a dozen nights alone—in civilization—in as many years." She now had "no sense of responsibility to anyone, not even Ruth." Bateson was now the central relationship in her life. Yet this was not yet clear to anyone except the two of them. Mead, after all, remained married to Fortune, and she had not yet informed the public, or even her mentor, Franz Boas, about their separation. "I have got to get out of my relationship with Reo before I can really be rightly happy with you," she told Bateson. "We are all caught now in this tangle."

Gregory Bateson (*left*), Margaret Mead, and Reo Fortune (*right*) arriving in Sydney Harbour from New Guinea, July 1933. *Courtesy Library of Congress.*

But the feeling about Bateson that she had shared with Ruth Benedict that spring—that "sense of the Heaven to be seen through the stained glass window"—was stronger than ever. And "if that blessed outcome ever arrives," she wrote to him on that day in October, making a promise they must have both known she could never keep, "I am all for cherishing my allegiance to you and for my part, doing no experimenting at all."

CHAPTER 3

# Artificial Paradises (1933–39)

> Shamanism remained very much abbreviated, with whole vision societies disappearing and new recruits obtained but rarely. The curing powers of the shaman were largely eclipsed by the curative properties of Peyote.
>
> —*Margaret Mead,* The Changing Culture of an Indian Tribe *(1932)*

## Spring 1933 to Summer 1935

THE SQUARES, MARGARET MEAD and Gregory Bateson both agreed later, was an embarrassing wrong turn. But the goals it represented—a utopian, interdisciplinary, applied science—never left them. It was a goal that centered on the idea of expanded consciousness. Mead believed that every individual, every society, follows patterns of behavior that "remain below the surface of consciousness." Taken together, they are that which we call "culture": how we walk, how we respond to pain, how we nurse our babies, how we have sex, how we sleep, how we dress, how we die. Mead was not alone in seeking to understand these subconscious patterns and, in surfacing them, to demonstrate that Western culture was in no way inherently superior. It was, as Charles King has documented, the larger intellectual project of Franz Boas and his students.

But it was Margaret Mead and Gregory Bateson who linked this project to their shared fascination with altered states of being. For the next four decades, Mead would return again and again to the idea of humanity transcending nationalism, racism, and polarization by establishing "a new kind of consciousness."

For this reason, those feverish first months together in the jungle of New Guinea, early in 1933, would define the rest of their lives. Margaret Mead and Gregory Bateson remained, at heart, representatives of the deeply idealistic moment in the history of science in which they came of age: the period in the late 1920s and early 1930s that witnessed a revolution in the scientific understandings of personality, culture, and sexuality that has been likened to modernism in the arts and literature. Suddenly, many of the old assumptions about human nature fell away. In their place was a new freedom to experiment and to dream.

We often remember the 1930s as a time of sepia-toned drabness, a troubled decade shadowed by a looming catastrophe in the form of the rise of Fascism in Europe. But this was not always what it felt like to *live through* the 1930s. The people of the decade, after all, did not realize that they were inhabiting an "interwar" era sandwiched between the two most destructive conflicts in human history. Many still believed they had instead lived through "the War to End All Wars." True, the Depression had erased life savings and harmed livelihoods. But the recovery from that global economic catastrophe was already beginning in the summer of 1933. And anyone who read the newspapers in the middle years of the decade knew that their world was witnessing an unprecedented pace of technological and scientific innovation. Headlines announced breakthroughs and inventions nearly every day, from electric guitars and trampolines to the radio telescope, nylon fabric, the jet engine, the digital computer, the discovery of nuclear fission, and the development of the first practical helicopter—not to mention the first widely available televisions, fluorescent lights, and Technicolor cameras.

Little wonder, then, that the idealistic tone of Mead and Bateson's anthropological work resonated with the public. In 1936, for instance, Margaret Mead's study of New Guinea's Arapesh people inspired a lengthy article in the *Dayton Daily News*. A decade earlier, articles such as this one would have borrowed concepts from the "salvage anthropology" of Franz Boas, which framed the work of anthropologists as a mission to recover cultural artifacts and ancient wisdom from vanishing tribes of so-called savages. In 1936, however, Mead's study of cooperation among the Arapesh seemed instead to point the way to a better future. "Any successful cooperative must be spiritually linked with that plane of consciousness," the article concluded, referring to the absence of dictatorial leaders in Arapesh society. "The cooperative idea is a hopeful indication of the development in civilization that may lie ahead of us." On the very same page was an article bearing the headline "Man Seen in Sight of the Goal of Brotherhood." It announced "the daybreak of a consciousness so much vaster and so much more glorious than the candle of the local self... the third stage of consciousness," following the evolution of the human species "perhaps 30,000 years ago." What was the source of that dawn? Science, specifically the science of psychology.

Similar claims appeared all throughout the newspapers of the United States and beyond in these years. The collective message was clear. Science would bring—*was bringing*—a new kind of utopia.

This was not just an American intellectual movement. In the Soviet Union, even as Stalin and his secret police were ramping up their campaign of mass murder and torture, idealistic science-fiction fans and proponents of "Russian Cosmism" were developing the first mass culture to imagine human space exploration. Across Europe and Latin America, Freudian psychoanalysis was reshaping understandings of the human mind and mental health. And British, Swiss, and German pharmaceutical firms released a host of new drugs with unprecedented properties, from the first widely available hormone supplements (1928) to methamphetamine, which was widely sold in Hitler's Germany as a legal, nonprescription stimulant. And it was in Germany, in the years between the decline of the Weimar Republic and the outbreak of World War II in 1939, that the darker implications of this era of rapid scientific progress were beginning to reveal themselves.

--------

In May 1934, as Margaret Mead, back in New York City, was finalizing her plans to divorce Reo Fortune, a neurologist was administering mescaline to the philosopher Walter Benjamin in Berlin. Long before similar claims were made by countercultural figures of the 1950s and 1960s, Benjamin contemplated, like Mead, the possibility that Dionysian drug experiences could transform modern society. They were a potential entryway, he thought, to "a profane illumination, a materialistic, anthropological inspiration" that offered an alternative to Western individualism. But Benjamin also wrote that "such intoxication has its dark side."

That dark side was already becoming clear to many of the mescaline researchers themselves. And it was not just nausea and headaches. The center of mescaline research in 1934 was in the Munich lab of Emil Kraepelin, a visionary experimental psychiatrist who had made it his mission to understand the nature of schizophrenia. If psychosis could be mimicked, Kraepelin reasoned, then it could be better understood, and the mescaline experience seemed to him and his lab to be a form of "artificial psychosis."

Many of Kraepelin's students were Jewish, and their earliest work with mescaline in 1920s Munich took place in a city that was increasingly dominated by a lank-haired artist turned propagandist who, as he foamed at the mouth in his vicious orations in beer halls, struck some as potentially psychotic himself. In early 1933, Adolf Hitler and the brown-shirted bigots who had inhabited Munich's seedy underworld were unexpectedly elevated to national power. Germany itself had become schizophrenic. And Kraepelin's students—including a young physician named Eric Guttmann, who had taken an interest in mescaline's relationship to creativity—began preparing to flee.

Funded by a grant for German Jewish refugees, Dr. Guttmann resettled in London. It was there, in 1934, that he launched a new mescaline experiment and began enlisting volunteers.

--------

Among those volunteers was a young Cambridge graduate who described himself as "relatively normal." In truth, Geoffrey Gorer was a deeply eccentric man who believed, among other things, that he could predict the future. Gorer volunteered for the mescaline experiment in the same spirit that had led him, the year before, to spend three months traveling through West Africa with a Senegalese dancer named Feral Benga who had become his lover in Paris.

"In the laboratory where [Dr. Guttmann] was working they were investigating the derangements of vision produced by schizophrenia," Gorer explained. As the drug took effect, with its characteristic tracers of vivid light and its ricocheting mental imagery, Gorer began scribbling his impressions on a sheet of paper. "Motor-bike revolving lighthouses," he wrote. "Aeroplane noise hurts my throat. Unpleasant associations—street accident... Noise like a band... Always artificial." At first, he found the experience upsetting. "Sounds—they cause colours," he scribbled. "Skin luminous... Frightfully cheap effect... As limited as if living in a colour film. I resent these visions."

Later, though, he reassessed his first impression. Mescaline, he decided, was what had helped him to become an anthropologist. And it was the post-mescaline Gorer who would shape a key moment in Mead's and Bateson's lives.

## Fall 1935 to Spring 1939

Geoffrey Gorer arrived in Manhattan in October 1935. He had spent the previous three months traveling in Indonesia and Cambodia. Somewhere along the way, he read Mead's *Sex and Temperament*, the book that emerged out of her time with Bateson and Fortune in New Guinea. Fascinated by it, Gorer made a point of seeking out both Mead and Ruth Benedict immediately upon his arrival in New York City.

During the fall of 1935, Benedict and Mead spent four months giving Gorer a crash course in anthropology. It marked the beginning of a lifelong friendship between Mead and Gorer. At the time, Gorer was writing a book about his recent travels through Southeast Asia. In Bali, he believed he had experienced "the nearest approach to Utopia that I was ever likely to see." His resulting book, *Bali and Angkor* (1936), was a lively travelogue, but it was also an unusual study of Balinese art. Gorer seemed particularly focused on describing individual Balinese carvings and sculptures with striking, almost hallucinatory precision.

The appendix of *Bali and Angkor* revealed that his mescaline experience in London had inspired the book. He believed that the trancelike state induced by mescaline had allowed him to access a mental plane similar to the "deliberate delusions" experienced by Buddhist monks or Balinese trance dancers. Gorer posited that these various altered states were linked, reflecting a common state of consciousness inherent in the human mind—*all* minds. In the weeks when he was befriending Margaret Mead, Gorer wrote that mescaline was a powerful new tool for a truly universal science of consciousness: "Everybody," he wrote, "irrespective of race, age, sex, or habits, would with a similar dose have had similar experiences."

Peyote had sometimes been portrayed as a relic of a vanishing culture, something to be swept up by ethnographers along with the artifacts that filled America's museums. But Gorer's account of mescaline, which resonated with Mead's earlier writings about the role of peyote in Omaha society, was destabilizing and experimental in a distinctly modern way. It suggested a path away from the colonial baggage of salvage anthropology, with its assumptions about the anthropologist as distinct from and superior to the culture they studied. A path toward a new kind of science that was participatory, experimental, future oriented.

It came at exactly the right time to shape Mead's new project. The same month she met Gorer, Mead and Bateson submitted a proposal for a trip to Bali of their own. And like Gorer's mescaline experiment with Dr. Guttmann, it was a proposal to study schizophrenia.

--------

Margaret Mead didn't tell her parents about her marriage to Bateson until after it took place on a detour to Singapore in March 1936. The couple's final destination, a village in the interior of Bali, was perfectly safe, she reassured them, "with doctors and motor buses and I will be able to work quite well there." The "official" culture of the island was Hindu. But Bateson and Mead were not interested in the forms of Hinduism practiced among Balinese elites. Instead, they dove into village religion, which centered on ceremonial events in which dancers fell into an extended trance state, speaking in altered voices and at times stabbing themselves with kris daggers without experiencing pain.

Later that year, Bateson's mother traveled by steamship to visit her son and his new wife in person. Mead found Beatrice Bateson "easy to please" because she was convinced that her son was doing important scientific work: "Her whole interest is that Gregory should represent his father as much as possible." It was perhaps a good thing that Mrs. Bateson stayed home when they went to document the Balinese dance ceremony that was their primary interest. Mead suspected that the key event in the dance, the stabbing of an eight-foot-tall witchlike monster with a flame-studded tongue, was an elaborate metaphor for killing one's mother.

Balinese trance seemed to involve a transformation of identity akin to both schizophrenia and the state induced by peyote. As Gregory Bateson put it, Balinese trance offered an escape "from the ego-organized world." It was, he said, a "trip... [into] the realms of altered consciousness." This made it an ideal field of study for the friend who joined them in Bali, Jane Belo, a Texan dancer turned ethnographer, who herself seemed not to occupy the ordinary consensus reality (she would later be diagnosed with schizophrenia). Belo interviewed ritual performers who spoke of moving into a different plane in which they "would become for a time an animal or an inanimate object—a monkey, a pig, a snake, or a potlid."

Mead remembered her time in Bali as a lifetime "condensed into a few short years." By 1938, though Mead, Bateson, and Belo did not realize it, those years were reaching an end. The insights gained from studying trance states and the fluidity of identity in Bali would soon be overshadowed by the darkness of a world at war.

--------

Toward the end of their time in Bali, Mead and Bateson took out a three-year lease on a half-ruined palace. They planned what Mead called "great interdisciplinary expeditions, complete with endocrinologists and psychiatrists." I Madé Kalér—Bateson and Mead's assistant in their two years of research, and their only close colleague who was Balinese—came to believe that they were deluding themselves. For all their theorizing, they were outsiders who could never fully understand Balinese identity. They "wanted to believe Bali was a paradise," a Balinese woman who knew the couple during this time remembered. In truth, this was "their flight from the West."

The spell broke somewhere on the Pacific Ocean in March 1939, as Mead and Bateson returned to Bali from a follow-up trip to the Iatmul in New Guinea. Japan's invasion of China had reached a new stage of destruction, and the empire's sights were turning to Southeast Asia. Meanwhile, Hitler had invaded Czechoslovakia. Abandoning their palace and their plans, Bateson and Mead prepared to flee. They carried with them a distinctive set of ideas involving hypnosis, altered states, sexuality, and an applied, utopian science that would help to inspire the field of psychedelic science.

A portrait of Mead and Bateson sailing between Bali and New Guinea, commissioned by their Balinese research assistant I Madé Kalér as a parting gift.

I Ketut Ngéndon (d. 1948), *Goodbye and Good Luck to Margaret Mead and Gregory Bateson*, ink on paper, 1938. *Courtesy the Asian Art Museum, San Francisco.*

By the time they reached Vancouver in early April 1939, it was clear to both Mead and Bateson that war was coming on the schedule Mead's father had predicted. The window had closed.

At sea, when Bateson filled out his landing card, he wrote, "TOURIST 6 MONTHS." Now he crossed it out. After they arrived in New York, he intended to sail for London and join the British war effort.

And Mead, at age thirty-eight, had just realized that she was pregnant.

CHAPTER 4

# "A Respectable Word for Hypnosis" (1939–43)

> With Lt. Ja[me]s Hamilton conduct experiment on T[ruth] Drug—Volunteer as subject—knock myself out cold with 1 gram Cig—5:30 pm... Dinner at Chinatown 7:30 pm.
>
> —*from the diary of George Hunter White, Monday, May 24, 1943*

## Late 1939 to Summer 1941

FROM PEYOTE IN NEBRASKA to trance in Bali, and from the Sepik River of New Guinea to the New York City of 1939, Margaret Mead sought not just to *understand* human minds and cultures but to *change* them. It was an explicitly utopian goal, tied up with ideals of pluralism and tolerance that were deeply personal to her. But in 1939, that goal became entangled with another: defeating Hitler. Germany's dictator once said that he acted "with the certainty of a sleepwalker." To Mead, Bateson, and many of their scientist peers, this seemed exactly right. Hitler was driven by his unconscious mind, they believed—and he was also tapping into the unconscious drives of his followers. This interpretation made the science of consciousness surprisingly important. If science could learn how to better integrate the conscious and unconscious minds, perhaps Hitler's hypnotic hold over the German people could be broken.

The philosopher Walter Benjamin had described the drug experience as a "loosening of the self," a form of "anthropological inspiration" that could reconnect a deracinated modernity to humanity's deep past. But Benjamin and others in the first generation of mescaline users were still seeing the drug as an outgrowth of peyote, with its redemptive history of colonial resistance in the form of the Native American Church. The rise of synthetic drugs like amphetamines created substances *without* histories, and without social norms governing their use. Under the Third Reich, tobacco use was discouraged—but housewives could buy methamphetamine-laced chocolates at corner stores. Shorn of any direct connection to older cultural or medical traditions, synthetic drugs were pure products of high modernity. No taboo or precedent guided them. There was no guarantee that their effects on the world would be positive.

And mescaline, as it turned out, was no more imbued with anthropological inspiration than a bullet.

--------

The final months of 1939 found Mead in her third trimester of pregnancy in New York, alone. Bateson was in London, having crossed the Atlantic that fall to join the British war effort.

Yet the only work Bateson managed to find was as a volunteer in a civilian organization known as the Mass Observation project. The goal of the interdisciplinary group—which included anthropologists, journalists, novelists, painters, and at least one spiritual medium—was to document the collective consciousness of Britain. In practice, this meant observing the mood on the street: jotting down conversations, interviewing strangers in pubs, inviting ordinary people to send the group their diary entries.

The Mass Observation project had obvious applications for the war. It gave important insights, for instance, into the impact of Churchill's speeches as prime minister. But it was not the position Bateson had hoped for.

In December 1939, Mead gave birth to a healthy daughter named Mary Catherine. When Bateson received the good news in England, according to a story told by Mary Catherine herself, her proudly atheistic father "tossed his pipe into the air and over the garden wall." He then rushed to the telegram office to send Mead a cable: "DO NOT CHRISTEN."

He also began making plans to leave. A few weeks later, in January 1940,

Bateson gave up on London—gave up, in fact, on any plans to resettle in his native country—and sailed back to New York City to meet his daughter.

Reunited in New York, Mead and Bateson focused their attention on a new organization known as the Committee for National Morale—a "pressure group" of scientists that sought to mobilize psychological and anthropological research in the war against Fascism. The committee, for which Bateson began working fifteen-hour days as a secretary, included familiar faces like Ruth Benedict and Geoffrey Gorer. But it also featured some of the world's leading psychiatrists and psychoanalysts: Erich Fromm, a German Jewish émigré who exerted a far-reaching influence on the intellectual history of the twentieth century; Henry Murray, an expert in psychological profiling who would later become involved in the 1961–62 psychedelic experiments at Harvard; Adolf Meyer at Johns Hopkins, a contemporary of Freud; and the influential psychiatrist Karl Menninger, founder of the Menninger Clinic in Topeka, Kansas.

The Committee for National Morale became such an obsession of both Mead and Bateson in the summer of 1941 that their professional and personal lives converged around it. They became especially close to Lawrence K. Frank, an executive at the Josiah Macy, Jr. Foundation who was among the most influential members of the committee. Trained in economics at Columbia, Frank had no formal education in psychiatry or medicine. Despite this, he had emerged in the 1930s as perhaps the most influential advocate for the new concept of "psychosomatic" medicine: the idea that healing physical and mental diseases was a matter of broad-based social change, not just targeted interventions to treat specific symptoms. In an influential essay written before the war, in 1936, Frank had proposed that society itself was now the "patient" that scientists possessed the power to cure—an idea that resonated profoundly with the utopian ambitions of both Bateson and Mead. Along with Cathy, who was now learning how to walk, Mead and Bateson spent the summer of 1941 living with Lawrence Frank and his family in his New Hampshire vacation cabin. The nucleus of the Macy circle was beginning to form.

Throughout 1941, Mead, Bateson, and this circle of scientists threw their energies into using science as an antidote to psychological manipulation. The members of the Committee for National Morale saw themselves as a shield protecting freedom, democracy, and diversity from the weaponized, manipulative forms of applied science emanating from Nazi Germany. They were "mobilizing attitudes," as Mead put it in a letter to Bateson's mother in the spring of 1941, to "avoid the paralyzing apathy-producing effects of propaganda."

And then the United States entered World War II, and everything changed. In the months after the Japanese Empire's December 7, 1941, attack on Pearl Harbor—months when the Third Reich reached its largest extent, and the combined population controlled by the Axis powers grew to encompass some 635 million people—American scientists came to see their role in distinctly more aggressive terms. These were months when a dreamy, poetry-loving Berkeley professor named J. Robert Oppenheimer joined the U.S. government's secret project to develop an atomic weapon. And they were also the months when the Macy circle emerged. It was the dawn of the Information Age—the era when *data*, even more than raw materials, first emerged as the royal road to wealth, knowledge, and power. The wielders of vast data-centric bureaucracies, such as J. Edgar Hoover in his filing-cabinet-filled FBI headquarters, rose to prominence. So, too, did new technologies such as the first programmable computers (1941–42). In Bali, Mead and Bateson had been at the forefront of the new data-driven science as they employed an innovative annotation system to document life in Bali with an almost hallucinatory level of detail. They had wedded this information-centric approach with a deep interest in altered states of consciousness.

Now, in the early months of 1942, those two interests made Mead, Bateson, and the intellectual circle forming around them in New York City important to one of the newest and most controversial methods of waging war: psychological warfare.

--------

The Office of Strategic Services (OSS) started with a push from Britain's foremost spy. On a spring evening in Lisbon in 1941, Admiral John Godfrey, the UK's director of naval intelligence, boarded a Boeing 314 flying boat and settled in for the twenty-seven-hour flight to New York. The day before, Godfrey and his assistant, Lieutenant Commander Ian Fleming, had spent an uneventful evening at a casino just outside of Lisbon. Watching clouds pass below as they flew over the Atlantic, Fleming entertained himself by imagining that a Portuguese man he'd gambled with at the casino was a Nazi agent in disguise. Fleming's passing fancy that day became the seed for the first James Bond novel, *Casino Royale*.

On arriving in Washington, DC, Godfrey and Fleming met with J. Edgar Hoover. They explained that they had been dispatched by Winston Churchill on a mission to help develop the intelligence apparatus of the United States, which still lacked a dedicated overseas spy service. Hoover, whom Fleming remembered as an "enigmatic man with slow eyes," disagreed. The FBI was, he said, already expanding into foreign intelligence, and didn't need any help from the British.

Their next meeting, on the thirty-sixth floor of Rockefeller Center's International Building, was more successful. It was with a man known as "the Quiet Canadian," Bill Stephenson, who, through his position as head of British Security Co-ordination, ran the nerve center of British intelligence in the Americas. Stephenson had solid connections to President Roosevelt, and he agreed to back their plan for a new intelligence service. For its chief, he suggested Bill Donovan, a Manhattan divorce lawyer and decorated World War I veteran. In June 1941, Roosevelt offered Donovan the intentionally vague post of "Coordinator of Information."

But a director was just the beginning. That summer, Fleming stayed behind in DC to advise on staffing the new organization. Recruits from the world of scientific research were among the most promising.

## 1942–43

As the OSS came into being, it started to explore the potential of hypnosis and truth drugs as methods for interrogating captured enemy soldiers and agents. And in New York City, it soon found a pool of experts on the topic. In May 1942, the Josiah Macy, Jr. Foundation hosted a conference on altered states of consciousness. Officially, it was "a small informal conference on cerebral inhibition" that had been organized by Lawrence Frank and the Macy Foundation's medical director, an ex–Harvard professor and expert on the nervous system named Frank Fremont-Smith. But Margaret Mead and Gregory Bateson anchored the group of attendees and had played an important role in planning the meeting. The title, Bateson later explained, was intentionally vague: "Cerebral inhibition," he said, was actually "a respectable word for hypnosis."

Today, the Cerebral Inhibition meeting is remembered as a milestone in the origins of the field of neuroscience. There was the Mexican neurologist Arturo Rosenblueth, the co-discoverer of "sympathin," a mysterious hormone that was thought to shape human behavior. And there was the visionary physician Warren McCulloch, looking more than a bit wizardly with his long white beard, who was the first person to apply Alan Turing's work on computing to the brain itself—one of the earliest steps toward artificial intelligence research.

For Mead, the highlight was a hypnotist named Milton H. Erickson. Born in a silver-mining camp in the scrubland of eastern Nevada, Erickson had an uncommon gift for observing elements of human behavior that others missed. As a child he had struggled with dyslexia and polio, and the adult Erickson believed that his remarkable ability to induce hypnosis grew out of these early challenges—a kind of superpower gleaned from his observation of hidden "tells" and unspoken cues. To Mead's delight, Erickson managed to hypnotize a Yale professor at the meeting itself.

In 1942, Mead and Bateson still saw science as primarily a defensive tool, a means of healing traumatized soldiers or analyzing enemy psychology, not as a weapon. "I am not really anxious to destroy this world in which we live but to build upon it," she wrote in a letter to Geoffrey Gorer. "I do not want my ideas—at least with my blessing—put in a setting of destructiveness instead of a setting of a fundamental love combined with a desire for change." But in the terrible summer of 1942, she began to change her mind. Many at this point were predicting that Nazi Germany and the Japanese Empire would prevail. Winston Churchill's warning, in a famous speech two years earlier, appeared to be coming true—it now seemed not just possible but likely that the entire world would fall "into the abyss of a new dark age made more sinister and perhaps more prolonged by the lights of a perverted science." If, as was widely imagined, science was allowing the Axis powers to win, then Allied scientists, including anthropologists, would need to match them or die. They were driven, writes Ellen Herman, by the belief that the war "was a fundamentally psychological conflict in which the psyche was a battlefield . . . and victory was measured in the capture of enemy minds."

Little wonder, then, that the U.S. military began to explore hypnosis as a tool of war in 1942. The work that Milton Erickson described at the Macy conference that year was of particular interest. Erickson claimed that his hypnotized subjects "remained unaware of a lighted cigarette placed in contact with their hand." He also claimed to be able to induce a "somnambulistic" trance in which a subject could walk, talk, and perform ordinary activities without any memory of being told to do so.

In mid-1942, Stanley Lovell, the chief of research for the OSS, met with Milton Erickson to discuss the military uses of hypnosis. Afterward, an excited Lovell began to dream up ideas for weaponized altered states. What if a Nazi POW could be put in a trance, he asked, then smuggled back into Germany and induced to "assassinate Hitler in that posthypnotic state"? It was perhaps the first formulation of the concept of a "Manchurian candidate." The idea was never put into action, but it was not considered fanciful, either.

Though the Cerebral Inhibition meeting centered on hypnosis, the participants also discussed other methods for inducing trancelike states. One of them was sodium amytal, the first substance to be given the "truth drug" label. Synthesized in 1923, sodium amytal was a traditional sedative, not a psychedelic. But by quickly inducing a "twilight state" between dreaming and waking, it had been found (so some claimed) to make it difficult or impossible for the user to avoid letting secrets slip.

In October 1942, Lovell and his team established a top-secret group—the "Truth Drugs committee"—to "investigate the feasibility of using drugs in the interrogation of Prisoners of War." In his memoirs, Lovell stated that the project was considered "ludicrous" and "unethical" by some. But not by everyone. Of the four members of the committee, three had attended the Cerebral Inhibition meeting a few months earlier.

--------

In May 1943, Lovell and the Truth Drugs committee decided that the time was right to begin real-world testing. The subjects selected for drug trials ranged from American soldiers (who volunteered as guinea pigs) to at least one captured German or Japanese prisoner of war, who did not.

To help, the committee selected a recent OSS recruit: a Federal Bureau of Narcotics officer from Pasadena named George Hunter White. The thirty-five-year-old White, who has been likened to "an extremely menacing bowling ball," had pale blue Siberian husky eyes set in a gin-blossomed face, a boundless appetite for intoxicants, and a lifelong fascination with Chinese culture. With his encyclopedic drug knowledge and a .45 pistol invariably strapped to a hand-tooled leather belt (he was a dedicated amateur leatherworker), White would be tasked with the dirty work of real-world drug testing. "There never was any officer in American uniform like Major White," Stanley Lovell of the OSS remembered of the man. "Behind his innocent, round face with the disarming smile was the most deadly and dedicated public servant I've ever met. Did we have a new hypnotic or narcotic to try? Major White would take it." Another OSS colleague described him as "compelling," "mysterious," and "almost mystical." White, he said, was someone who carefully concealed his true nature.

Photo of George Hunter White from a vaccination certificate relating to his investigation of opium smugglers in Rome, Cairo, Tehran, Baghdad, Beirut, and Istanbul in the spring of 1948. *Courtesy Stanford University Special Collections and Archives.*

Joining him was James Alexander Hamilton, a man who would become close friends with both Mead and Bateson. It would have been difficult to find a starker contrast with the Falstaffian figure of White. Trimly handsome, cheerful, and witty, Hamilton was a young UC Berkeley psychology professor who had grown up on a farm in western Illinois. By late 1942, Hamilton had joined the OSS and been assigned to help oversee the truth drug experiments. White and Hamilton were set up with abundant funding, a large supply of experimental drugs, and a safe house in New York City.

After exploring mescaline as a potential truth drug, the team settled on another experimental drug—THC—that had been isolated from cannabis plants just three years earlier. It was at the time almost entirely unknown. Even the adventurous Major White handled it with care due to its potential dangers.

--------

Meanwhile, drug-induced hypnosis was being put into action in the war effort as a curative for traumatized soldiers. Two future members of the Macy circle—Roy R. Grinker and John Spiegel—were the most important catalysts.

Grinker and Spiegel met while working as U.S. Army psychiatrists during the hard-fought Tunisian campaign of 1942 and 1943. As they described in their book *Men Under Stress*, the bitter desert fighting against Rommel's forces in North Africa demanded new methods for rapidly treating trauma. Both were committed Freudians. As a medical student, Grinker even journeyed to Vienna in 1933 to seek out psychoanalysis from Sigmund Freud himself, managing to obtain a discount rate ("My fee is $25 an hour," Freud wrote, "but in consideration of my special interest in your case I would agree to a reduction"). But they also recognized that traditional Freudian talk therapy simply did not work in a battlefront context: it was far too slow.

In a classified 1943 report published by the Macy Foundation, Grinker and Spiegel argued that injected pentothal, a potent hypnotic drug, acted as a shortcut for obtaining what they called "a therapeutically effective discharge of emotions associated with a recall of a repressed memory." Grinker and Spiegel harnessed the same nascent concepts to describe the effects of narcosynthesis. Traumatized soldiers were stuck in thought loops that reinforced a painful memory each time it was relived. Injection of pentothal *with* talk therapy, they said, speedily transformed this "malignant cycle" into a "benign cycle," restoring order to a disordered mind. They called their new technique "narcosynthesis."

--------

Throughout the rest of the war, Mead did her best to remain on the side of what was known as "white" (defensive) psychological warfare. Gorer and Ruth Benedict, too, spent the war in the Office of War Information, working with Mead in Washington to apply anthropological insights to analyzing the character and motives of the enemy. Their war was fought with paper, deskbound, always with an eye toward the postwar period when true scientific research could resume.

But not all their peers stayed behind desks. Other anthropologists and psychiatrists enlisted in the Office of Strategic Services, with its emphasis on secret actions behind enemy lines, covert dosing with drugs and poisons, and the tradecraft of espionage: dead drops, encrypted messages, so-called black propaganda.

In 1943, Gregory Bateson became one of them.

CHAPTER 5

# The Office of Strategic Services (1943–44)

> In battle, in forest, at the precipice in the mountains,
> On the dark great sea, in the midst of javelins and arrows,
> In sleep, in confusion, in the depths of shame,
> The good deeds a man has done before defend him.
>
> —*J. Robert Oppenheimer, 1945*

## 1943

THROUGHOUT JANUARY 1943, AN unusually tall man of about forty could be observed shuffling past the victory gardens that had sprouted in the shadow of Carnegie Hall. Each day, he disappeared into the employees' entrance at the Museum of Modern Art on West 53rd Street. And each evening he emerged with tired eyes.

Gregory Bateson had talked his way into becoming the director and sole staffer of an obscure project funded by a government agency tasked with combating German propaganda in the Americas. From the perspective of the museum's employees, Bateson was a quiet man who spent his time in a tiny room that smelled of 16-millimeter film, mimeographed paper, and pipe tobacco ash. All he appeared to be doing was obsessively rewatching a single Nazi propaganda film from 1933. Later, he would tell a friend that he had watched it in reverse so many times that he had learned how to speak backward.

The film, *Hitlerjunge Quex*, told the story of a teenage boy, Heine, who joins the Hitler Youth and dies a martyr on the streets of Berlin. Bateson believed it revealed the heart of Nazi psychology: the worship of death as a pathway to a false utopia. Bateson wrote of "the hypnotic fascination" of the film's electrified world of "bright lights, the whirl of circuses, and city life." But the forward movement implied by the film's wavelike imagery was, he wrote, only the "illusion of progress." It seemed to him to reveal the technological inhumanity of Hitler's regime as a whole—the willingness to treat human beings as replaceable parts in a vast machine. Perhaps, Bateson speculated, this was the future that awaited us all. "Man, the mechanical puppet," Bateson wrote, buffeted by psychological manipulation, "is endlessly liable to be swept off his feet into some sort of whirlpool."

The bright lights of technology were as false as the glowing lure on an undersea anglerfish. Those who followed them, Bateson wrote, would be "sucked down into a swamp."

--------

Iatmul houses were built on tall stilts over swampy fishing areas; whirlpools in the Sepik River were frequent. The Iatmul cosmology, as Bateson understood it, grew out of their contemplation of "ripples and waves on the surface of water." Among the Iatmul, he once wrote, "it is said secretly that men, pigs, trees, grass—all the objects in the world—are only patterns of waves." Now Bateson seemed to be applying the cosmology he'd picked up in New Guinea to Nazi propaganda.

Perhaps, too, he was applying his own trauma to the task. His brother Martin's suicide had been in the heart of central London, surrounded by the bright lights and whirling automobiles of Piccadilly Circus. The moment was mirrored back to him in Heine's death in central Berlin.

Bateson left the temporary posting convinced that the film displayed a frighteningly sophisticated understanding of mass psychology. And he concluded that if the Nazis were using science for manipulation, scientists on the other side were duty bound to contribute their own expertise to combat it.

The Bateson that emerged from that room was different from the man who had described the war as a battle of freedom versus weaponized science. Never again would he be so naïve.

But he would also never know a period of more heightened life than the two years that followed. They were years of training with jungle commandos; years of befriending a young Julia Child (who, improbable as it may sound, became one of Bateson's closest companions in the OSS); and years of serving as the direct superior to a suspected Soviet double agent. In London, Bateson had been a stranger at the window, peeking into the British war effort via the passive watchfulness of the Mass Observation project. Now, in the summer of 1943, he finally had the chance to take an active role.

--------

It was not until afterward that Bateson came to regret it. At the time that he joined the OSS, in July 1943, it was considered a thrill, and an honor, to be recruited into the United States' newest and most elite spy agency.

Staying true to the OSS's oath of secrecy, Bateson never wrote publicly about his recruitment or his training. In her 2011 history of Julia Child's time in the OSS, journalist Jennet Conant described the OSS as having "the lenient, idiosyncratic atmosphere of a small college, with the same tolerance for campus radicals, zealots and oddballs." Gregory Bateson had found a natural institutional home.

In mid-March 1943, Bateson underwent a physical exam for "possible appointment to the Army of the United States." On March 25, he was summoned to Washington, DC. This was, apparently, the beginning of his recruitment to the Office of Strategic Services. A hastily scribbled note in Bateson's archives records that he was told to wear street clothes to his interview, and that "Dr. Abrams" would contact Margaret Mead—a reference to Dr. Allen Abrams, the OSS scientist who at that time was busy collaborating with James Hamilton and George Hunter White on their truth drugs research.

By May 11, 1943, Bateson was writing to a friend that "the OSS has begun trying to pull me in." Mead eagerly encouraged the effort. She met directly with an OSS officer involved in recruiting Bateson; afterward, she wrote a careful memo for the absentminded Bateson reminding him of his passport number and explaining how he should go about obtaining an FBI clearance. Bateson traveled to Washington, DC, for a final interview. (When he was asked to list his aliases and pseudonyms, he wrote down "Mevembelawan," explaining that this was a "name given me by New Guinea natives.") Not long afterward, he made his way to OSS headquarters and enlisted in the secretive agency.

Located on Navy Hill just north of the Lincoln Memorial, the location code-named Q Building would later become the first headquarters of the Central Intelligence Agency. It was not so much a single building as a patchwork of structures from different eras, some dating back to the War of 1812. The complex adjoined a defunct hospital and chemical weapons laboratory that had been operated by the National Institutes of Health. Across the street was a brewery, filling the headquarters buildings with the scent of roasted barley.

After Q Building, recruits proceeded via unmarked military vehicles to a secret location for weapons training. Exiting their blacked-out trucks, some were surprised to discover that the mysterious site known as "Area F" was in fact the Congressional Country Club in Bethesda, Maryland. Elizabeth MacDonald, who deployed with Bateson during the war, remembered the enormous range of people she encountered at the country club: "a Chinese artist, a Thai missionary, a newspaper reporter, a Shanghai businessman, a private detective, the producer of the Lucky Strike Hit parade," not to mention a Japanese American veteran of World War I, a dog trainer, and a patent-medicine salesman.

The recruits learned how to fire machine guns on the fairways of the golf courses, smashing the pristine green turf into mud. They learned how to disguise their identity, plant bombs, and evade capture behind enemy lines. "I teach what is called 'gutter fighting,'" the British martial arts expert William Fairbairn lectured them, drawing on his experience as a police officer in 1910s Shanghai. "There's no fair play; no rules except one: kill or be killed." One student remembered that the grizzled Fairbairn "taught us how to roll an ordinary newspaper into a lethal dagger." As the narrator of a classified OSS training film put it: "In this phase of the instruction period, the student is taught the gentle art of murder." Other arts were taught as well. In July 1943, during Bateson's training, George Hunter White took time out from his work testing truth drugs to visit the new recruits and share his encyclopedic knowledge of drug-assisted interrogation techniques.

After weapons training came the psychological assessment school: Area S. With Henry Murray as its director and James Alexander Hamilton as his lieutenant, the school was a kind of living laboratory for the experimental techniques of psychological testing that the two men had been developing at Harvard and Berkeley in the years before the war. "Everything about Area 'S' was very sneaky, even going down there," one recruit remembered. "You went there, stayed three days, took all manner of tests—physical, mental, and moral—then the school powers made out your 'S' Report and sent it back to Washington. An 'S' Report was terribly top secret. It said what was wrong with you, and what was right." Everyone involved was given a fake name and backstory. A key part of the assessment process involved a trainee's ability to cause their peers to slip up and divulge their real name. Everyone, civilian or military, wore standard-issue fatigues without any identifying information. If you "broke cover" in any way, you were kicked out. If you induced someone else to share information about their identity, you got points.

It was the perfect environment for an anthropologist.

--------

By the fall of 1943, Bateson had completed his training and begun his first assignment assisting the OSS in hypnosis-assisted interrogations. Only fragments of this work are available to researchers. Taken together, however, archival documents reveal the outlines of an effort to utilize hypnosis and hypnotic drugs in interrogations of Japanese and German POWs. Clearly, a second step had been taken in the path toward weaponizing science, a path Mead and Bateson had both claimed to detest at the outset of the war. The explanation was in the news every day: 1942 had been a year of unmitigated horrors, and, although the tide of the war had decisively turned by the final months of 1943, there was no guarantee of victory, much less a speedy one.

It was in these anxious months that Bateson, Mead, their hypnotist friend Milton Erickson, and an OSS intelligence officer named Arnold Meadow began working on a program that complemented the Truth Drugs committee. Their goal was to use hypnosis in place of drugs as an interrogation technique.

A mimeographed transcript of one unsettling experiment, conducted just before Christmas in 1943, lay forgotten in Bateson's papers until the day he died. In it, Bateson questions a man whom Mead described as a "Japanese psychologist" who was "working with GB as an informant." They speak about inane things such as whether he likes cigars—while Erickson slowly brings him into a state of deep trance.

The man, according to Erickson, was soon entirely under his control. In fact, Erickson claimed, the man was so lulled by Erickson's quietly commanding voice that the hypnotist had been able to make Bateson completely invisible to him.

--------

For the rest of his life, Bateson struggled to integrate what he did in the service of the U.S. military with his goals as a person and a scientist. One of his collaborators in San Francisco, where he moved in 1948, complained that he was "completely alienated" from the world, a broken man who "can't really engage in anything." Bateson himself would refer obliquely to traumatic memories from the war until the year he died.

Bateson's role in the disturbing hypnosis experiments of 1943 suggests one reason for his ambivalence about his OSS service. And so, too, does his later work in the psychological warfare unit known as Detachment 101 in the China-Burma-India theater of World War II—the poorly documented missions that won Bateson a medal for bravery, put him back into contact with James Alexander Hamilton, and, in the aftermath of Hiroshima, led him to write a letter to General Donovan that would influence the creation of the CIA.

CHAPTER 6

# Operation Bittersweet (1944–45)

> In wartime there is a tendency to say that all people with premises different from our own are mentally sick and that this is especially the case when they happen to be our enemies. But in a time of comparative peace I think it is more honest to say that the propagandic messages aimed at an enemy country during war were a sort of anti-therapy attempted on a mass scale.
>
> —*Gregory Bateson, 1958*

## December 1944 to July 1945

BATESON'S FIRST MONTHS WITH the Office of Strategic Services were a continuation of his work as a deskbound researcher. This began to change in February 1944. The OSS had promised him an assignment at "an undetermined destination for a trip of undetermined duration," Bateson wrote to an ex-girlfriend that month. "The machinery has not made up its mind about what it wants to do with me," he added. "What a life."

In early March 1944, Bateson and his Morale Operations unit received orders to report to a dock in Los Angeles and ship out to a military base called South East Asia Command (SEAC), which was located in Kandy, the ancient capital of Sri Lanka. Along with the orders came the standard-issue kit for a Morale Operations officer in the field: an Abercrombie and Fitch duffel bag packed with a bedroll, a pith helmet draped in mosquito netting, a compass, a wristwatch, two thermometers, two dog tags, and a lightweight typewriter. There was also a booklet entitled "This Is No Picnic."

SEAC was housed in a mansion that had once belonged to an imperial tea merchant. With its wide verandas and pith-helmeted army officers, it looked like—and in a sense was—a faded outpost of the British Raj. But it was also the site of one of the newest techniques of modern conflict: psychological warfare. After a brief training and orientation session in Kandy, Bateson and a group of OSS colleagues (including Julia Child and the anthropologist Cora Du Bois) headed into the Sri Lankan highlands. They would be living and working in a secret outpost of Morale Operations amid a former coconut plantation in the mountains.

The setting was picturesque, the conditions difficult. There were enormous neon-blue bees, vivid geckos, and eight-foot-long monitor lizards. There was also dengue fever, which sickened several of Bateson's colleagues. Continual mist and rain made it difficult to keep paper intact—not to mention the industrious termites, which could eat clothing locked in a steel trunk. Someone before their arrival had trapped a baby cobra and stashed it in a large glass bottle, to which small frogs were added each day. The civilian director of the Morale Operations unit, Carlton Scofield, started work at 7 a.m. and ended it at 2:30 p.m. in time for a siesta. Scofield assigned Bateson to duty as a "desk head," tasked with developing plans for psychological warfare operations to be carried out by agents in the field. The team looked more like the faculty club of a research university than a cadre of intelligence agents. One new arrival, the Burmese journalist Edward Law-Yone, was surprised to find not hardened military types but "anthropologists, ornithologists, psychoanalysts."

Meanwhile, Bateson went from writing about the science of survival to living it. His unit underwent instruction in how to identify poisonous snakes, smoke a porcupine for valuable protein in the field, and steer a kayak through high waves. After years of fieldwork, this was not terribly challenging training for Bateson, although it may well have been for some of the other scientists Law-Yone mentioned. "Mr. Bateson is the only mature person on hand," an internal report complained.

By the time of the D-Day landings in Normandy, in June 1944, it seemed increasingly likely that Hitler would be defeated. But in the Pacific, Japanese forces remained dug in on a string of fortified islands, and the outcome of the war was still uncertain. The Allies were making major gains in the Pacific, but they had also suffered significant losses. The Battle of Saipan, fought for control of a key foothold on the Mariana Island chain, claimed over three thousand American lives in a three-week campaign that ended in the suicides of the Japanese commanders, as well as some one thousand Japanese civilians. It was not just the bloodiness of the battles that made the "island-hopping" campaign of the Pacific theater so daunting. It was also the fact that the Japanese Empire covered nearly three *million* square miles of sea and land inhabited by hundreds of distinct cultures. All of these regions, all of these cultures, needed to be understood and navigated in order to effectively fight the war.

This is where anthropologists played a crucial role. From the Philippines to the Chinese mainland to New Guinea—where over one hundred thousand Japanese troops lay in wait behind massive fortifications—anthropologists had developed deep knowledge of the region's cultures. By leveraging their expertise and their contacts, they could develop effective propaganda campaigns, recruit local agents, and gain insight into the motivations of both enemy and friendly populations. Bateson, with his years in New Guinea and Bali and his deep knowledge of the social, political, and religious structures of the region, was among the most valuable of these assets.

As Bateson's importance to the OSS grew, so did the risks he took. He began spending more and more time in the company of a mysterious figure who would become a close friend—a man who would later grow so close to Margaret Mead that she spent months at a time sharing her home with him. Corporal James Mysbergh was tall, well built, twenty-five, and still a Dutch citizen when he enlisted in the U.S. military in 1941. Described in OSS documents as a capable intelligence agent who "as a boy was brought up in the Far East"—he was born in Java and spoke five languages—Jimmy Mysbergh reveled in risky missions behind enemy lines. Bateson and Mysbergh began spending long hours together hatching unusual plans, such as their proposal to release yellow dye into a Burmese river because the color supposedly symbolized "political independence" in Burmese culture. The younger man seems to have brought out something dark in Bateson's nature.

Bateson began showing signs of strain as he devised increasingly strange psychological warfare scenarios. One especially macabre plan involved scattering the ashes of enemy pilots over Japan by airplane "with appropriate accompanying texts and prayers." This was promptly rejected by the Army Air Force, whose own pilots refused to participate. One of Bateson's superiors around this time authored a memo warning that the "cool and scientific" strategy Bateson developed for one operation would result in "a frenzy of torture and killing" by the Japanese. For all Bateson's eloquence, the colleague considered it a "paranoid fantasy, disguised as a military plan."

Then, toward the end of 1944, Bateson volunteered for a series of dangerous missions behind Japanese lines in the Arakan Mountains of Burma.

The shades of both of his brothers were with him then, and they were leading him down a dark path. John died in a pointless infantry charge in the final days of World War I, long after the fate of the war had already been settled. Martin had set off on a series of quixotic quests—a motorcycle tour across Europe, a failed effort to become a playwright, an equally failed attempt to romance a young actress who was already engaged—before killing himself in one last, futile romantic gesture. As the historian Peter Mandler writes, Bateson "felt he could not back off from his dirty tricks, he could only dive further in, as if he were in one of his feedback loops, without knowing whether it was ultimately for good or for ill—perhaps both: at this stage he was thinking that even if the enterprise ended in the sacrifice of his own life, it might yet prove a fruitful suicide."

--------

In February 1945, Mead waited for a letter from the husband she had not seen for nearly a year. For weeks at a time, the only signal that he was still alive came from photographs addressed not to Mead, but to their daughter, Cathy—and the only content that could be gleaned from these was that he now lived in the jungle and appeared to have adopted a pet mongoose. It was during these missions in the backcountry of Burma that Bateson reunited with his friend from OSS assessment school, James Alexander Hamilton.

Hamilton's experience working with George White on truth drugs had left him with an abiding interest in the concept of "psychochemical warfare": the use of mind-altering agents to incapacitate soldiers or otherwise influence the course of battles. Documents declassified in 2015 show that Stanley Lovell, the OSS's head of research and development, corresponded in June 1944 with Hamilton about the military applications of a toxic Mexican fungus, *Coccidioides immitis*—which can cause extreme fatigue, lung damage, and a painful skin rash—then being studied at Stanford University. The documentation is fragmentary at best, but George White's diary offers a startling clue that weaponized fungi may have been part of Hamilton's official mission to the China-India-Burma theater. In 1945, Hamilton visited White in Calcutta, India, where White was serving as the OSS's security officer for the region. White's diary contains a cryptic entry relating to Hamilton's visit: "Hospital with Hamilton re fungus." Quite possibly, Hamilton was seeking samples of tropical infectious agents to bring back to Lovell and the OSS medical team. One other piece of evidence, recently declassified, lends credence to this conjecture: a U.S. military report on biological weapons from 1946 mentioned an infectious agent that reached the OSS labs via Calcutta.

Just before meeting with George White in Calcutta, Hamilton had been living with Bateson and parts of the Morale Operations team on Ramree, a sparsely populated island off the coast of Burma. Ramree Island occupied a strategic choke point for Japanese resupply lines and was bitterly contested between Allied and Axis forces. By February 1945, the actual fighting for the island had ended, but a contingent of Japanese troops remained. With Bateson's help, Hamilton wrote an eight-page memorandum on how to convince Japanese forces to surrender peacefully. Instead, many of the troops stationed on the island committed mass suicide.

Depressed and at loose ends, Bateson followed James Hamilton to the OSS field station in Calcutta. On May 8, when news of Hitler's death and Germany's surrender reached them, Bateson celebrated by drinking until midnight with Hamilton, Jane Foster, and George Hunter White.

## Summer to Fall 1945

On the day that the first atomic bomb was dropped on Hiroshima, Jane Foster was seated beside Bateson underneath the palm-thatched roof of one of the

Morale Operations huts. They both had typewriters in front of them and were writing up field reports. Then a tinny voice interrupted their work. "We were all stunned," remembered Foster, when "the Army loudspeakers blared out the news that an 'atomic device' had been dropped."

Foster, who would later be charged with espionage as a Soviet double agent, was an amateur painter with a taste for surrealism. Her response to the news was to imagine a scene that resembled a Salvador Dalí painting: "a vast eyeball-searing yellow desert with huge orange suns blazing and, in a corner, a small steel structure, twisted, crumpled and torn, and not a living thing around."

The sound of a typewriter broke her reverie. She looked around to find her other colleagues in a state much like her: either lost in thought as they contemplated the magnitude of the moment, or talking and shouting, eager to learn exactly what had happened. But not Gregory Bateson. He seemed to be writing something.

"Gregory, what in God's name are you typing?" Foster asked at last.

As she remembered it, Bateson said he was "writing about the future of life insurance in the atomic age." It sounded to her like some kind of dry British joke. But Bateson was deadly serious. And the memo he began that day would become perhaps the most influential thing he ever wrote.

--------

In 1953, the Central Intelligence Agency's official historian, Arthur Darling, wrote a classified history of the Agency's origins. It put Bateson's post-Hiroshima memo front and center. "Within the week" in which the Hiroshima bomb fell, Darling wrote, OSS director General Donovan received "a report from Gregory Bateson concerning the effect of the atomic bomb upon 'indirect methods of warfare.'" Darling framed this as a key moment in the decision-making process that led to the formation of the CIA. "Mr. Bateson forecast changes in psychological warfare, clandestine operations, and strategic intelligence," the historian wrote.

Specifically, Bateson predicted that international relations for "the next ten or twenty years" would be fundamentally reshaped by the combination of "rocket aircraft and atomic bombs" because they tipped "the balance of advantage on the side of attack" and away from defense. As a result, warfare would be reconfigured away from traditional methods and toward information-based, "indirect methods" against which "the atomic bomb is powerless." Bateson re-ranked the importance of U.S. agencies. First were the technicians manning "the physical laboratories and industrial plants concerned with atomic weapons." Second was the Air Force "and its research and production industries." Third was the new intelligence agency Bateson proposed. Only after those would come the traditional "land and naval forces." A follow-up memo from another member of the OSS to Donovan gave crucial support to Bateson's "analysis of the situation." The wheels of government bureaucracy began turning. And Gregory Bateson began preparing for the long trip home.

~~CONFIDENTIAL~~

16. Rating

| | Superior | Excellent | Very Satisfactory | Satisfactory | Unsatisfactory |
|---|---|---|---|---|---|
| MOTIVATION, energy, effort, initiative, interest in assignment. | | X | | | |
| PRACTICAL INTELLIGENCE, speed and accuracy of judgement, resourcefulness in solving problems. | | X | | | |
| STABILITY, emotional control and maturity, absence of nervous symptoms. | | X | | | |
| ABILITY TO WORK WITH OTHERS, teamwork, tact, absence of annoying traits. | | X | | | |
| LEADERSHIP, organizing ability, ability to win cooperation. | | X | | | |
| PHYSICAL ABILITY, agility, daring, ruggedness, stamina. | | | X | | |

17. REMARKS (Be sure to record outstanding accomplishments and weaknesses)

Mr. Bateson is a person of outstanding intellectual ability. He has shown personal bravery and devotion to the Mission. He does not, however, have much interest in subjects not related to his areas of specialization.

Bateson's OSS service rating, which notes his "personal bravery" and "devotion to the Mission" but also his low interest in "subjects not related to his areas of specialization." *Courtesy National Archives and Records Administration.*

--------

Bateson resurfaced in New York City four months later. He refused to speak in detail about his military service. When asked, he would tell friends only that he had been away on an intelligence mission in the Far East. It didn't seem to have gone very well. He told one friend that his tour of duty was "interesting in patches," but in the end "a total waste of time." Even Margaret Mead was not able to get much more out of him than this. There was something closed off in him now, a door she could not open for the rest of their lives. In his twenty-two months overseas with the OSS, Mead complained to another friend, Bateson had written her only four letters. One of them was fifteen single-spaced pages explaining an Indian film he had seen. It broke off midsentence, without a single word about himself—or a single question about how his wife and daughter were faring half a world away.

As the historian Paul Boyer has written, the raw recognition of the human devastation in Hiroshima and Nagasaki "gradually sank, unconfronted and unresolved, into the deeper recesses of American awareness," leaving behind "thick layers of psychic keloid tissue." But in the weeks immediately after the bombs fell, we also find a vivid, intense grappling with the moral costs of technology.

Those debates and questions linked Bateson and Mead, once again, in a common cause: building a science of human potential that would allow humanity to survive our species' vastly increased capacity for violence.

Meanwhile, the unintended consequences of Bateson's push for a new intelligence agency specializing in unconventional warfare continued to play out. A month after Bateson's memo, Lovell's medical laboratory at OSS headquarters itemized its "toxic materials." The list ranged from "poison bullets" to potential truth serums tested by the Truth Drugs committee to the pathogenic Mexican fungus that Hamilton had researched before going to Burma. It also included over a hundred more esoteric drugs, including anthrax, adrenaline, "Malta fever toxin," rattlesnake venom, and muscarine, the compound found in mushrooms used by Siberian shamans.

As the intelligence agencies that succeeded the OSS began casting about for alternative methods of atomic age warfare, they couldn't fail to notice the work on weaponized altered states that figures like Hamilton, White, and even Bateson himself had begun to explore during the war. And they began to take it further.

The final months of 1945 were marked by growing tensions between the United States and the Soviet Union. But in addition to the traditional military and political dimensions of what some were already calling the Cold War, a new kind of conflict was taking shape, one that would be fought through the unconventional means Bateson had specified. These means included weaponized psychology and experimental drugs.

This distinctly *psychedelic* dimension of the Cold War—in the sense that the conflict was, in part, a battle involving drugs thought to expand or transform consciousness—would eventually contribute to the fracturing of the Macy circle. But before that, it helped to popularize one of the most promising pathways for consciousness expansion yet discovered: lysergic acid diethylamide.

CHAPTER 7

# "A Not Unpleasant, Intoxicated-Like Condition" (1945–46)

> Ergot is uncertain in its action...Dr. Engelmann of St. Louis emphasized the evils of ergot in the hands of midwives and ignorant practitioners.
>
> —*"The Proper Use of Ergot in Obstetrics,"* Boston Medical and Surgical Journal, *September 28, 1882*

## Fall 1945 to Fall 1946

IN APRIL 1943, WHILE Gregory Bateson was preparing to enlist in the Office of Strategic Services and Margaret Mead was writing about the fate of Pacific islanders in the "new world society" that would follow World War II, a bookish Swiss chemist was tripping on LSD for the first time in world history.

The story would not become famous until many years later. But by 1946, a young psychiatrist named Werner Stoll had already heard it many times, both from his father, Arthur, and from the man himself, Hofmann. It was a darkly comic tale. There, in his laboratory on a spring day, sits his father's friend, Dr. Albert Hofmann. He wears a lab coat and is an earnest, patient, careful father of three. He is handling a minute pile of semi-translucent crystals. Then Dr. Hofmann begins feeling dizzy. A restless energy pulses through him. It is a Friday afternoon in spring—the fields outside the lab are filled with wild-flowers. Fearing he is getting sick, he mounts his bicycle and unsteadily cycles home. In a letter to Stoll the next day, Hofmann describes the experience as "a not unpleasant intoxicated-like condition, characterized by an extremely stimulated imagination. In a dreamlike state, with eyes closed (I found the daylight to be unpleasantly glaring), I perceived an uninterrupted stream of fantastic pictures, extraordinary shapes with intense, kaleidoscopic play of colors." Hofmann realizes that he must have unwittingly exposed himself to the substance he had just synthesized, the twenty-fifth in a series of compounds derived from a fungus called ergot: LSD-25. If the drug had managed to affect him despite his precautions, it must be extraordinarily potent.

A few days later, Hofmann tries again—this time taking what he assumed was a "threshold dose" of the drug intentionally, to test it. It leaves him incapacitated, lying on his couch at home. *Am I dying?* Hofmann wonders. Is this "the transition"? Then a doctor arrives, summoned by his wife. Hofmann's pulse is normal. His heartbeat is fine. He is in fact entirely healthy—just incredibly high. The next morning, Hofmann writes, "Breakfast tasted delicious and gave me extraordinary pleasure."

The dose makes the poison, the German alchemist Paracelsus famously wrote: any healing substance is also a potentially toxic one. But the reverse can be true, too. In ancient times, ergot, the drug Hofmann had been investigating when he discovered LSD, was known as both a destructive blight that poisoned grain crops *and* as a medicine used to aid childbirth. Since at least the sixteenth century, midwives had been scraping off the reddish powder of ergot from barley and administering it to pregnant women to induce labor. This potential medical use had in fact been the reason for Hofmann's research into ergot-derived drugs in the first place. Later, when he discussed the role of psychedelics in human society, Hofmann would claim that LSD had put him in mind of childbirth in a distinctly different way. "I had the feeling that I saw the earth and the beauty of nature as it had been when it was created, at the first day of creation," the chemist said in one interview. "I was reborn."

The Sandoz chemists, secure in neutral Switzerland, tried their best to ignore the war raging around them, and Hofmann's account of his first experience with lysergic acid diethylamide is no exception. But it was impossible to ignore. Hofmann's trip in April 1943 is today commemorated as "bicycle day." It is less commonly remembered that he was only riding a bike in the first place due to wartime automobile shortages.

LSD, it is worth remembering, was born in one of the most sinister moments in human history. The haunted ambience of central Europe at the end of World War II soon made its way into Hofmann's acid trips, though he rarely, if ever, mentioned the fact. In one trip report from later in 1943—which Hofmann never published and which only recently resurfaced in his archive—the chemist described a disquieting episode when he took LSD while on military reserve duty in a Swiss mountain fortress on the Italian border. That night, he experienced "horrible dreams" of a "deranged mutilated woman with severed arms." His comrades, he wrote, "think I am insane."

Throughout 1944 and 1945, Hofmann and Arthur Stoll—Hofmann's supervisor at Sandoz—continued sampling the drug in a series of self-experiments. They also tested it on animals. ("A caged community of chimpanzees reacts very sensitively if a member of the tribe has received LSD," Hoffman noted.) But Albert Hofmann and Arthur Stoll were research chemists, and they had little sense of what to actually *do* with the bizarre new substance. Arthur Stoll's psychiatrist son Werner did, however. By the end of 1945, Werner Stoll was ready to see if this oddly euphoric, time-dilating, unpredictable, and incredibly potent drug had any value in psychiatry.

--------

Werner Stoll's notes from his clinic at the University of Zurich in late 1945 record the earliest known uses of LSD in psychotherapy. They are painful to read. "Complaints that the soul has died," Stoll scribbled in the case notes for a young woman who walked through his clinic's doors in November 1945. "Feelings of being hypnotized... condition worse than appalling." Stoll's notes continued: "Often blank facial expressions. No hallucinations. Diagnosis: incipient schizophrenia." Electroshock and insulin therapy, the state of the art at the time, were "ineffective." So Stoll prescribed "LSD for 2 consecutive days in December," followed by "continuation of the insulin." There is no record of whether it worked.

Stoll himself, however, had begun taking LSD on his own, and he was increasingly fascinated by its potential. "Besides the hallucinations," he said cryptically of his first trip, "I could now see the *real* as well." He described another LSD session as a "consciously euphoric" experience of kaleidoscopic, firework-like visions during which "I felt myself one with all romanticists and dreamers." But a follow-up trip was different. "I longed to imagine bright fire as in the first dark experiment," he wrote. "And I did see fires; however, they were sacrificial fires on the gloomy battlement of a citadel on a remote, autumnal heath." Even in the earlier trip, as it wore off, Stoll had momentarily seen the city around him as a post-apocalyptic landscape at some point in the future, a "landscape of ruins... inspired undoubtedly by the memory of war events."

This link between psychedelics and war trauma is not new. It stretched back to some of the earliest accounts of psychedelic experiences, such as the Aztec man who told an early Spanish missionary that peyote allowed him to see his own death in battle. In the mescaline experiments of the 1920s and 1930s, too, it had appeared. An early mescaline patient, who had served in World War I and lost a brother in the conflict, spent his mescaline trip seeing "wavy lines" of "brilliant colors." This standard visual hallucination reminded him, he wrote, of "coloured lines of trenches on a war trench map—the red, blue, brown, purple lines which used to represent various objectives to be reached at successive stages of an attack."

But this was also something truly different. LSD had no precedent in prior psychedelic traditions. Mescaline could draw on the store of wisdom embodied in thousands of years of Indigenous American peyote usage. LSD was a synthetic drug.

What, then, to make of this high-tech by-product of an ancient poison that could induce such strange mixtures of euphoria and despair, insight and comical misunderstanding? Like the atom bomb, it was something new under the sun, something born from the flames of the Second World War.

--------

It came at the right time for new things. In a 1945 speech, aware that the United States would emerge from World War II as a global hegemon, Margaret Mead described a future for science as a tool for enlarging human consciousness. By this time, it had become commonplace to think of humanity's destiny as entangled with the destiny of machines. Since the 1920s and 1930s, as the historian Richard Wilson has argued, the perception that humanity had entered a "Machine Age" was "the fundamental fact of modernism." But the mechanized brutality of the 1940s brought a distinctly darker perspective to bear on this observation. The writer J. R. R. Tolkien, in a letter to his son, called the global conflict that came to an end in 1945 "the first War of the Machines." And, he predicted, "as the servants of the Machines are becoming a privileged class, the Machines are going to be enormously more powerful" in the years to come. For many, it was self-evident that this was a bad thing. In a widely read essay from October 1945, George Orwell predicted that the age of atomic weapons would split the world between "two or three monstrous super-states," each dominated by a "machine civilisation" overseen by a "self-elected oligarchy": the United States and its European allies, the Soviet bloc, and "the third of the three super-states—East Asia, dominated by China." Orwell predicted a future in which science was permanently weaponized in the service of a shadow conflict that depended on the mobilization of vast reservoirs of technical expertise—for, as he put it, "a complex weapon makes the strong stronger." In this eerily prescient essay, Orwell coined a term that would define an era. The future, Orwell believed, would be locked in "a permanent state of 'cold war.'"

Margaret Mead fundamentally disagreed. Rather than envisioning the postwar world as split between three hermetically sealed empires, she began describing it as an interconnected whole. Human society, she said, was now "a great complicated machine" in which every part was mutually dependent. "We," she declared, apparently meaning not just scientists but everyone who cared what scientists like her had to say, had a "responsibility for keeping it in good repair and improving it." The machine was not, as more skeptical commentators like George Orwell saw it, an inhuman juggernaut poised to trample the human spirit. The machine *was* us.

Mead saw not despair but enormous opportunity in this fact. She believed that there was now a chance to institute a new global culture that would "tap the whole range" of human potential, drawing together "the infinitely diverse expressions of the human spirit" that survive within thousands of different cultures. The task of the postwar world, Mead believed, was for the scientists and engineers who had made the defeat of Fascism possible to learn humility as they confronted a fundamental truth: science created the conditions for winning the war, but science alone could not create the conditions for a lasting peace. The only solution, she believed, was a new world order built not out of conformity but out of vast diversity. And yet, even as she warned that science did not have all the answers, Mead was open about the fact that she believed scientists would have a central part to play in "planning" a "world order... that would appeal to millions of people."

But any attempt to create a "synthetic culture"—a society wholly invented by scientists, to match the era's synthetic drugs and chemicals—would fail. What was needed was an explosive, radical, vivid diversity, one that interwove scientific modernity with hundreds of other living cultural traditions so that human nature could be realized to its fullest possible extent. In the past, Mead wrote, humans had "unconsciously" forced everyone within a society into a limiting set of arbitrary cultural forms: "America developed one kind of person, England another, France another." What made her optimistic about the future was precisely this recognition. For what had once been an unconscious pattern was now being pushed into the light—and now that pattern, that range of possibility, could be consciously expanded.

"This," Mead concluded, "seems to me a possible ethical platform" for the postwar era. It was something to stand on, this platform of a science with utopian yet practical goals. A way to build new worlds not on the ruins of the old, but on their firm foundations.

--------

During one of his first LSD trips, Stoll claimed to have seen a "constant flow" of "motifs from all kinds of foreign cultures." He saw everything from the jewel-like colors of the medieval Isenheim Altarpiece to motifs from pre-Columbian Mexico and the Indian subcontinent. He moved through "desert landscapes, terraces, flickering fire, starry skies of unbelievable splendor." There were "serene systems of arches... idols, masks, strangely mixed all of a sudden with childish drawings."

Stoll was experiencing one of the most commonly reported characteristics of psychedelics. Although the neuroscience of psychedelic drugs remains poorly understood, one theory is that substances like LSD and mescaline influence the brain's "default mode network," a system of linked brain regions associated with selfhood, memory, and a state of mind commonly described as daydreaming or mind-wandering. Psychedelics appear to encourage a more chaotic or "entropic" mental state, leading to more free-form, unexpected, emotive—and possibly more creative—ways of thinking. At the trip's euphoric heights, Stoll experienced what he called "a peculiar feeling of anticipation." He saw a network of ropes like the inside of a circus tent. These morphed into a vision of a Spanish city at dusk, the sky an indescribable pale blue.

A feeling of mysterious joy welled in his chest. And then the city dissolved in a "dense rain" of sizzling sparks.

There was another image from the trip that lodged in Stoll's memory. He saw "a succession of towering, Gothic vaults, an endless choir." It was like something from a medieval cathedral. But, he noticed, it was curiously ungrounded. Where the bases of the arches should have stood, Stoll could see nothing at all.

Next came a "landscape of skyscrapers." The scene reminded Stoll of "pictures of the entrance to New York harbor." The graceful art deco towers of that city were still new in 1946; many were still being built. In Stoll's vision they were endless, teeming, "innumerable."

He felt he was moving toward these artifacts of high technology like an angel, or a rocket.

And then he saw it. The skyscrapers had no foundations either.

CHAPTER 8

# Deviants Like Myself (1946–48)

> So I was right in presuming, on their second visit, that M. and G. were, like myself, deviants.
>
> —*from the diary of Maya Deren, March 1, 1947*

## Winter 1946 to Summer 1947

NINETEEN FORTY-SIX WAS MARGARET Mead and Gregory Bateson's year of ambition. It seemed the vast danger of the new atomic era had provoked in them a corresponding determination to do *something*—something concrete and effective and universal—not just to understand, but to *prevent* that danger.

They were not alone. The year after World War II's end marked a period of uncertainty and introspection as nations sought to redefine their roles in a new world order. But it was also a period when truly big ideas seemed achievable. New international organizations like the United Nations inspired hopes of a genuinely effective global federation of states, and despite rising antagonism between the United States under President Truman and the USSR under an aging Stalin, the geopolitical battle lines between the war's victors had not yet been fully drawn.

Mead and Bateson were not the only scientists who saw this period as an opportunity to intervene directly in the world before the Cold War solidified the looming standoff between the two superpowers. But they *were* among the most active. One notable example was their involvement in an ill-fated Hollywood film about the atom bomb. By this time, Margaret Mead had a powerful connection in the film industry: her polymathic brother-in-law Leo Rosten. Remembered as both a chronicler of Yiddish humor and as the deputy director of the OSS's public-facing sister organization, the Office of War Information, Rosten was also a successful film noir screenwriter. In February 1946, Mead arranged a meeting between Rosten and an atomic scientist who was working with the Hollywood studio MGM to develop a film called *Manhattan Project*. Rosten, in turn, put the scientist in touch with Hal Wallis, the producer of *Casablanca*—but Wallis was already working on a competing film with a screenplay by the libertarian novelist Ayn Rand. In the end, the project fizzled out into a tepid docudrama called *The Beginning or the End*.

No matter. Mead and Bateson were already moving on. It was also in 1946 that they joined with Albert Einstein and other atomic physicists in pushing for a world state. Mead and Bateson met with a group of physicists and other scientists, the Federation of American Scientists, who believed that only a unified planetary government could halt nuclear war. At one meeting, attended by Einstein, Bateson circulated an essay that had grown out of his conversations with Mead. Its original title was "Anthropologists vs Atomic War." The "competitive and anxious culture" of postwar society, he warned, would be vulnerable to civilization-destroying shocks. He foresaw "a sudden plunge from organized life into social entropy...the spread of destruction, famine, disease and chaos to the whole human species."

But Bateson also identified a solution. He urged scientists to back the creation of a world federation with "large scale propagandic and educational functions...from flags and parades to postal services and epidemic control." The first step was developing "higher orders of awareness": new, "more rapid" ways of diagnosing and healing the pathologies that lead to conflict. Shortcuts to consciousness expansion.

Bateson was evoking the widely held belief that there would soon be breakthroughs in the human sciences to rival the Manhattan Project. After all, scientists in those years were celebrating not just a new era of wonder drugs such as penicillin, but a genuinely revolutionary approach to thinking about consciousness and communication. Starting with a series of Macy conferences that began in March 1946—later dubbed the "cybernetics conferences"—Mead,

Bateson, and colleagues such as Norbert Wiener of MIT and Claude Shannon at Bell Labs were reimagining the brain itself as a computer.

History's first general-purpose digital computer, ENIAC, had been announced at the beginning of that year. ENIAC was a weapon of war, used to calculate atomic bomb yields. Crucially, however, it could be *reprogrammed*—altered for peacetime use, gifted with new potentials. What if the human mind was the same? Perhaps assuring human survival meant developing a new science of sanity, a new set of tools for reprogramming minds and societies stuck in self-destructive feedback loops.

It was against this backdrop that Mead and Bateson began retraining themselves as experts in mental health. Bateson set about researching methods to cure the world's "addiction" to war. Mead became a founding member of the World Federation for Mental Health, a first step toward creating a global psychiatric organization. Bateson underwent intensive Jungian psychoanalysis, while Mead—who declared herself "unanalyzable"—began helping her psychoanalyst friends with difficult cases.

The Mead and Bateson of the early postwar years were more overtly experimental than at any other time in their lives. In this, they were not alone. Around the world, as colonized peoples fought for independence, gender roles shifted, mind-altering drugs multiplied, the American civil rights movement and early gay liberation struggled into existence, and global population jagged upward at an unprecedented rate, the period from 1945 to 1960 was among the most radical eras of social experimentation in history.

This fact has, in popular consciousness, been overshadowed by the attention lavished on the 1960s and 1970s. But looked at another way, these decades were an aftereffect, an echo. It was in the late 1940s and 1950s that the forces of change, stored up during the war like a coiled spring, most rapidly unwound themselves.

--------

Maya Deren, the experimental filmmaker whom Gregory Bateson fell in love with in 1946, was a case in point.

Born Eleonora Derenkowska to a secular Jewish family in Kiev, Ukraine, Maya Deren escaped pogroms with her mother and psychiatrist father to settle in Syracuse, New York, in 1922, when she was five. She died in 1966, at forty-four, from a brain aneurysm that was probably linked to her habitual use of injected amphetamines. Her films, according to one critic, explored "bisexuality, promiscuous love, and erotic fantasy" as they performed "self-ethnographies" of America in the 1940s. The characters Deren plays often fragment into multiple simultaneous realities, like Schrödinger's cat. Two or three independent Maya Derens appear to move through space at the same time, pursuing different goals.

Much the same could be said for her life itself. At various points in her twenties, Deren was a Trotskyite activist, a graduate student studying symbolist poetry, a photographer based in Hollywood, a trainee occultist, and the touring manager of one of the earliest African American dance companies. Her social circle was no less eclectic. She threw raucous, amphetamine-fueled parties at her apartment at 61 Morton Street in Greenwich Village, dancing barefoot to calypso music. Her guest list included Anaïs Nin, Ralph Ellison, Gore Vidal, and Marcel Duchamp.

And, starting in December 1946, it grew to include Bateson and Mead.

That month marked the high point of Maya Deren's career as an artist. A tiny profile in *Esquire* magazine had come out just a few days earlier, one that trumpeted her recent award of a Guggenheim grant—the first ever given to an experimental filmmaker. It quoted her as saying that she made entire films for the same amount of money that Hollywood spent on lipstick. It was the first volley in a wave of press attention that would culminate in Deren winning the experimental film prize at the Cannes Film Festival in September 1947.

By that time, however, Maya Deren was living out of a suitcase in a cheap hotel in Port-au-Prince, Haiti. She had divorced her spouse—a sad-eyed Czech refugee filmmaker named Alexander "Sasha" Hammid—mere weeks earlier. Her chief possessions were a film camera, two letters of introduction declaring her intent to make a documentary about voodoo, and a diary overflowing with regret.

"Until the very last minute I expect Gregory to have wired," one entry in this diary reads. "Even perhaps flowers to the boat."

Deren and Bateson, after all, were supposed to have gone to Haiti *together*. They were supposed to have married.

Maya Deren in her 1944 experimental film *At Land. Public domain.*

--------

Margaret Mead had always been tolerant of her third husband's affairs. She even had a word for those Bateson pursued: "Steves," after his nickname for an older paramour with whom he had once lived on the Sepik River. Mead wrote of various Steves during the war itself, and several more before it. Their ranks even included Jane Belo herself, the Texas-born dancer turned anthropologist who accompanied them to Bali, and whom Bateson once told his daughter he had been in love with. These were not just tolerated affairs, but elements in what today would be described as an open practice of polyamory. Sexual jealousy was simply not something Margaret Mead trafficked in.

But Deren, who began spending increasing amounts of time with Bateson in the spring of 1947, watching and rewatching the footage of Balinese trance dancers that Bateson had shot in the 1930s, was something different. Whereas Mead saw herself as Bateson's mentor—the older, more experienced scientist who would bring rigor and clarity to his self-defeating and elliptical mind—Deren led him in the opposite direction. She reveled in the mystical and the irrational. And in Bateson, she saw a kindred spirit.

Bateson seems to have agreed. In July 1947, he moved out of the Greenwich Village flat he shared with Mead, their daughter, Cathy, and the Frank family. Living alone in an attic room in Staten Island, without permanent employment, Bateson became markedly more eccentric. The two gifts Cathy remembered receiving from her father that year were a Picasso print of a boy dressed as a clown and a dead horseshoe crab.

It was around this time that Mead wrote a letter to a friend about Bateson's failure—not as a husband, but as a scientist. Bateson had devoted his life to two opposing goals, she said. On the one hand, there was "the part of him that wanted to be a great scientist." On the other, a kind of conceptual artist, someone who wanted to live his entire life as "some strange romantic gesture."

Mead pointedly used the past tense to describe Bateson's former goal. He had *wanted* to become a great scientist like his father. But that potentiality had now passed. And there was another implication in Mead's letters from this time. Their relationship *had*, in the end, produced a scientist of greatness. But it was Mead, not Bateson, who wore that mantle.

Things came to a head in August 1947, when Deren planned a research trip to Haiti. She asked Bateson to join her, and he said yes.

Bateson began filling a reddish-brown notebook with plans for their trip. Would it be better to book a steamer from New York City to Martinique, or to take the train down to New Orleans and sail direct to Haiti from there? Could his bank in London wire $5,000 immediately to cover costs for "eight to ten months scientific anthropological studies in Haiti"? He listed the necessities of a long voyage: quinine and glycerin, bandages and boric powder, scissors and safety pins—not to mention a substantial trove of drugs, including amphetamine pills, cocaine, and opiates. He had complicated dreams in which Maya Deren appeared alongside his father, or alongside Margaret Mead, in different guises ("using Maya as my attorney," he scribbled in an attempt to explain one). He researched locations in which he might easily file for divorce: Puerto Vallarta, Ciudad Juárez, Jamaica. He planned a new life.

Maya Deren believed they would get married along the way. Bateson appears, for a time, to have believed they would, too.

In the end, though, he pulled back. One day in August, Deren stood waiting on a platform at Penn Station for a night train to New Orleans. She lingered on the platform until the train was about to leave the station, then boarded it alone. Bateson never appeared.

In the insomniac late night of a train cabin rattling across the Midwest, Deren took two sips from her silver flask. One for her, and one for Bateson. "It goes up and down," she scribbled in her journal. "At first a sense of freedom and that it is right and has to be this way—and then, because in the street the beggars play 'Peg o' my Heart,' and I see in the distance a tall figure in khaki—there is a sense of anguish and loss... Oh Gregory, Gregory, there is a hunger in me for all we had."

Then she thought about the inner life of an older woman she had spotted in the train's restroom. The woman was entirely ordinary, respectable, unmemorable... until Deren noticed her using a hypodermic needle to inject herself with cocaine. "It has become increasingly clear to me that I have been infinitely destructive in my stupidities," Deren wrote.

After her death, Deren's possessions were slowly cataloged by film scholars. It was a tricky process due to the complexity of her various collections, which ranged from voodoo priestess robes to Japanese drums belonging to her third and final husband, Teiji Itō. And although Bateson never joined Deren on her experimental wanderings through the first half of the twentieth century, one small part of him did. It was a relic from his past—a relic that, perhaps, he had hoped to exorcise along with his marriage to Margaret Mead.

There among Deren's belongings, researchers found two tiny pieces of metal: "Gregory's dog tags."

## Fall 1947 to Summer 1948

The man who provided Maya Deren with amphetamines, Dr. Max Jacobson, exemplified the truly experimental and even dangerous approach to pharmacology that defined the late 1940s. Jacobson injected his patients at will with a secret formula of vitamins, hormones, and stimulants. His eye-popping client list included Thelonious Monk, Niels Bohr, Winston Churchill, Ingrid

Bergman, Humphrey Bogart, and Marilyn Monroe. In 1960, just before his famous televised debate with Richard Nixon, John F. Kennedy joined Jacobson's patient roster. From that point onward, Kennedy's long-standing injections of testosterone and barbiturates would be regularly augmented by Jacobson's hormones, enzymes, and amphetamines.

Jacobson's story has been told many times, usually in accounts of celebrity excess—a not unsurprising fate for a doctor whose clients included Judy Garland and Elvis Presley. "Ideas come at the speed of light," Truman Capote once said of Jacobson's injections. "You go 722 hours straight without so much as a coffee break. You don't need sleep, you don't need nourishment... Then you crash—it's like falling down a well, like parachuting without a parachute."

Yet aside from his celebrity-oriented practice and the aura of mystery he maintained around his proprietary drug blends, Jacobson was actually a fairly *typical* representative of a truly adventurous, and often reckless, era in consumer pharmacology.

At the time, few pointed out the use of stimulants of the famous names mentioned above. After all, it was not unusual. Amphetamines and other novel pharmaceuticals—including wartime narcosynthesis drugs that, like amphetamines, were repackaged as consumer-facing products—were being massively employed in postwar America for such things as athletic performance enhancement, weight loss, and the treatment of depression, anxiety, and trauma. This was, in short, an age of "do it yourself" pharmacology. Drug experimentation after the war was self-directed, individualistic, and unmoored from professional opinion.

--------

Bateson's love affair with Maya Deren flamed out as quickly as it had ignited, and they do not appear to have spoken again after her solo trip to Haiti in August 1947. For a few months afterward, Bateson and Mead attempted, half-heartedly, to rekindle their own relationship.

But then, at the end of 1947, Bateson made a decision that forced a permanent split. Surprising everyone close to him, he accepted a position in a field in which he had no practical experience, and in a city thousands of miles away:

he was going to become a sort of in-house anthropologist at the Langley Porter Clinic, a psychiatric hospital in San Francisco. The idea of moving west appears to have been planted by none other than James A. Hamilton, who wrote a letter to Bateson saying that he was thinking about working at the Palo Alto Veterans Administration hospital, encouraging Bateson to consider applying there as well. Hamilton never did take the job, instead becoming a professor at Stanford. But by 1949, Bateson had relocated to the Palo Alto VA—the inspiration for *One Flew Over the Cuckoo's Nest*. While there, he would witness the origins of Silicon Valley and help to inspire the field of psychedelic therapy.

Meanwhile, Mead was moving with characteristic individuality in a different direction. As society jagged toward the experimental, she pulled back. By this time, she was beginning work on a new book on human sexuality that would be published in 1949 under the title *Male and Female*. In it, she drew on cutting-edge hormone research to make the point (which at the time was not just novel but radical) that "each sex depends for full functioning, upon both male and female hormones." But she also she pushed back on "the extreme advocates of a varying bisexual balance," reframing her approach as one of seeing the full potential of two sexes rather than jettisoning the concept of binary sex entirely.

A poll of American newspaper editors at the end of 1949 declared Margaret Mead to be one of the "women of the year" (*Charlotte News*, December 23, 1949). *Public domain.*

Writing of the possibility that she might make Geoffrey Gorer her fourth husband, Mead told Benedict in 1947 that "one of my blocks is that I can't conceive of my life not oriented to a man." Marrying Gorer was not appealing to her, she said, but was "an entertainable idea" given the circumstances. Meanwhile, she wrote to Benedict, "I miss you very much." The more or less openly gay Gorer was, in other words, a potential beard. He was also a new partner in her work with Benedict on a project officially known as Research in Contemporary Cultures. Benedict and Mead informally called the work "culture cracking"—picking the lock of a society's psyche like a safe. It boiled down to trying to decode, at a distance, the conscious and subconscious motives of entire nations. That work required security clearances and government approval. And *that* required a clean public reputation, no suspicion of deviance, no radical opinions.

Which was why the trouble for Mead, in 1948, could be spelled out in a single word: Kinsey.

CHAPTER 9

# "The Scientist Is *Not* Outside" (1948–49)

> I believe that the world's leaders will discover that infant conditioning and narcohypnosis are more efficient, as instruments of government, than clubs and prisons.
>
> —*Aldous Huxley to George Orwell, 1949*

MAYA DEREN GOT HER drugs from a high-priced Manhattan doctor. But for a young Allen Ginsberg and his friend Herbert Huncke—a merchant mariner turned Times Square hustler—the portals to mind expansion were usually dirty doorways near 42nd Street. And it was in Times Square, where Huncke turned tricks, that Ginsberg and his friend first encountered a sex researcher who was not yet famous but within a matter of years would become an iconic figure in American popular culture.

Born to a religious, working-class family in New Jersey, the young Alfred Kinsey was one of the first Eagle Scouts. He graduated from college in two years, with two degrees. Kinsey's Harvard doctoral work reflected his love of the outdoors—and his obsessive focus. Roaming through the fields of rural New England, he became fascinated by the behavior of the gall wasp. By one count, Kinsey collected up to five million wasps between the years 1916 and 1920.

It was in the mid-1930s, as breakthroughs in hormone science repeatedly made the news, that Kinsey first began to study the science of sex. Now a professor of biology at the University of Indiana, Kinsey befriended a younger colleague, Robert Kroc, who was studying the effects of hormone therapy on lab rats. At one of their lunches, Kroc remembered, he and Kinsey began "trading information" about one another's sexual histories. Soon after, Kinsey surprised his neighbors and colleagues by adopting a daring new outfit for working in his beloved garden: "a loincloth type of thing," as his daughter put it. By 1940, he was advocating for sexology in urgent language that echoed the wartime writings of Mead and Bateson. "It is desperately strategic," Kinsey wrote, "that our civilization realize something of the diversity in human sex behavior, and acquire some sympathetic understanding of that which is different from one's own."

Throughout World War II, Kinsey dove into the study of male sexuality with the same restless energy that had made him the world's leading expert on gall wasps. He interviewed soldiers and prisoners, outcasts and executives, roaming American cities in search of informants who could lead him to undiscovered sources of knowledge. In March 1946, while Mead and Bateson were attending the very first Macy conference on cybernetics (Bateson spoke on gender fluidity in New Guinea), Kinsey was exploring the dive bars and all-night diners around Times Square. It was in one such diner that he first met Herbert Huncke. One night, Huncke assembled three of his friends for Kinsey to interview. Their names were Allen Ginsberg, Jack Kerouac, and William S. Burroughs.

"We were part of Kinsey's statistics," Ginsberg remembered of that evening. In anonymized form, the sexual histories of these founding members of the Beat Generation appeared in Kinsey's famous report.

--------

Leaving the drugs, poets, and sex workers of Times Square behind, Kinsey returned to his suburban home and his gardens in Bloomington, Indiana. And then he wrote a letter to Margaret Mead.

"I am very glad to have met you in New York," Kinsey wrote. "I am sorry my stay in New York was not long enough to contact all the persons who offered histories. I should very much like to have you go through a history with us someday to see how the material is handled."

Had Mead herself agreed to submit a personal sexual history to Kinsey's study? Without additional context, it's impossible to say. What *is* clear is that Kinsey sought to enlist Mead as a collaborator, offering to share his data and making plans to meet again. (Bateson, too, had met Kinsey during his trip to New York City; the two men spoke together on a panel on human sexuality.) Kinsey even began speaking of his own research as a form of anthropology. He hired a young anthropology PhD student at Harvard whom he promptly dispatched to the American Museum of Natural History to work with Mead on cross-cultural sexuality.

Both sexologists and drug researchers were realizing that anthropology offered a wellspring of inspiration. Mead's ambition to expand human potential through science had acquired a sinister cast during World War II, but as memories of the war faded, her science again took on a utopian character. She argued that an expanded awareness of the range of human sexuality could prevent individuals from being forced into "an ill-fitting mold." Kinsey and his team (which in 1947 won a large grant from the Rockefeller Foundation) seemed poised to contribute to that noble goal.

And then Kinsey's report came out—and Mead became Kinsey's most powerful opponent.

--------

Drawing on a raft of statistics, Kinsey argued that around one tenth of all men were "more or less exclusively homosexual." Almost immediately, the explosive claim and its author seemed to become part of mass consciousness in the United States and beyond.

When Mead was asked to comment on Kinsey's report at a symposium, she accused Kinsey of puritanism. By supposedly describing sex in purely biological terms, stripped of its cultural significance, Mead said he had contributed to a larger trend of framing sexuality in impersonal, scientific language that was "extraordinarily destructive of both intra-psychic and interpersonal relationships." Though it wasn't quite what she had said, newspapers nationwide proceeded to run her comments under headlines like "Kinsey Sex Report Is Called Destructive and Dangerous."

Kinsey was shocked by this intensely negative reaction from a colleague he had seen as an ally. Behind the scenes, it is possible that rumors of disquieting behavior at Kinsey's research institute had reached her by this time. Kinsey's biographer notes that despite his vision of a "sexual utopia" of uninhibited people, there was by the late 1940s "something grim in the way Kinsey was approaching sex, not only in his private life but in his research." When Kinsey demanded that his employees have sex on film to provide additional data for his studies, they complied—but felt, in the words of one female colleague, a "sickening pressure." After all, she remembered, "I felt like my husband's career at the institute depended on it." His researchers supposedly had to "clear" their extramarital affairs with Kinsey in advance.

Mead, in at least one respect, was right. There *was* something puritanical about Kinsey. He policed the sexual lives of the men and women who worked alongside him with an unabashedly patriarchal power. Mead may have recognized that tyrannical mentality in Kinsey and been repelled by it.

But she also had a more personal problem. When she had first met Kinsey, Mead was a married woman with a daughter and a government security clearance. By the end of 1948, she was estranged from her husband, who now lived across the country. Moreover, she was now beginning a romance with a younger female colleague, a former OSS officer turned anthropologist named Rhoda Métraux. The two women began spending enough time together that those around them noticed.

The timing of Kinsey's efforts to make Mead a comrade in arms could hardly have been worse. In August 1947, for instance, during the summer when Kinsey was trying to work directly with Mead on the anthropology of sexuality, Mead was in Paris, leading an American delegation of scientists. Rhoda Métraux was among them. This was all taking place at the precise moment that the State Department, urged onward by a July 1947 warning from a Senate committee, began investigating supposed Communist infiltration of the American government. The committee's letter singled out "the extensive employment in highly classified positions of admitted homosexuals, who are historically known to be security risks." Within months, the FBI uncovered several high-profile cases of "disloyalty" at the State Department. These cases involved not suspected Communists, but suspected homosexuals. What has come to be known as the Lavender Scare had begun.

--------

Ruth Benedict, who had worked herself into a state of ill health during the war, died of heart failure on September 17, 1948. Margaret Mead visited her oldest lover and greatest mentor in the hospital just before she passed. Benedict looked to her "almost supernatural" at the moment of her death, she recalled: "like a figure in the ceiling of the Sistine chapel." Mead had loved the older woman for over twenty-five years, loved her with a devotion that Mead's daughter, Mary Catherine, remembered as forming an "axis of her life" that intertwined with her love for Bateson.

The FBI began a loyalty investigation of Margaret Mead a few days later. This was no simple background check. It was an extraordinarily thorough search, overseen by J. Edgar Hoover himself. It meant poring over Mead's college yearbooks, collecting everything she ever published, and interviewing everyone from a sorority girl who met her in Indiana when she was eighteen to her Barnard College PE teacher to her neighbors in Greenwich Village. But though there was ample opportunity for Mead to break down under the strain, she carried on as always. And, as always, she disguised any sign that she was not a heterosexual, happily married woman. The closest the FBI came to discovering the truth of her separation and her bisexuality was when they began asking Mead's friends where her husband was. But now her ties to Lawrence Frank, her mentor at the Macy Foundation and longtime landlord, paid dividends. When the FBI questioned Frank's wife, Mary (who, as Mead's close friend, knew all about both her romance with Benedict and her split with Bateson), she assured them that the two remained happily married. Bateson, Mary Frank said, was simply out of town.

The night after Benedict's funeral, the cord attaching a mirror to Mead's wall snapped. As it fell, it knocked over the carvings that she and Bateson had gathered in Bali before the war. It was easy to read as symbolic: the two great loves of her life, both now fallen out of it. But Mead was both sentimental and an optimist. She prided herself on the sturdiness of her relationships, if not her home furnishings. Back on the wall went the mirror. Back to their rightful place went the carvings.

For the rest of her life, Mead kept a photograph of Ruth Benedict on her mantel, and one of Bateson on her bureau.

--------

By this time, the fall of 1948, Bateson was settling into his new life in San Francisco, reporting for work at the Langley Porter Psychiatric Clinic, a modernist building located on a misty, eucalyptus-scented avenue near Golden Gate Park. After work, he found time to reconnect with his psychiatrist friend James Alexander Hamilton. Bateson's former OSS colleague was happy to show Bateson around San Francisco's bars and cabarets. On September 29, there was even an Office of Strategic Services reunion. George Hunter White was in town for a narcotics case, and Hamilton and Bateson met up with their old wartime comrade at a downtown nightclub called Shanghai Lil's.

A month later, White was back in town to raid a suspected opium den. This time, Hamilton, White, and Bateson spent no less than seven hours drinking cocktails. It is impossible to know what they spoke about or why their drinking bout ran so long, but Bateson's archive offers one clue: two days earlier, the anthropologist had observed his first electroshock therapy session. He found it unsettling.

The Langley Porter Clinic in the late 1940s. *Courtesy UCSF Archives and Special Collections.*

Bateson's decision to switch careers midstream—to give up on being an anthropologist and instead to begin working at a psychiatric hospital in a city where he knew almost no one besides Hamilton—baffled his ex-wife. Mead saw it as a sign of his lack of willpower, a flight from his responsibilities in New York, and proof that he could not do serious science without her there to help him. From Bateson's perspective, though, the move made sense. At the time that Bateson joined Langley Porter, the clinic's director, Karl Bowman, was transforming it into one of the world's most important hubs for research into transgender sexual identities. The crucial contribution of the Langley Porter Clinic was that rather than treating trans men and women as patients, Bowman—and Kinsey, too, who visited Langley Porter repeatedly in these years—worked alongside transgender people as colleagues.

The Langley Porter Clinic would also become an important early center for the development of psychedelic therapy. This unique intersection of interests made it a perfect fit for Bateson, who since his early days studying the *naven* ritual among the Iatmul people of New Guinea had been intrigued by the relationship between altered states of consciousness and an expanded conception of human sexual difference. These had been considered eccentric interests in the 1930s. But now, in the postwar era, they were the cutting edge of science. Sex hormones and psychedelic drugs were linked in the midcentury scientific imagination due to the perception that these two categories of drugs, uniquely, had the ability to reshape one's core identity. This was not a transient alteration of consciousness; it was not the treatment of a symptom. It was an explosive, self-directed transformation, an atom bomb for the self.

--------

The ability to fundamentally alter one's sense of self was not just a pathway to personal fulfillment, however. It was, and remained, a tool used to control and manipulate.

A conference bringing together researchers involved in the study of consciousness, psychiatry, and sexuality in 1948 made these stakes clear. Margaret Mead and Alfred Kinsey both attended. Despite their public disagreement and personal dislike, the conference crystallized the extent to which they (and Bateson, too) occupied the same side when it came to their thinking about mental health. On the other side stood the conference's organizer, Paul H. Hoch, who bore the title of New York State commissioner of mental hygiene. Hoch's views on sexuality were regressive even for his time. For him, schizophrenia, addiction, and homosexual or transgender identity were all part of a continuum of deviance.

At the time, Hoch was researching mescaline and sexuality. But rather than seeing mescaline as a therapeutic tool for surfacing repressed trauma or encouraging self-reflection—in short, as a far less destructive *replacement* for older methods like lobotomies—Hoch was exploring psychedelic drugs as a *supplement* to lobotomy. He discussed one unfortunate patient, a young man who had suffered from depression for six years and also (in Hoch's words) exhibited "deviant sexuality, compulsive masturbation, homosexuality, exhibitionism, and transvestism." Mescaline therapy induced what Hoch called "severe silliness" in the patient. And then he proceeded to what he apparently considered the natural next step: frontal lobotomy. Afterward, the commissioner reported, his patient displayed "intellectual and emotional fragmentation." From that point on, nothing more about this appalling experiment is known.

For all Kinsey's considerable flaws, the contrast between his own practices and those of Paul Hoch pointed to how truly revolutionary his approach was. Kinsey—and Bateson, too, in his psychiatric work at Langley Porter—treated his patients with friendship and empathy. Kinsey's letters to trans women like Louise Lawrence and Christine Jorgensen show him enlisting them as colleagues and fellow researchers. Hoch saw only a disease to be eradicated.

These were the true stakes for Margaret Mead as the government began its probe of her private life in 1948. Homosexuality and other forms of what Hoch called "aberrant" behavior might go unnoticed, unremarked upon—but in the hands of the wrong doctor could also lead to a hideous fate.

--------

As the winter of 1948 blossomed into spring in San Francisco, Gregory Bateson was coming into his own as a mental health professional. Amid the experimental culture of the Langley Porter Clinic, he began to see his work with psychiatric patients as a chance to atone for his role in weaponizing the science of consciousness during World War II. He was beginning to form the identity that would shape the rest of his life—beginning to see himself as an early warning sign of the vast dangers of science in the twentieth and twenty-first centuries.

In World War II, Bateson believed he had seen the earliest stirrings of a new science of "social manipulation" in which "our tools are people." Not just seen, in fact, but directly participated in it, helped to *inspire* it. Like a hero from the Greek tragedies he loved, he now saw himself as someone who had been allowed to survive a terrifying ordeal so that he might serve as a warning.

He rarely put these thoughts into words. But when he did, they came out in a torrent.

In April 1949, Bateson participated in a "Roundtable on Modern Art" at the San Francisco Museum of Modern Art. Maya Deren's friend the legendary conceptual artist Marcel Duchamp was one of the attendees. And so was a notably grumpy Frank Lloyd Wright, who was at that time the world's most celebrated living architect. At one point in the proceedings, Wright began goading Duchamp by insulting his painting *Nude Descending a Staircase*, which hung outside the room. When Bateson jumped in to defend Duchamp, Wright aimed his fire at Bateson himself. Someday, centuries hence, the two fields of art and science might merge, Wright said. But the scientist—and Bateson was at that moment the only scientist in the room—was "the enemy at the present time of all that the artist would represent." Religion was declining, Wright said, and the false prophets now came from science. "We have all been *had* by scientists," the architect fulminated.

Bateson had once hoped to be a great scientist—*still* hoped to, deep down. But when Wright spoke of science as a dangerous force of manipulation, Bateson agreed. This, he said, was not the sort of science he wanted anything to do with.

*But—*

Before Bateson could finish, Wright interrupted again. Scientists, he said, stood outside the pattern of life, "trying to take the thing apart and see what makes it click." Wright's words recalled the visual language of Rivera's mural. Scientists might think their hands were on the controls of the great mechanism of the world, but in seeking to control society, they had also placed themselves apart from it.

This appears to have genuinely angered Bateson.

"*No*," he said, his voice rising, "the scientist is *not* outside."

Perhaps they *thought* they were, once, Bateson said. But not anymore. The scientist—or at least the kind of scientist he now hoped to be—was part of the world, "as much as the artist."

William Bateson had taught his sons that the arts were "the great thing in the world but that no Bateson would ever be capable of contributing to them." Batesons knew their limits. "When Martin switched from science to drama," Bateson remembered, "his sin as WB saw it was of presumption that he, a Bateson, might contribute at that higher level."

Now, in April 1949, the ghost of Bateson's father had reappeared in an excruciatingly public setting. And Bateson, at long last, was fighting back. Deren was gone from his life. Martin Bateson was dead. But Bateson now carried on their legacy—his sense that, like an artist, his job was to see the world askance. To pull out beauty from it, even from the terrible parts. *Especially* those.

In the years to come, Bateson existed in that ambiguity. He spent months making a film about the nature of play among river otters—refusing to stop filming, day after day, even as weeks went by with the otters steadfastly refusing to play. And by the decade's end he had found his way into the most outlandish, perhaps the most artistic, corner of science available at the time: psychedelic science.

--------

A few days after Bateson's run-in with Wright, at the height of a New York spring in April 1949, Allen Ginsberg found himself in the backseat of a stolen car piled high with his own diaries. It was being driven in the wrong direction down a one-way street in Queens.

After the car crashed, the police used Ginsberg's diaries to trace the four occupants, who fled the scene, back to an apartment at 1401 York Avenue, near 74th Street. Allen Ginsberg's apartment. They found it to be full of drugs and stolen goods. Ginsberg claimed, according to one newspaper article at the time, that "he was a copy boy for a news service" who had joined a gang "to obtain 'realism' he needed to write a story."

A court didn't buy it. That May, Ginsberg was sent to Columbia University Hospital for psychiatric observation. He spent most of the rest of the year institutionalized, with a tentative diagnosis of "Schizophrenia—Pseudoneurotic type."

--------

Just before his arrest, Ginsberg remembered a vision that he had experienced the summer before. At first it was disturbing: an "apparition of an evil, sick, unconscious wild city." But then it resolved into something else.

"As I was walking down 125th Street, I suddenly stopped and stared around me in amazement," Ginsberg wrote. He was seeing something he had never noticed before, something that now inspired awe: the accumulated "intelligence and care" that had gone into molding the cornices and rooftops of the buildings in Harlem. All around him were these relics of the consciousness of people long dead. He had a sudden awareness of a "vast endless space reaching back into time and reaching forward into the future." And he felt a parallel awareness that he was also one of the "artifacts of that long evolution." It was a glimpse of something "longer than my own life, extending beyond my life and my former consciousness."

If Ginsberg had walked long enough down the streets of Manhattan and Harlem in April 1949, he might have passed L. Ron Hubbard and the science-fiction editor John Campbell, talking through the modified version of narcosynthesis that would become Dianetics and Scientology. Or Maya Deren in Greenwich Village, back from Haiti and looking skinnier, about to premiere a new experimental film called *A Meditation on Violence*. On Central Park West, he could have passed Margaret Mead carrying offprints of an article in which she argued that science would propel society into a "a new state" by revealing unforeseen "potentialities" in human nature.

All of them in the same time and place, all grappling with the same question. The question of potential: how to expand it, how to live up to it, how to harness it to save the world—or perhaps just themselves.

Werner Stoll, trying LSD for the first time just after World War II, had seen a city much like Manhattan, with towers rising upward above a bustling harbor. Yet each tower lacked a foundation. Ginsberg's vision was similar, but it was focused not only on buildings, but on *people* without foundations: unmoored, lonely, psychically sick, adrift in a vast city. The vision was at times sinister, but it unlocked something beautiful inside him. For a week afterward he felt "glimmerings, hints of possibility, secret amazements."

It was not until later that Ginsberg realized that his vision had been like an acid trip.

He had been a young man in those experimental days of 1949, unsure of himself, his future, and his decision to share an apartment with a Times Square hustler who worked on the side for Alfred Kinsey. But after his first experience with LSD, in a building just outside Stanford University in May 1959, Ginsberg felt something like certainty. He wrote a letter to his father, Louis, a high school teacher in New Jersey, and urged him to try the drug, too.

"It was astounding," Ginsberg said. "I lay back, listening to music, & went into a sort of trance state...and in a fantasy much like a Coleridge World of Kubla Khan saw a vision of that part of my consciousness which seemed to be permanent, transcendent, and identical with the origin of the universe—a sort of identity common to everything—but a clear & coherent sight of it."

The scientist who set up Allen Ginsberg with his first LSD trip was Gregory Bateson.

## Part II

# THE PROJECT

*(1949–59)*

The world is at an exceedingly acute stage—as crucial a phase of history as we've been in for the last 100,000 years. The moment you try to integrate the whole world—that's the most difficult time. This is the first period in history where we might destroy everything at once. The job is to get through the next half century.

—*Margaret Mead, quoted in the* Los Angeles Daily News, *May 4, 1950*

Note that I am tending to label any new or good ideas as associated with the [LSD] project.

—*Margaret Mead, "Preliminary LSD Memo," September 25, 1954* coming to the conclusion that I would try the drug, especially after reading Abrahmason's third paper, a case of removing fear of homosex. in one theapeutic 4 hour session.

September 22. Second LDS session, present as before plus two femela subjects, Schwartz - old subject, and Weltman ( one before 25). and John Eberhartd as an interested observer. ( Had been talking to Jarvik about it. I asked him). Both givgn Caligors before, then had dinner, and did Caligor's again with same O.l. drawing( this was duggested to me by Abrahamson after Sept.22, I called Caligor, he agreed).Then Caligor got Koretsky to to a TAT type interrogation on them. (This I havent seen yet) Saw the drawing before, though Schwartz, potentially sick, paranoid and stiff, Weltman probably a dancer, or interested in body rythms, and not sick. No comments from anyone.

During evening both girls around( I had dinner early with Abrahamson, hiw sec. who in a state of giggles from another drug, her husband and Korhetsky. Jarvik, the 2 subj cts Caligor and the Polish named chemist had dinner later. Orig. plan had been one placebo, but changed to be sure enough of interest, with Eerberhardt coming too, so we had one experienced subject, and one with second increased doe.

Schwartz - giggling uncontrollably and meaingingly, short bursts, no depth, reported sense of focussing on one parson then others would blendinto a circle, seemed farway(? Jarvik), yes, when in groups had felt this way, distantiated, not realy listening. Stiffness in face, made gestures of opening mouth, stretching face oddly. When questioned eyes would slew about. Reported drugs was going away, was back. Reported no affect to dug, had it so often. Questionned me about my trip, had been an anthro major, said difficult to concentrate, but did.

Weltman, gracefuldark little German Jewish girl, neurological seretary in a studying to me an opera singer with an Italian who feels everything starts with soul, never mentions body, soul must grow and voice come out. Wants to BE this, but wonders it its wrong to try to be a dead and gone culture, feelsteacher is right "soul is a little material thing for the voice to cling to, likes her job but feels that the people given shock aren't really psychotic, talks to them when acting as assistant and feels contact with them. Maybe the doctors wouldn't like this. Wants to be a singer, but would not sing under LST. Is concerned with keeping control, will keep it, brought up rigid German, her parents now know better, look what happened, she likes Eastern Eur pean Jews. (gestures beome more Jewish after excursions away in talking. Asked about Caligor, no ádisturbance of precision" I drew well, I never draw, isn't it a pity - we often confuse ends and means... I get so interested in other things during the way. (All this in an interview out in the hall, alone with me, had wanted to get out, felt nauseated( no fear of vomiting), stiffness in chest, ( she thought what everyone had boys said not), dryness ofthroat, stiff neck, misery, fear of laughing, had laughed a lot before..xxxxx

3 conversations with her, first in group people sitting around, she reported no distortion of vision, audirtory distortion, own voice too loud, people distant, harmonization of voices not attened to, sense of need for more control.

The second page of Margaret Mead's LSD Memo, September 23, 1954. *Mead Papers, Library of Congress.*

CHAPTER 10

# Lights Coming on in White Rooms (1949)

## Summer to Fall 1949

THEY LOOKED A BIT like tiny lightbulbs. The tops were thin tips of glass, the bottoms cylinders. Inside was liquid LSD. In 1949, thousands of these bottles, now labeled with the newly copyrighted brand name "Delysid," began rolling off the production line of the Sandoz chemical factory in Basel, Switzerland. The experimental compound known as lysergic acid diethylamide was on sale.

Though Sandoz-branded LSD was available only to licensed physicians engaged in psychiatric research, the drug traveled widely. Delysid bottles made their way to Poland, Austria, and Germany, and then to places farther afield. In Baghdad, Iraq, ten men and two women volunteered to try it. "The air is full of shimmer like a heat wave," one of the participants reported.

One of the earliest recorded uses of LSD in the Western Hemisphere occurred in Boston, at the Boston Psychopathic Hospital, where neuropsychiatrist Robert Hyde volunteered to be the first to test the drug on himself. At the time, Hyde and his colleague Max Rinkel, a Jewish refugee from Nazi Germany, were seeking effective therapies for their patients. For the most part, they were failing. One of Hyde's efforts included exploring the use of Chinese checkers and jigsaw puzzles as tools for breaking down the communication barrier; another involved developing a new notation system for transcribing rapid conversations between psychotic patients. Rinkel experimented with administering methamphetamine to counteract the sleep-inducing effects of drugs used in narcosynthesis, like sodium amytal. The tools at hand for treating their patients were simply inadequate. It was as if a cook had been put in a kitchen with only two ingredients—a tub of ice cubes and a pot of boiling water—and told to make a meal.

And so when a Viennese psychiatrist named Otto Kauders visited the hospital in May 1949 and told the psychiatric staff about a new drug called LSD, Rinkel and Hyde took an immediate interest.

Robert Hyde put in an order for several vials of the drug. And then he waited, until the day he was greeted by the arrival of a package from Basel. After his injection, Hyde seemed to forget that he had taken LSD at all. He insisted on making his usual rounds, accusing his colleagues of having given him a placebo. "That was not Dr. Hyde's normal behavior," Rinkel reported. Hyde and Rinkel and their colleagues began assembling a group of test subjects to further study the drug's effects.

--------

As it happened, one of the assistants at Boston Psychopathic was Arnold Meadow, the OSS officer who had collaborated with Mead, Bateson, and the hypnotist Milton Erickson in their study of how trance states could be used in wartime interrogations of Japanese and German prisoners. Meadow had left the OSS three years earlier and entered Harvard's PhD program in psychology. To gain some experience on the side, he began working as a research assistant at Boston Psychopathic—and, in particular, as an assistant in Rinkel and Hyde's new trials of LSD.

Throughout the summer of 1949, as the trials began in earnest, Mead was trying and failing to get in touch with Meadow. "What's happening to you? Why don't you answer letters?" she implored at one point. Mead did not hear back from her old colleague in the wartime hypnosis experiments until December 1949, some eight months after her original letters. Meadow apologized. He had been trying to finish his Harvard psychology thesis, he said. But progress had been slow because he had also taken a "stimulating" new job on the side—working with Hyde and Rinkel. He made plans to meet with Mead in the new year. It was at this meeting that he apparently first told her about LSD. In 1950, virtually no one outside the psychiatric and medical community had yet heard of the drug—but now, thanks to her connections, Margaret Mead had. True to her self-description as a "listening post," she had noticed that something was happening at Boston, something she wanted to learn more about.

But what *was* happening? In those early days, it was not clear to psychiatrists how substances like mescaline and LSD fit into their fluorescent-lit wards and their drug cabinets. Today, nearly one in five American adults has a prescription for a psychiatric drug. And although the United States ranks near the top of per capita psychiatric drug consumption, it is not alone. Internationally, psychiatric drug sales account for tens of billions in annual revenues. These are truly striking numbers considering the industry as we know it began in 1949. The reason for the enormous faith placed in this age of wonder drugs was simple: the postwar United States was witnessing an unprecedented surge in psychiatric hospitalization. By the end of 1949, over half a million Americans were living in mental hospitals, facing crowded conditions and utterly inadequate treatments.

At first, LSD followed the path of mescaline in Weimar Germany: it was studied as a substance for experimentally modeling hallucinations. But given the conditions of the world in 1949, it was inevitable that LSD would not stay bottled up in one category for long.

A photograph of Max Rinkel (*left*) and two successive drawings of him created by the painter Hyman Bloom, in 1954, under LSD. When presenting these images at a 1955 Macy conference, Rinkel described Bloom's LSD reaction as "mixed feelings of curiosity and hostility" and explained, "He wrote in the upper corner 'Hindu religion' " and then "fell into a trance like state." (Max Rinkel, "Experimentally Induced Psychoses in Man, in *Neuropharmacology: Transactions of the Second Conference* [New York: Josiah Macy, Jr. Foundation, 1956], 235–58.) *Courtesy of the Macy Foundation.*

--------

As Ido Hartogsohn writes, the earliest psychedelic researchers stood at an "interpretive crossroads" at the end of the 1940s. On one side was the perspective that these substances simply mimicked schizophrenia. This made them interesting and useful diagnostic tools but not significant medicines, let alone vehicles for social transformation. On the other side was the belief that LSD and other psychedelics could open a new world of experience that would help redirect global culture away from its pathological addictions to nationalism and violence.

For those who pursued the second path, exciting revelations were in store. The 1950s would prove to be a golden age for psychedelics, the decade that witnessed the birth of psychedelic therapy. A sense of optimism prevailed among drug researchers, who were buoyed by the introduction of drugs such as lithium and rauwolfia that were potential cures for disabling illnesses like depression and schizophrenia. Even Methedrine and Desoxyn—two brands of methamphetamine—were hailed as miracle drugs that could amplify the abilities of "normals."

As postwar wonder drugs became a big business, however, they also became a growing factor in military strategy. If narcosynthesis could "clear" the mind of trauma, could a modified version of the same technique be used to *induce* it? If a drug like LSD was useful for modeling psychosis as part of psychiatric research, might it also be useful for causing psychosis among an enemy army battalion—or an enemy city? Such questions were on the minds of drug researchers in 1949 for a simple reason: this was a key moment in the Cold War, the year that NATO was founded, the Soviet Union tested its first atomic bomb, and the Chinese Communist Party came to power. It was also the year that Margaret Mead became a consultant for the National Institute of Mental Health, where she began thinking deeply about how the new age of psychoanalysis and wonder drugs could reshape mass consciousness.

As she did so, she continued her collaborations with the U.S. military and the Department of State. Mead's leadership of the Research in Contemporary Cultures project at Columbia—which employed both a former Soviet spy and James Mysbergh, the ex–OSS operative who would soon become a career

CIA officer—placed her near the center of a unique intersection between the worlds of social science, espionage, and experimental pharmacology. It was a world she had moved in during World War II. Now she was returning to its heart.

One example was Mead's ties to Irving Janis, a policy analyst who had a small but important role to play in the early Cold War. During World War II, Janis had worked closely with Mead and Bateson. Afterward, Janis joined Project RAND—a new military-funded think tank that was later renamed the

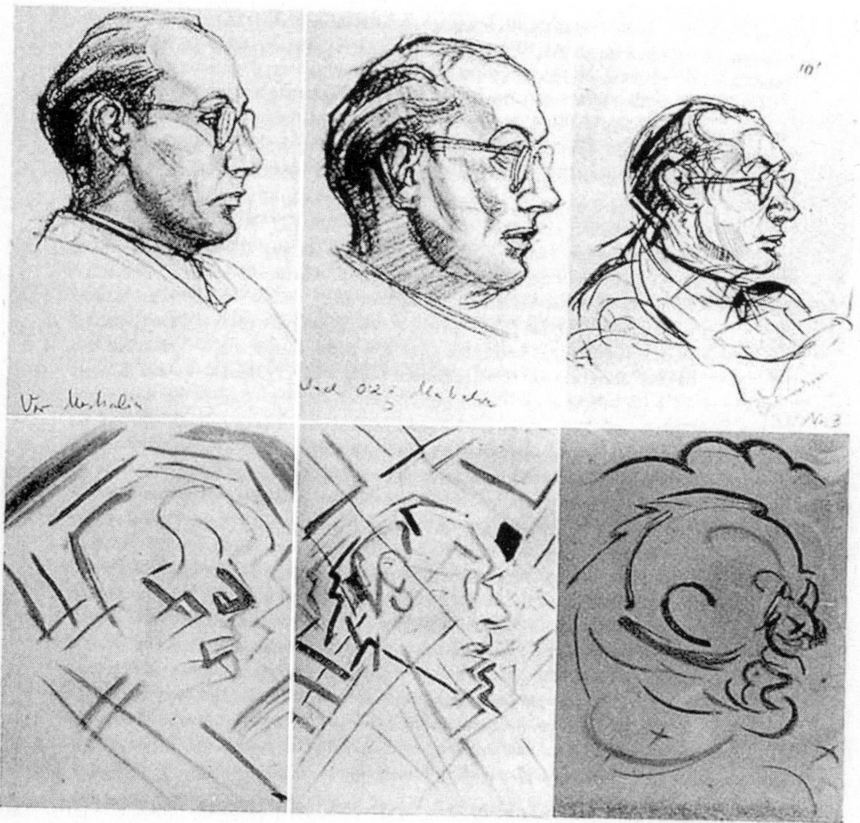

"The trip permeates me and makes me dissolve in it...I have a pronounced sense of impending death, but again, without the urge to fight against it or complain about it." Notes from an experiment with an anonymous artist who sketched a series of portraits under the influence of LSD and mescaline in January 1951. (Laszlo Mátéfi, "Mezcalin- und Lysergsäurediäthylamid-Rausch," *Confina Neurologica* 12 [1952], n. 3.) *Public domain.*

RAND Corporation—as an analyst of Soviet psychology. In April 1949, Janis issued a RAND study that helped trigger an entirely new concept, one intimately bound up with the concept of a weaponized science of consciousness: the idea of "brainwashing." The show trial of the Hungarian dissident Cardinal József Mindszenty, which took place in Hungary early in 1949, seemed to Janis decisive proof of the use of drugs in Soviet interrogations. "Somehow they took his soul apart," agreed intelligence adviser Paul Linebarger, noting the cardinal's glaze-eyed confession to misdeeds that were physically impossible. "Nobody is going to make me believe that logic or persuasion did this to Mindszenty. I think that this was quite literally *psychiatric warfare*."

Janis believed that such confessions had been the result of a "special psychological technique," which he conjectured involved "a series of electroshock convulsions" combined with drugs and "hypnotic suggestion." This combination, he argued, could potentially "induce a somnambulistic trance... in perhaps 90 percent or more of all defendants." Janis called for a parallel American program testing "the effectiveness of various drugs" for "weakening resistance." He warned that this research could "represent a serious threat to democratic values" and produce "unintended consequences." Nevertheless, he thought it needed to happen. In May 1949, Janis attended a secret meeting of scientists and military officers tasked with investigating the use of drugs and hypnosis in interrogations. According to an FBI observer, the group discussed such things as "the eradication of information from the conscious memory," "the extraction of classified information from an individual through subconscious isolation," and "the re-creation of personality and character traits" via psychological and chemical manipulation. "People had quite a lot of fears," one CIA official recalled of this time, citing the widespread belief that Soviet and Chinese scientists had perfected brainwashing techniques such as these.

In his report, Janis directly cited the hypnotism research of Milton Erickson to support his points—as well as themes and concepts that Janis had developed during his wartime work with Mead and Bateson (though, as that work remained classified, he was unable to cite it). While Janis was researching and writing his report, Mead was working alongside him as a RAND consultant on Soviet psychology.

--------

These debates about the social roles of science, the possibility of healing mental trauma with drugs, and the question of "reprogramming" the brain ran through many scientific circles in 1949. But they converged on a series of conferences sponsored by the Macy Foundation, reaching a kind of high-energy state that threw different scientists from different fields into exciting and sometimes bizarre contacts.

Today, only one series of Macy conferences from this era is widely remembered: the Macy cybernetics conferences, which appear frequently in the early pages of books about the origins of artificial intelligence research, neuroscience, and the digital era. The participants included figures who were directly responsible for creating some of the earliest programmable computers. They knew that the world was heading toward a revolutionary moment—the birth of the Information Age. Each meeting promised profound insights. In one, the emerging field of neuroscience was transformed by the presentation of the legendary paper "What the Frog's Eye Tells the Frog's Brain." In another, the information theorist (and ardent amateur juggler) Claude Shannon shared a new coinage: the word "bit," for measuring units of data.

And then there was Harold Abramson. On the surface, Abramson, a physician specializing in allergies at Mt. Sinai Hospital in Manhattan, was an impressive but unmemorable man. Indeed, one of his only published writings outside of academic journals is a letter to the editor of the *New York Times* explaining that the paper had gotten him mixed up with a *different* Dr. Harold Alexander Abramson, who happened also to be a New York City physician and Columbia alumnus. Beneath the surface, however, a roiling cauldron of bizarre stories about Abramson has been bubbling for over sixty years. Abramson was, no less than Mead, energized by the idea that science could expand human potential. Unlike Mead, Abramson seems to have had few qualms about weaponizing his research. During World War II, he had worked on methods for delivering chemical weapons in the form of aerosol gas. By June 1949, he was back at the same task, consulting for the U.S. Army Chemical Center at Edgewood Arsenal in Maryland, an elite and highly secret facility specializing in chemical and biological weapons research.

Given this record, it would be easy to see Abramson entirely as a sinister and amoral figure. But this was certainly not how he saw himself.

At the 1949 Macy conference on cybernetics, Abramson, who was usually soft-spoken, delivered an impassioned speech about the survival of the human species. Improved communication between scientists working in different fields, he said, was critical to the fate of life on earth: "Those who control the weapons of hostility and those who understand the psychomotive forces originating hostility," he said, "must meet on common ground as soon as possible... if man is not to destroy himself." Abramson was talking about nothing less than an entirely new model of science as a transformative social project. He concluded by saying that he believed his vision of applied science was "necessary for the achievement of world peace."

In the summer of 1949, while Mead was starting to plan the new Macy conference on altered states of consciousness, Abramson was undergoing an FBI loyalty investigation to clear him for continued work as a consultant at Edgewood Arsenal, a government research facility at the center of a highly classified initiative to test the effects of chemical and biological weapons on enlisted volunteers from the U.S. armed forces.

That August, L. Wilson Greene, Abramson's superior officer, authored a report on "psychochemical warfare." Abramson appears to have started to work with psychedelic drugs soon after.

CHAPTER 11

# The Psychedelic Cold War (1950–51)

Signs of desiccation, but very little pain.
I followed Raines and Kolb, in that dark backward,
Seeking a clue.

—*Weldon Kees, "The Clinic (To Gregory Bateson)," 1949*

Patients kept asleep longer than 48 hours will need special nursing care.

—*George N. Raines and Lawrence C. Kolb, "Combat Fatigue and War Neurosis (Concluded),"* United States Naval Medical Bulletin *(September 1943)*

## 1950

THE 1950S WERE A decade of contradiction, and so it is little wonder that the psychedelic science that decade fostered was contradictory from the start. In laboratories and clinics, in basements and bedrooms, the science of consciousness was becoming part of mass culture. But it was split between opposing goals. In 1950, the year that witnessed the outbreak of the Korean War, the human mind was reimagined as something *programmable*—by drugs, by hormones, by electricity, by words, and in ways both transcendent and oppressive.

Thus began the era of what might be called the psychedelic Cold War: a war fought not with bombs and bullets but with pills and wires, suggestions and commands. Some sought to heal their minds by using new techniques of mental health that involved taking charge of their own therapy with novel drugs. Others sought to manipulate minds by using those same substances without consent. As an anonymous CIA scientist put it, any techniques aimed at the "building up of a personality" could also be used to do just the opposite: "A personality could be changed, intense mental confusion could be produced by deliberately attacking an individual along psychological lines."

It was this power of science to manipulate that continued to haunt Gregory Bateson five years after his stint as a psychological warfare expert. As he wrote to the MIT scientist Norbert Wiener, science had become the weapon of "the advertiser who manipulates the public into buying some particular product; the colonial administrator who uses anthropological insights to maintain law and order in a native population; our friends in Rand who apply the theory of games to the strategies of war." Bateson repeated a version of this comment to Mead herself, in an angry letter that all but accused her of helping lead military-funded scientists into World War III.

"No doubt the Rand people are right and have 'useful' advice to offer on such subjects as intercepting hostile aircraft or bamboozling hostile diplomats... but still it stinks," he wrote bitterly. He began predicting that "15 to 30 percent" of humanity would die due to imminent nuclear or biological war.

By this time, old friends and colleagues of Bateson like Irving Janis—and to an extent, Mead herself—had taken up the challenge he set in his letter to General Donovan in the days after Hiroshima. In that letter, Bateson had effectively said that the new battlefield of the atomic era would be the mind. He had suggested that manipulative science, a science of *narrowed* consciousness, might be a weapon more powerful than any conventional army. He had watched as the idea of brainwashing emerged out of the inner policy sanctums of the Cold War to become a national preoccupation.

Bateson's prediction of a dark future of unconventional warfare, written in the aftermath of Hiroshima, appeared to be coming true. And with it, a troubling fact was again made apparent to him. He had helped to *create* the covert world of Cold War science. The thing he now loathed most was also a thing he had helped inspire.

## 1950–51: Control of the Mind

When, in their argument at the San Francisco Museum of Art, Frank Lloyd Wright had attacked science, Bateson partially conceded the point. The kind of applied science that Wright had in mind really *was* indefensible, he agreed. It was part of the machine, the circus with its spinning lights, the whirlpool that sucks humans into a Pavlovian system of addiction and reward. Bateson reacted to Wright, and to his own past, by pushing into theory, avoiding anything that could be misapplied.

Margaret Mead could not have been more different. By 1950, she had a clear channel to the uncanny core of power in the early Cold War, and she intended to use it.

One particularly vivid example of this was the CIA officer who lived in Mead's basement. He wasn't there on a permanent basis. But the room was set aside for Jimmy Mysbergh, the Javanese-born former OSS intelligence officer who had befriended Gregory Bateson in Burma, whenever he needed it. The room he occupied had in fact once been Bateson's own: in 1947, when Mead's husband moved out, Mysbergh moved in. And in the years that followed he became what Mead called "a member of the family."

After the end of World War II, Mysbergh had continued for a time as an intelligence officer working under official cover in the U.S. embassy in Kuala Lumpur. He left this position to work directly for Mead, who employed him along with several other ex–OSS spies as part of the Research in Contemporary Cultures project at Columbia University. Around the same time, in the spring of 1949, he began serving on a Council on Foreign Relations committee devoted to psychological warfare. One of his colleagues there was a fellow OSS officer, Allen Dulles, who became a senior leader of the CIA late in 1950. It is possible that this was around the time when Mysbergh, too, joined the Agency. Throughout the 1950s, Mysbergh would maintain his close friendship with Mead, even as he became, as their shared friend James Alexander Hamilton put it, "a deep cover CIA agent all over the world."

Mysbergh was operating on the front lines of the Cold War policy that came to be known as "containment": the global effort, played out in the shadows, that sought to prevent the spread of Communism. But as the historian Elaine May has argued, there were really *two* versions of containment at work in the United States of the 1950s. The Cold War was felt not just in foreign policy, but in the inner lives of individuals and families. In 1950s America, the domestic world of the home became a figurative (and in many cases a literal) bomb shelter: a refuge not just from all the familiar prewar stresses of work and public life, but also from the threat of World War III. This double meaning of "containment" defined Mead's life and work at the beginning of the 1950s. She was the only public figure of the era who claimed expertise both in the domestic containment of midcentury family life and in the shadow struggle of Cold War grand strategy.

Her dreams of the future had once been of a post-nationalist world state and a liberated, sexually adventurous society built on an interweaving of thousands of past and present cultures. But as with her detour to Washington during World War II, the onset of the Korean War put this utopian mission on hold. The Mead of 1950 and 1951 was more concerned with repurposing science to defeat the Reds.

One of her allies in this goal was Harold Abramson, the allergist turned chemical weapons researcher, who was Mead's invited guest at the 1950 Macy cybernetics conference. Abramson's life was outwardly conventional, a model of midcentury domesticity. He and his wife, Virginia "Beaver" Abramson, had four children. They collected Japanese netsuke carvings, carefully cultivated the lawn of their palatial home in suburban Long Island, and played bridge with their neighbors once a week. But a case could be made that Harold Abramson was the single most influential—and infamous—psychedelic researcher of the twentieth century. He came to see the drug as a potential pathway to a union of the sciences, and as a powerful new tool for dissolving psychic trauma. Yet today he is mostly known for violating medical ethics as a drug researcher for the CIA.

Before Bateson moved to California in 1948, he had bought his daughter a present to remember him by: a small aquarium containing two Siamese fighting fish. He said it would teach her about feedback loops of aggression and the difficulty of maintaining homeostasis, or order within a closed system. "When two males meet," Bateson told his daughter as they walked through a pet store in lower Manhattan, "they engage in competitive display that rapidly escalates into conflict; one threatens and this stimulates the other to a further exaggerated threat." This was an interest that Bateson and Abramson shared. By 1950, Abramson was tending to dozens of Siamese fighting fish at his laboratory at Cold Spring Harbor. Once, Mary Catherine remembered her father giving the fish "a tonic, a strange smelly medicine" that made them "swim around as if they were drunk."

Abramson, too, poured medicine into his Siamese fish tanks. But the parallels ended there. Bateson bought his fish tonic at a pet store. Abramson's Siamese fighting fish were being dosed with LSD.

--------

At the spring 1951 Macy conference on Problems of Consciousness, Abramson mentioned LSD in public for the first time. In the process, he strongly implied that he had begun taking the drug himself.

"Does anyone know about the diethyl amide of lysergic acid and its effects in producing experimental psychoses?" Abramson asked abruptly, changing the subject from a discussion Mead was having with Roy Grinker, the narcosynthesis pioneer. He described it as a substance producing hallucinations of "sprinkling, sparkling, glittering spark whirls," "green and red fogs," and "arcs and colored circles" alongside a "loosened up" thought process. Abramson then returned to the case Mead was discussing—a young woman diagnosed as schizophrenic in her twenties. LSD, he said, could have allowed him to detect her illness before it manifested, by bringing the latent psychosis to the surface. Mead seemed to disagree. "To isolate one drug," she said, wasn't enough. It had to be part of a larger pattern of scientific inference.

But at this point in her relationship with Abramson, Mead was seeing only the visible part of an iceberg. By the spring of 1951, Abramson was already employed as a consultant for the U.S. military at a secret chemical weapons lab at Edgewood Arsenal in Maryland; soon afterward, he began consulting as well for the CIA, on a new top-secret project known as Project Artichoke.

The Macy circle's deepening CIA ties came at a time when the United States was facing multiple challenges and conflicts at home and abroad. The Cold War with the Soviet Union was intensifying, with the "hot" conflict of the Korean War entering its second year. President Truman was about to fire the man in command of American forces in Korea, General MacArthur, for insubordination, and intelligence services around the world were expanding rapidly in preparation for the onset of a global conflict that now seemed inevitable following China's entry into the Korean War. Perhaps, then, it did not seem strange to Margaret Mead that so many people close to her were forming direct ties to the Central Intelligence Agency. Fighting Communism during the height of the Korean War, before the worst abuses of McCarthyism but after the revelations of Stalin's Great Purge, was still a largely bipartisan affair that united people across a wide range of the political spectrum. Nor did the CIA, which had existed for only four years in 1951, have the negative public reputation it would earn later in the Cold War. Instead, many scientists considered collaboration with the CIA to be a patriotic duty and a professional honor.

Still, though, it is not much of an exaggeration to say that Mead, in 1951, was positively *surrounded* by spooks. There was James Mysbergh, one of her closest friends. There was Harold Abramson. And there was James Alexander Hamilton, another of Bateson's wartime colleagues, with whom Mead was now fast friends. For by 1951, Hamilton, too, had become a CIA consultant.

--------

Gregory Bateson did not follow his OSS colleagues back into war. Instead, he busied himself with an eccentric series of collaborations with his Langley Porter colleague, the poet and painter Weldon Kees.

With his small frame, elegant sense of dress, and tendency to roam city streets in search of adventure, Kees resembled a human version of Stuart Little. Like Bateson, he tried to wear his ambition lightly. Kees's diary from this period records a life of martini lunches and aimless afternoons in bookstores and jazz clubs. In fact, though, Kees was one of the most polymathic American artists of the 1950s. He directed films and theater. He wrote and sold pop songs with titles like "Anaïs Nin Stomp" and "Television Baby," practicing the piano for hours each day. His poetry was published regularly in the *New Yorker*, his criticism in *The Nation*. And his paintings were remarkable. Kees's murky, mysterious abstractions—which bore a distinct resemblance to New Guinea tribal art—hung in galleries alongside works by Jackson Pollock and Willem de Kooning.

Weldon and Ann Kees at their home in Richmond, California, in 1952. One of Kees's paintings hangs on the wall. Beside it is a tribal mask that may have been a gift of Gregory Bateson, who owned a large collection of similar objects. © *Dan Wynn Archive and Farmani Group, Co LTD.*

Kees was amassing a body of work that had the potential to make him a household name... and likely would have, if he had learned to stop burning so many bridges. Just before he left New York City, *Life* magazine ran a group photo shoot of a dozen or so abstract painters who came to be known as "The Irascibles"—including Pollock, de Kooning, and Mark Rothko. The group had originally included Kees, but he alienated too many with his mordant wit. Kees did not appear in the photos.

It was Kees who became Bateson's closest partner in his new life in San

Francisco. They first met in 1950, when Kees was hired as a filmographer for a Langley Porter research project about communication that Bateson was conducting alongside an eccentric Freudian psychiatrist named Jurgen Ruesch ("that loveable exponent of Viennese charm and a phallic symbol in every home," as Kees described him). Bateson and Kees, both recent transplants from New York City, soon found that they shared a darkly comic approach to life. Kees's diaries record a friendship between the artist and the anthropologist that had deepened into a close bond by early 1951:

> **Friday, Feb 2, 1951:** To Bateson's 11:00 seminar: cybernetics. With him to lunch at Roselli's. Looking at film we shot.
> **Friday, Feb 16:** Ate with Greg [Bateson] at Simpson's. In the afternoon, looking at films.
> **Friday, Feb 23:** To Gregory's seminar: on the configurating influence of 2 people in a dominance-dependence situation.
> **Friday, March 2, 1951:** With Gregory to Italian restaurant. Picked up his wife at Susan West's place, puppy along with us.

Bateson had married Betty Sumner, his second wife, at the beginning of the year. She gave birth to their son John at the end of April. That same month, Kees visited the Batesons at home in Menlo Park to watch "Gregory's Nazi picture." Surrounded by his new family and his West Coast friends, Bateson softened his English accent into something more American. He began inviting over neighbors for backyard garden parties, trying out new recipes for salad dressings and barbecue sauce. He bought a television and a car (a military-grade jeep that he was barely able to drive). He was becoming a 1950s Californian.

--------

But he was not becoming conventional. One sign of this was Bateson's surprising endorsement of the work of a man who had, throughout 1950 and 1951, made it his business to tear down the psychiatric establishment and convince the world that he was a genius. L. Ron Hubbard also claimed to be a nuclear physicist, an explorer, a happily married father, and a decorated World War II naval captain. None of these claims were even remotely true. In reality, L. Ron Hubbard was a science-fiction writer. He typed his stories in a closet, with a single red lightbulb for illumination, on giant scrolls of butcher paper that he tore off with a straight-edge ruler. His mailing address was in suburban New Jersey (PO Box 666). He prided himself on being able to write well over ten thousand words in a single day. Perhaps unsurprisingly, he was also a longtime user of amphetamines.

In a 1950 book, Bateson cautiously commended L. Ron Hubbard's Dianetics, a so-called science of the mind that sought to replace traditional psychiatry. Though the two never met, Hubbard's disdain for mainstream psychiatry and society apparently resonated with Bateson. It may also have helped that throughout 1950, Hubbard had been doing his best to ingratiate himself into the Macy circle, befriending Claude Shannon and attempting to attend the Macy cybernetics conferences in person. Hubbard exemplified the era of DIY pharmacology in the early 1950s, a time when expertise was highly valued, but also a time when expertise could be faked more easily than ever before. Figures like Hubbard were able to bypass traditional academic publishing by publicizing their work in mass-market magazines, capturing the attention of audiences worldwide. As a self-proclaimed mental health expert who claimed he had developed a unique combination of self-hypnosis and narcosynthesis therapy, Hubbard tapped into the growing disillusionment with mainstream psychiatry, offering an alternative vision that promised adherents they could find inner peace—"going clear"—in the comfort of their own homes. And Hubbard's messaging about the dangers of a technique like his in the wrong hands (he darkly hinted that the Soviets studied his methods) was just as perfectly tuned to the mood of the age. He embodied a potent mixture of dread and technological optimism regarding the power of science to reprogram the human mind.

Hubbard and his ideas formed a shaky, treacherous rope bridge between the two sides of the Macy circle, the techno-pessimist and the techno-utopian. Like Bateson, he had a truly jaundiced view of psychiatry and of mainstream science and a well-developed fear of the Cold War military-industrial complex.

But like Mead, he argued that anthropology and cybernetics could be welded together to forge a transformative "science of survival," and that drugs had a part to play in it.

Hubbard embodied many of Mead's worst qualities, offering up a particularly simplistic and exaggerated form of techno-utopianism in his "shortcut" to mental health through drugs and hypnosis. In his distorted popularization of the Macy circle's thinking, he showed how potent the allure of a new science of expanded consciousness could be in the era of the early Cold War—and how dangerously cultlike it could become.

CHAPTER 12

# "Narcosynthesis Is Not Very Difficult" (1951–52)

## Summer 1951 to Spring 1952

THE UNITED STATES AFTER World War II has been called "the era of the expert." Experts—and certainly Margaret Mead, typing serenely behind her huge desk in her turret at the American Museum of Natural History, deserved that title—formed a protective bulwark around an increasingly precarious society. For this reason, there was something almost mystical about the cultural role of elite scientists as the 1950s dawned. The men and women who attended the Macy conferences or who gamed out atomic warfare scenarios in Washington were becoming high priests of a technological sacrament. It was they who, in the words of another study of Cold War culture, "took over the role of psychic healer" for an era in which science seemed to have swept away all other modes of explanation.

But how was expertise doled out? Who deserved the title? Who enforced it? The truth was that the era of the expert was also an age when it was becoming increasingly difficult to distinguish between knowledge and nonsense.

And it was not just public figures like L. Ron Hubbard who capitalized on these openings. In the halls of the Central Intelligence Agency, from about 1951 onward, a group of intelligence officers began conducting their own amateurish experiments with consciousness-expanding drugs, testing them on themselves and on poorly informed "volunteers" from a host of settings, including hospitals, military bases, and psychiatrists' offices. They were not experts in drug safety (far from it), and the physicians they hired to assist them held subordinate positions within the hierarchy. The CIA was the citadel of Cold War expertise par excellence. But the truth is that much of the Agency's work with psychedelics was a toxic mishmash of amateurism, unchecked megalomania, and simple incompetence.

The entangled stories of L. Ron Hubbard's Dianetics and the early psychedelic testing of the CIA reveal the limits of scientific idealism in a world where the lines between expertise, opportunism, and manipulation were increasingly blurred.

--------

"The goal of Dianetics," L. Ron Hubbard wrote in 1951, "is a sane world—a world without insanity, without criminals and without war." It was a "workable technology" built on "thousands of years" of human knowledge. This rhetoric was not the only thing he borrowed from the Macy circle. Instead of Milton Erickson and Lawrence Kubie's "hypnagogic reverie," Hubbard put his patients into "Dianetic reverie." In an amateurish attempt to co-opt the cybernetics theories of the Macy circle, he likened the human mind to a computer that could store and retrieve data.

And, to a surprising degree, L. Ron Hubbard positioned himself as an ally to Boasian anthropologists like Margaret Mead and Gregory Bateson.

"Psychology—no, I'm afraid we don't have anything to do with psychology," Hubbard once said in an interview, when asked about the ideas that influenced Dianetics. Instead, he remembered, "I had been studying ethnology and anthropology." Hubbard's closest collaborator on Dianetics was his editor at the pulp science-fiction magazine *Astounding*: John W. Campbell. A strong hint about the anthropologists Hubbard and Campbell were reading at the time came in *Astounding*'s December 1950 issue. Alongside a flurry of commentary from readers about Dianetics ("It sounds very much like the kind of talk I hear from some of the patients in the mental hospital where I work"), Campbell ran a long article on the research of Margaret Mead, Reo Fortune, Ruth Benedict, and Gregory Bateson. It is safe to guess that when, in his Dianetics lectures, Hubbard referenced such things as the "absence of jealousy among the Polynesians" and his "hobby" of reading ethnology, he had Mead in mind.

And he had also been doing drugs. Early Dianetics supporters spoke of it quite openly as a modified narcosynthesis. John Campbell told Robert Heinlein that Dianetics improved on "the sodium pentothal system of clearing battle neuroses." In June 1950, Hubbard told an audience that "working a patient with narcosynthesis is not very difficult when the narcosynthesis works," but went on to argue that his hypnotic techniques (and his drug regimen) were both superior.

L. Ron Hubbard (*foreground*) conducting a Dianetics "audit" of a woman in Los Angeles in September 1950. *Courtesy* Los Angeles Times *photographic archive, Library Special Collections, Charles E. Young Research Library, UCLA.*

In an October 1950 lecture, Hubbard argued that drugs were valuable as shortcuts. They could reduce by 75 percent the time it took to "clear" a traumatic memory, or "engram." Hubbard said he was searching for "the one-shot clear, whereby a person walks in, you take a hypodermic syringe and shoot it in his arm, he goes up against the ceiling and comes down clear." He even mentioned working with "a chemical engineer" at a drug company on "an atropine derivative" that might fit the bill. But for the time being, he had settled on a substance he called GUK, a blend of vitamins, an amino acid, and amphetamines.

Hubbard was a tuning fork for the era, resonating back the frequencies of a newly experimental, pharmacologically adventurous postwar culture. Dianetics was not just influenced by the Macy circle. It also anticipated many of its failings, notably a persistent tendency to understate risks and overstate the potential benefits of radical behavioral change driven by technology. Dianetics, in other words, was both an offspring and an omen of its time: an age of experts in which no one seemed to be able to agree on whose expertise to follow.

## Summer and Fall 1952

The secret, CIA-sponsored drug experimentation program known as MKULTRA was launched in 1952 (the name may have been inspired by the codebreaking project "Ultra," overseen by Alan Turing in World War II). It resembled Dianetics in two important respects. First, it, too, was an outgrowth of the postwar culture of free-form drug experimentation, when motivated amateurs felt inspired to claim expertise in a field so new that no one really knew what the limits were, ethical and otherwise. And second, it was, like L. Ron Hubbard's teachings, a by-product of the applied science of consciousness advocated by the Macy circle.

The origins of MKULTRA lay in a lingering fascination with hypnosis among several mid-level CIA officers. One of them, Morse Allen, met in 1952 with none other than Milton Erickson, the hypnotist friend and collaborator of Bateson and Mead. Erickson was a skeptic regarding the idea of a "Manchurian candidate" scenario, and after a series of bizarre hypnosis experiments at CIA headquarters—including one in which it was claimed that a hypnotized secretary was nearly induced to commit murder—Allen gave up on the idea. But as with the OSS research, the CIA program had two strands. One was devoted to hypnosis. The other, to hypnosis-inducing *chemicals*. This was MKULTRA, one of the most notorious intelligence programs in the history of the Cold War.

A young chemist named Sidney Gottlieb, newly hired by the Agency, became a leader of the second research project, and he dove into it with great enthusiasm. The son of Hungarian Jewish immigrants, born in the Bronx with a clubfoot that did not stop him from pursuing a lifelong passion for folk dancing, Gottlieb was handsome, charming, highly intelligent, and driven by the conviction that, in the fight against Communism, the ends justified almost any means. Starting in 1951, when he was hired by the CIA at the age of thirty-three, Gottlieb began to oversee a covert program of experiments in hypnosis, drug interrogations, and other forms of mind control. One of Gottlieb's first assignments in the fall of 1951 was to dig through the CIA's files relating to the 1943 "truth drug" experiments. On orders from above, he then devised a plan to bring back George Hunter White to reprise the experiments he had done with THC-laced cigarettes on unwitting New Yorkers. Only this time, White would be tasked with giving his subjects LSD, and recording the results.

An internal CIA report on MKULTRA summarized the project as a "comprehensive" hunt for "biochemical controls of human behavior." This included "the search for and procurement of botanical and chemical substances," their analysis in laboratories, and a real-world testing program that "progresses gradually to more and more realistic operational simulations." In addition to drugs, the project had a wide latitude to study and implement "additional avenues to the control of human behavior" ranging from radiation to electroshock therapy to techniques from "psychology, psychiatry, sociology and anthropology." Although this work was potentially "distasteful and unethical," the report argued, it was also a field in which "opposition intelligence services are active and highly proficient," and therefore defensible—within reason.

Harold Abramson was among the first researchers enlisted in the program, part of a group of four that also included the first American to take LSD, Robert Hyde of Boston Psychopathic. Explaining the CIA's outreach to expert researchers, the report stated that the "key individuals" possessed top-secret clearance and an awareness of the CIA's role; "the system in effect 'buys a piece' of a specialist in order to enlist his aid in persuing [*sic*] the intelligence implications."

Gottlieb recalled being asked to set up a meeting with George Hunter White, the former OSS operative who had been tasked with investigating "truth drugs" during World War II, in January or February 1952. But it was not until June 9, 1952, that White and Gottlieb met for the first time. "I drove up to Boston with him," Gottlieb remembered. "[White] was working on an interesting narcotics case around there that he thought I might enjoy experiencing up close, as they say. Riding up together, we had a chance to really discuss matters of interest to OSO [Office of Special Operations]. Most of what we talked about was his work with the Division 19 truth drug program and other, related matters... George... was always armed to the teeth with all sorts of weapons; he could be gruff and loutish, vulgar even, but then turn urbane to a point of eloquence."

When Gottlieb asked White if he would like to be a consultant for the CIA, White immediately said yes. It is unclear when White's old partner James Hamilton joined the project as well, but it may have been a package deal: by this time White and Hamilton had become fast friends, celebrating birthdays and wedding anniversaries together with their wives. White was elated to join the Agency. He had hoped for the opportunity for years, and he nurtured a simmering resentment against those he assumed had kept him out: elitist Ivy Leaguers lacking street smarts. Now he would show them how the job was done.

But all was not well in White's personal life. The following morning, he received a distressing phone call from his wife, Albertine. Their beloved pet parrot had died. "Poor little bastard just couldn't make it—tried hard," White wrote in his diary. "I don't know if I'll ever get another bird—or pet. It's tough on everyone when they die."

The next day, White began working with the Central Intelligence Agency. Ten days after that, he drove to Queens and bought a new parrot.

--------

White's job was to perform what the CIA called "realistic testing" of a scenario that haunted the imagination of many Agency officials at the time. How would undercover operatives behind the Iron Curtain react if Soviet spies spiked their food or drink with a tasteless, odorless, and incredibly potent drug like LSD? Could they be manipulated into becoming double agents?

There was also an offensive component. If psychedelics like LSD really *did* function as "truth drugs," the CIA was interested in doing the same thing to Soviet spies. As Sidney Gottlieb put it, "Both interrogation and provoking erratic behavior were of interest."

In October 1952, on the day before Halloween, White visited CIA headquarters to meet with Gottlieb. He also reconnected over lunch with an OSS colleague who had risen to become the Agency's head of counterintelligence, the orchid-collecting, poetry-writing, alcoholic arch-spy James Jesus Angleton. White was truly in the inner sanctum now. In the weeks afterward, White began experimentally dosing friends and acquaintances with LSD—without telling them. White's cryptic diary entry about a "delayed reaction" following one evening with Albertine and four of their friends was later fleshed out by the recollections of one of those friends, a man named Gil Fox. "We were all boozing and smoking pot in those days, including George," Fox recalled, "and one night George gave us LSD. He slipped it to us secretly... Afterwards we went slumming around the Lower Village. It was snowing. We stopped the car on Cornelius Street and the snow was red and green and blue—a thousand beautiful colors—and we were dancing in the street... Then we went into a lesbian bar."

In hindsight, Gil was angry at the violation. "It turned out to be a bad thing to do to people," he concluded, "but we didn't realize it at the time."

The closing months of 1952 bore witness to the palpable paranoia that defined the apex of the Cold War. They were also evidence that sometimes, that paranoia was justified. George Hunter White's safe house, where he performed secretive experiments on unwitting subjects with the assistance of the psychiatrist James Alexander Hamilton, was a real-world scenario that seemed to be ripped straight from Hubbard's fantasies. And it was a scenario that centered on two men whom Gregory Bateson knew well. In the year that followed, the links between the covert side of Cold War drug experimentation and Mead and Bateson's circle only deepened, drawing the Macy Foundation itself into the heart of the CIA's mind control program.

The very *term* "psychedelic," coined a few years later, encapsulated the tension between the promise and the peril of these substances. Literally, it means "revealing the mind." Humphry Osmond, the civilian scientist who invented the word, assumed that this was an inherently benevolent goal. But in 1953, key members of the Macy circle were learning that the project of revealing the mind resembled the ancient Roman god Janus. Janus was the god of transitions, passageways, new ideas, and new historical epochs. And his two faces looked out on both war and peace.

CHAPTER 13

# The Telephone at the End of the World (1952–53)

> We were literally terrified, because [LSD] was the one material that we had ever been able to locate that really had potential fantastic possibilities if used wrongly.
>
> —*testimony of an anonymous CIA officer recalling the origins of the MKULTRA program in 1952*

## Fall 1952 to Spring 1953

FREEZING WINDS HOWLED ACROSS the battlefield, biting at the exposed skin of Lieutenant A. Robert Abboud. Then he spotted him—the soldier lying motionless in the no-man's-land between the American and North Korean front lines. Abboud skidded to a halt beside the young officer, who had just been struck by shrapnel. The sight that greeted him left Abboud doubtful that anyone could survive such an injury. A portion of the man's brain was now cradled within his ruined helmet. Abboud heaved the grievously wounded man onto his shoulder, ignoring the weight bearing down on him as he raced back toward American lines. The medics there were struck not only by the severity of his injury, but by a sudden realization about the identity of the victim. It dawned on them, as the wounded soldier's name was repeated, that this was no ordinary casualty of the Korean War. This was *Allen*

*Dulles Jr.*—the nephew of the incoming secretary of state, and the only son of the man who was about to become the head of the CIA.

The date was November 15, 1952, less than two weeks after Dwight D. Eisenhower's election. Those two weeks had been a period of professional triumph for Allen Dulles Sr., deputy director of the CIA, to whom Eisenhower had promised an appointment to the top job at the Agency in the new year. And it was in this moment of triumph that the news of his son's devastating brain injury reached Dulles and his wife, Clover.

Dulles worked as a diplomat and lawyer before joining the Office of Strategic Services in 1941. He had been fascinated by the science of the mind for almost as long as he had been a spy. As an OSS operative in Zurich, Switzerland, during the height of World War II, he had formed an unlikely bond with that city's most celebrated citizen: the psychiatrist and psychoanalyst Dr. Carl Jung, great friend and rival of Sigmund Freud. Though Jung was approaching seventy, Dulles managed to enlist him as an OSS agent. "Nobody will probably ever know how much Professor Jung contributed to the Allied cause during the war," he later claimed, citing Jung's "highly classified" work, which involved constructing psychological profiles of Nazi leaders. By November 1952, then, Dulles had already been thinking for a decade about what he called "brain warfare." But now he was not just thinking about the science of consciousness as a professional tool—he was desperately seeking to enlist it to save his son.

For help, Allen and Clover Dulles turned to a respected neurologist named Dr. Harold Wolff. Clover came to look on Dr. Wolff as an almost godlike figure, a genius capable of miraculous cures. This was the age of experts, after all, and Wolff's research had given him nearly unmatched insight into the workings of consciousness. In the end, Wolff made little progress. Allen Dulles Jr. had permanently lost his short-term memory. He had gained a tendency to concoct paranoid delusions (such as the belief that his father was a Nazi double agent), which later led to an unusual diagnosis of injury-induced schizophrenia. What Harold Wolff *did* achieve was a lasting and lucrative bond with a man who, following his formal appointment as director of the CIA in 1953, numbered among the most powerful people on earth—a man who believed the Cold War was being fought in the fields of consciousness itself.

In the years to come, the Dulleses' personal tragedy would ripple outward in unexpected ways, intertwining their fate and that of the doctor they looked to for salvation with the destinies of Margaret Mead, Gregory Bateson, and the larger Macy circle.

What Daniel Immerwahr has called the "hidden empire" of the United States began before World War II, but it was in the early 1950s that it truly became global in scope. From islands in the Pacific to secret bases in Greenland, and from New Guinea airstrips to the modified German V-2 rockets that American engineers were now shooting into space from New Mexico, the United States was moving aggressively into a transformed world. And this expansion was not limited to geographical territories; America's postwar empire also infiltrated the realm of ideas. Throughout 1952 and 1953, the CIA extended its covert research funding into a range of scientific activities, from the work of legendary psychologists like B. F. Skinner to esoteric chemistry and ethnobotanical projects. As a result, the lines between science and espionage blurred, connecting researchers and the clandestine operations that sought to exploit their work for the advancement of American interests during the Cold War.

--------

It appears that the CIA began channeling covert funding to Harold Wolff not long after he began his treatments with Allen Jr. He soon pushed his research away from migraine headaches and toward new work on the consciousness-altering effects of LSD, mescaline, and issues relating to the newly coined topic of "brainwashing." Harold Wolff was, after all, an old hand in such matters. As far back as 1942, he had attended the Macy conference on Cerebral Inhibition (a "respectable word for hypnosis," as Bateson put it). He was also an old friend of Margaret Mead's. Wolff was a prickly character, a man invariably described as "intense" by subordinates and colleagues alike. But he seemed to have liked Mead quite a bit. They often spoke about aspects of Cold War science, from their mentorship of Mark Zborowski, a Soviet spy turned New York–based anthropologist and pain researcher they had both taken under their wing, to their shared work on a research group funded by the National Institute of Mental Health (NIMH).

Meanwhile, other key members of the Macy circle were making their own inroads with the Agency. In the final months of 1952, Harold Abramson—the wartime chemical weapons researcher and peacetime specialist in allergies and asthma—was hard at work preparing for the upcoming Macy conference on the "Problems of Consciousness," whose proceedings he edited with Margaret Mead. In the same period, he was first contacted by Sidney Gottlieb, the chief of the CIA's new "chemical division." By October 1952, Gottlieb had enlisted Abramson's assistance in what would become one of the original MKULTRA subprojects: George Hunter White's real-world testing of LSD and other drugs.

To a surprising degree, the MKULTRA project covered the same ground as the "Problems of Consciousness" conferences. MKULTRA was a profoundly ambitious attempt to learn how to alter human consciousness in ways that had implications not just for individuals, but for the fates of nations. In this respect, it mirrored Mead's long-standing dream of a science of expanded consciousness that would reshape human society—but with the important distinction that it intended to use this science as a *weapon*, not as a peaceful tool of what Mead called "cultural evolution."

--------

It is important to clarify the nature of the CIA's contacts with these scientists. The CIA funded an eclectic range of projects during the Cold War's height, and those who received this funding typically had no idea of its true source. In addition to researching psychedelics, for instance, the Agency anonymously bankrolled the launch of the *Paris Review* literary journal (in 1953), sponsored an international touring exhibit of paintings by leading abstract artists (1950), and printed thousands of copies of the novel *Doctor Zhivago* (1958).

Yet the CIA's role in midcentury psychedelic research was something altogether different. The Macy Foundation became what intelligence professionals call a "cut-out": a public-facing organization that serves as a conduit for covert money, information, and expertise. By early 1953, serious money was flowing directly from the CIA to the Macy Foundation and onward to psychedelic researchers. In that year alone, Sidney Gottlieb used the Macy Foundation as a cut-out to funnel at least $85,000 ($973,000 in 2023 dollars) toward

Abramson's research on the effects of LSD on snails, Siamese fighting fish, and his human psychiatric patients—not to mention himself. Macy Foundation research director Frank Fremont-Smith (an expert in the central nervous system who had been collaborating with Harold Wolff since the 1920s) appears to have been the primary middleman linking the CIA to the foundation's projects. Fremont-Smith took a personal interest in LSD, using it himself and collaborating with Harold Abramson at his laboratory in Cold Spring Harbor, New York. Meanwhile, members of the CIA's Technical Services Staff began quietly attending Macy Foundation conferences and taking careful notes.

It is unclear how much Mead knew about the early preparations for MKULTRA. But given the fact of her long-standing relationships with men like Wolff and Fremont-Smith—not to mention her familial closeness with James Mysbergh, the undercover CIA operative who spent months at a time living with Mead and her daughter, Cathy, in their Greenwich Village apartment—it is highly likely that some word of it reached her.

By the spring of 1953, moreover, it was not just her professional and social life but her personal geography that the MKULTRA project was beginning to push on. George Hunter White had set up shop just five blocks south of that very apartment. Posing as a louche merchant seaman named Morgan Hall, White rented a unit in a brownstone on Bedford Street in Greenwich Village. An aesthete with a taste for East Asian antiques, he spent considerable time perfecting the interior design of what he began calling his "pad." After this, he opened it up to a CIA security team that installed hidden recording equipment and a two-way mirror. And then, with occasional assistance from James Hamilton and White's mysterious sidekick, a petty criminal and police informant named Pierre Lafitte, White began his project of "real-world" testing of psychedelic drugs on civilians. A CIA memo described White's modus operandi. "For one reason or another," the memo stated, "he is able to get these people to his apartment at 81 Bedford Street." White then "gives drugs to these unaware individuals through drinks, cigarettes, and perhaps food," after which he "tries to elicit information."

History was truly repeating itself. Almost exactly ten years earlier, Hamilton and White had been tasked by the OSS with conducting real-world trials of THC cigarettes on unsuspecting New Yorkers, then recording them to determine whether the narcotic functioned as a "truth drug." The substance had changed, but the team and the motive were almost exactly the same.

--------

It was against this backdrop that Mead decided to travel about as far as it was possible to get from New York City.

James Mysbergh was living in Burma during this period, tracking Communist ideologies among the Burmese elites, and it was with him that she shared her plans for a return to the field and an escape from the Cold War. In the final months of 1952, he and Mead exchanged notes about what they called the "communist utopia" being planned there. Mead found utopias fascinating—after all, at some level she hoped to create one herself. And though she was no Communist, she was intrigued by what Mysbergh reported. The Communists, he said, did not just dream of a revolution in the material circumstances of life. They wanted a revolution in *consciousness*—a fundamental reshaping of what it felt like to be human. Mead began to wonder about the places she had studied in the 1920s and early 1930s, places like Bali and New Guinea, which were now the sites of some of the most rapid cultural and technological change in recorded history due to the massive upheavals caused by World War II.

Was the revolution in consciousness already taking place? After three years of disappointment, Mead hungered to find something positive about recent history: some sign that rapid change could *enrich* human potential, not just diminish it. She had spent 1952 watching as the Cold War grew hotter, and her own ability to shape events toward a hopeful, utopian course faded. And she responded by choosing to spend six months on an island off the coast of New Guinea that was thousands of miles away from her closest friends and family (including Cathy, then twelve, who stayed behind in New York), an island that numbered among the world's most remote places.

Her plan was to return to the village on the island of Manus where she had lived with Reo Fortune in 1928. At the time, Manus had been much like the other islands in the archipelagos northeast of New Guinea—isolated societies reliant on age-old practices of spearfishing and small-scale agriculture. But that old social order was said to have transformed during World War II, when the world's most advanced militaries suddenly converged on the strategically located Admiralty Islands and the neighboring Solomon Islands chain. Over a million American GIs passed through this region between 1943 and 1945. Their numbers included John F. Kennedy (who nearly died commanding a patrol torpedo boat in the coral seas south of Rabaul), and Lieutenant Richard Nixon, who launched his first political campaign with proceeds from his time playing poker at a naval supply depot off the coast of New Guinea. As Mead put it, the American GIs represented

> every sort and kind of American male, fighting a war with the most highly developed technical equipment the world had ever seen. Some fourteen thousand Admiralty Island people were exposed . . . to this tremendous spectacle, as miles and miles were packed with barracks, built on the spot from wood sawed in saw mills set up in the bush. The Americans knocked down mountains, blasted channels, smoothed islands for airstrips, tore up miles of bush . . . [and] the Americans took no responsibility for the preservation of the caste relationships which existed between Europeans and natives.

It seemed to her that Manus was the perfect place to observe how rapid technological change could lead to the creation of new cultures, new forms of consciousness and of being human—a world in miniature that could help prepare all the planet's societies for similar changes to come.

--------

Reaching the village of Pere from Manhattan took weeks. There was no electricity. The nearest telephone was a full day's journey away by boat. But there was beauty in abundance. In 1928, Pere had been a group of bamboo stilt houses perched over a white sand beach. Those houses had been torn down and new, "American-style" dwellings erected in their place, all of the same design. The crystal-clear ocean, however, had not changed. The waters still teemed with iridescent reef fish in shades of electric blue, greenish yellow, and rich violet. They fought for life amid pods of dolphins, sharks, and manta rays.

Mead recruited two anthropologists in their midtwenties to join her on the trip. They were a married couple: the man, Theodore Schwartz, tall, skinny, cerebral, and interested in photography; the woman, Lenora, short and outgoing, fascinated with dance and interested in the science of personality, given to wearing pigtails and white muslin dresses. In other words, the Schwartzes strongly resembled herself and Bateson some twenty years earlier. She was blunt about the fact that they were cogs in a machine. Or perhaps blocks of stone in a cathedral. "Science is a cooperative task," Mead told the Schwartzes, who regarded her with a mixture of awe and fear. "Individual geniuses advance IT, not themselves," she wrote, capitalizing the word. "It is not like art, in which each work of art would never have occurred at all if its individual creator had not painted or modeled or written. The ego motivations of the silly petty little academic world of credits and rivalries have nothing to do with SCIENCE except to muddle and mar it."

Night had fallen by the time Mead arrived at Pere in early June 1953. Soon after she reached shore, a man wearing well-ironed white clothes and a white tie approached her. He explained that he was a newly elected representative of a newly created body—the village council. He handed her a letter written in Tok Pisin, the creole language used across the region. Looking out at the faces lit by kerosene lanterns in the darkness around her, Mead felt a shock of recognition. Their appearance had changed, as had their clothing, but many of these men and women were people who knew her. Their names came back to her: Kilipak, Pokenau, Manuwai. They had been children in 1928 but were now among the leaders of the village, and they were expecting her. "I realized for the first time the relative emptiness" of her earlier fieldwork, Mead wrote later. She and Reo Fortune and all the other prewar salvage anthologists had been illegible to the people they studied, arriving and disappearing like UFOs, extracting information but never providing meaning. Now, however, Mead and the people of Pere "lived in the same world." Where before she had seen "natives" to be studied, Mead now saw the people of Pere as fellow travelers into an uncertain future, "facing decisions as complex as any that face us, and facing them with full consciousness." There had indeed been a revolution in consciousness in Manus. And from the beginning, Mead was determined to see it in positive terms. "They had taken their old culture apart piece by piece and put it together in a new way... to achieve their new goals," she wrote.

A few days later, a volcano erupted, sending a massive plume of smoke skyward and necessitating the evacuation of the village to higher ground. For days, the villagers gathered on a hilltop, discussing what to do. As they spoke, Mead began noticing references to a cataclysmic event a few years earlier, an event that was not a natural disaster but a homegrown religious movement that attempted to attain transcendence through the destruction of "everything belonging to the past." Some on Manus called it the Noise.

In 1947, a prophet had emerged who predicted a coming age of abundance, even immortality. But first, the old ways had to be cast out. Reports poured in of visionary experiences, trance states, even seizures. Census records were destroyed, the hats of colonial officials ritually burned. The tossing out of remnants of colonial life had a deep resonance for the people of Manus. Mead once explained their cosmology as one in which human souls experienced a powerful but short-lived afterlife, making their wishes known "through diviners and mediums." But after any of the deceased's male family members died, the old soul "was deposed, his skull was thrown out, and he became a homeless, malicious inhabitant of the lagoon, first a member of a set of anonymous kin ghosts, then a sea slug and finally, after a few generations when all memory of his period on earth was gone also, he ceased to exist."

The Noise was, in some sense, an acting out of this belief on a grand scale. Now it was colonialism itself whose powerful ghost was being pushed into the lagoon. Though the first prophet of the new religion was killed, his movement survived under a cannier, younger man, Paliau. By the time of Mead's arrival, Paliau had emerged as the mysterious political leader of the island, an éminence grise who was rarely seen in person.

The Noise fascinated Mead. In the aftermath of World War II, the people of Manus had been forced to develop a new social order in a matter of months, not decades or centuries. Mead believed that a similar dilemma—the dilemma of how to adapt to rapid technological and social change while retaining at least *some* cultural traditions—would in time confront all human societies in the twentieth and twenty-first centuries. Would adaptation to rapid change be mystical and sudden, trancelike, potentially violent? Or could it be measured, reasoned, led by scientists rather than prophets?

Or might it be some combination of the two?

## Late 1953

One day on Manus, Mead was urgently summoned to the nearest telephone, which was located a full day's journey away from her village. It was from New York City, she was told, and sounded serious.

After spending the day in a motorized canoe crossing choppy seas, Mead returned the call. The line crackled with the sound of technology strained to its limit, the sound of a long chain of operators and distant machines struggling to link the two sides of the earth. Through the static, a voice reached her. It was a man calling from an ad agency on Madison Avenue. He explained that he was compiling a list of famous smokers for an advertisement. The important phone call turned out to be the following question: "What brand of cigarette does Margaret Mead smoke?"

The adman's call was a reminder of how interconnected the world was becoming. But it was also a reminder of how potentially meaningless—how noisy—those connections were.

--------

Margaret Mead sought out one of the most remote places on earth to distance herself from the things that troubled her at the beginning of 1953. Gregory Bateson did not have that option. For parents who have just lost two infant children, no escape is possible.

In early December 1952, Bateson's wife, Betty, gave birth to twins, a boy and a girl. Both were sickly; by Christmas, both had died. The physical and emotional toll Betty must have endured is beyond description.

As for Gregory, he pushed the grief downward, never mentioning it in his published writings. The enormous trauma he experienced surfaces only in the long-lost diaries of Weldon Kees, to whom he grew even closer in the weeks and months following the twins' deaths. These entries suggest that both Batesons were in the depths of a months-long personal crisis throughout the first half of 1953. "Betty looking like hell and scarcely speaking to Greg," Kees wrote in his diary following one especially "depressing" visit in May, when the Batesons and Kees were supposed to see a Disney nature documentary together but Bateson "told me to go on alone."

Yet it was during this period of darkness that Bateson began to develop a new set of groundbreaking ideas. These ideas, born partly out of the trauma he faced, would eventually become literally field-defining. The group of eclectic researchers Bateson began to gather around him in 1953 is mostly remembered today in connection with the concept of a "double bind," a form of "no-win situation" that Bateson came to believe was related to schizophrenia. But this group was also one of the seedbeds for modern psychedelic therapy. What set the "Bateson group" apart was its emphasis on the importance of set and setting, its focus on personal transformation and expanded consciousness, and its skepticism about the purely scientific approach to understanding the mind. Though deeply flawed, these scientists were also pioneers who laid the groundwork for a new way of understanding and using psychedelic substances that still shapes the field today.

The interests of the Bateson group dated back to Mead and Bateson's time in 1930s Bali with their friend and research partner Jane Belo. Their stated aim then had been to understand schizophrenia. But they had somehow forgotten their original goal, returning instead with steamer trunks stuffed with notes on trance states and gender reversals. Now Bateson returned to this older aim, building on his ex-wife's writings on the interaction between symptoms of illness in an individual and the pathologies of the culture they inhabited. He was thinking in terms of a science of expanded consciousness, identifying the causes of mental illness in the "ecology" of an individual's mental experience—the intricate pattern of social relationships, culture, and family in which we all move—and not just in the mind itself.

The other influence on Bateson's revived interest in schizophrenia was more personal: Jane Belo, whom Bateson had once been in love with, now suffered from this devastating disease. The Dallas-born Belo, a dancer turned anthropologist, had always been an unusual person with an extremely vivid imagination. In 1952, she suffered a major episode of depression and checked herself into a mental hospital, where she was given a schizophrenia diagnosis. The new "tranquilizer" drugs were tried, but nothing worked. She spent the rest of the decade in and out of institutions. In this period, Bateson began keeping careful notes on these same drugs. But they failed, again and again, to heal the schizophrenic patients—or "schizophrenic families," as he now called them—that he was treating in Palo Alto.

By the late 1950s, it was relatively commonplace to refer to the world as a "schizophrenic society." Science textbooks and neuroscientists alike imagined the human mind as something like a telephone switchboard: a vast apparatus of electronic relays that could be "reprogrammed"—and that could also short-circuit. Reprogramming the mind to cure the ills of society remained a preoccupation of the era, from L. Ron Hubbard's Scientology to the Macy circle's "Problems of Consciousness" conference series. Bateson and Mead had both reoriented themselves toward communications and mental health in the years after World War II because they genuinely believed that achieving such a goal might prevent a nuclear apocalypse. But in the years that followed, their own scientific and personal goals began to seem schizophrenic, too.

CHAPTER 14

# Nembutal and Siamese Fighting Fish (Late 1953)

Do you feel drowsy? Do you feel as if in a dream? Are you anxious?

*—questions from a patient survey distributed by Harold Abramson, as recorded in H. A. Abramson et al., "Lysergic Acid Diethylamide (LSD-25): I. Physiological and Perceptual Responses," 1955*

## Thanksgiving Week 1953

IN THE EARLY MORNING of November 28, 1953, Dr. Harold Abramson of Cold Spring Harbor, New York, fielded what was possibly the most disturbing phone call of his life.

His phone rang around 3 a.m. An operator in the switchboard room of the Hotel Statler, an enormous brick structure just across from Penn Station, asked if Abramson could please hold. Then the voice of a thirty-five-year-old man came on the line. It was Robert Lashbrook, a junior CIA officer working under Sid Gottlieb in the Agency's chemical division. Lashbrook was calling from a hotel room with two beds and a shattered window, its curtains billowing into the freezing night air.

A crowd was starting to gather on the sidewalk thirteen stories below, where one of Abramson's psychiatric patients lay crumpled on the pavement.

--------

Within hours, this phone call would become an item of interest in a police investigation of a possible murder. New York City police detective David Mullee arrived first. The dead man's skull was fractured in several places, his leg twisted underneath him. There was no identification on the body, but a search of the hotel's registry revealed that his name was Frank Olson, and he was sharing a hotel room with another man, one Robert Lashbrook.

Mullee and his partner had no difficulty locating Lashbrook, though he stubbornly resisted questioning. In Lashbrook's suit pockets, they found airline tickets and hotel bills, a piece of paper with the address of the Chestnut Lodge psychiatric institution, another with Harold Abramson's home and office addresses, and a cryptic slip of paper bearing the initials G.W., M.H., and J.M. and the date November 25. When the detectives asked him about these names, Lashbrook "indicated that he preferred not to identify them because of security reasons." All Lashbrook would tell the detectives was that he was in New York on top-secret Department of Defense business, and that Olson was known to suffer from stomach ulcers.

Detective Mullee's best guess was that Lashbrook and Olson were "engaged in some homosexual affair" that had turned sour. If so, "the case was actually a homicide." He also pondered an alternative theory. Mullee had heard that Senator Joe McCarthy's House Un-American Activities Committee was in town. Perhaps the dead man was a Communist who had committed suicide under the strain. But to get anywhere, they needed Lashbrook to cooperate, and he continued to stonewall. Mullee and his partner were so incredulous about his claim to be involved in a top-secret national security operation that they consulted an FBI agent, who said "he did not know of any government work so confidential as to justify a lack of cooperation with police officials." At any rate, there was little they could do. Olson's wallet was missing, and nothing could be learned about him through public records. And because of a newspaper strike, reporters were not following the story closely, removing any hint of public pressure for a resolution. The case ground to a halt.

A few days later, the CIA dispatched its own team of investigators. They interviewed Mullee and his partner. In the process, they convinced the police to close the case and rule Olson's death a suicide.

The CIA team knew very well that Robert Lashbrook didn't work for the Pentagon. And they knew that Frank Olson was a leading chemist at Edgewood Arsenal—he specialized in the dispersal of poison gas—as well as a fully witting consultant for the Agency. To an agency insider, it was clear what Lashbrook was doing with Olson: he had been sent by his bosses to clean up a potential security threat. If Olson truly had lost his mind, he could not be allowed to wander around Washington, DC, or New York telling strangers highly classified secrets about chemical weapons and interrogation drugs. But even *they* could not understand how Harold Abramson had become linked to the death of a chemical weapons researcher. One CIA officer spent a fruitless day in the library looking up Abramson's name in old editions of *Who's Important in Medicine*. Each yielded the same drab fact: "Dr. Abramson's practice is limited to immunology and allergy." Why, then, did Lashbrook rush Olson to Abramson's office when he began to show signs of psychosis? And why was it Abramson whom Lashbrook called after the death?

Digging deeper, the same officer found an internal CIA report from the year earlier which stated that Abramson was "engaged in psychiatric research testing the efficacy of a new drug." This seemed to offer a clue. By that time, though, it didn't matter. Nearly two weeks had passed since Olson's death, and the press had virtually ignored it. Their work finished, the CIA's investigative team let the matter rest.

--------

Abramson, too, spoke to the CIA investigators. He told them that he had been shocked by Olson's decline. After all, Frank Olson was not an anonymous government employee to Abramson. He was a colleague from World War II. They had known one another for over a decade, having worked together on aerosol delivery methods for chemical weapons.

In the golden light of a November afternoon three days before Olson's death, Abramson and Olson reunited in a different context: not as peers, but as a psychiatrist with a psychotic patient.

They were sitting in Harold Abramson's office in an art deco tower on East 58th Street. Olson looked terrible. He was pale and shaking, his thinning hair disheveled, his careworn eyes weighed down by dark bags. Self-loathing choked his voice. Until recently, he had been next in line to lead his division at Edgewood. Now that was impossible, Olson explained. His mental collapse began, he said, when he was dosed with LSD that had been slipped into a glass of Cointreau during a weekend gathering of chemical weapons researchers at a cabin in Maryland. Robert Lashbrook was there. Sidney Gottlieb, too. Ever since the weekend at the cabin, he hadn't been the same.

On that afternoon in November, Robert Lashbrook, accompanied by Olson's direct superior at Edgewood, Vincent Ruwet, was sitting in the waiting room of Dr. Abramson's office. As Olson's official minders, they had been tasked with ensuring that the chemist didn't spill government secrets to strangers. By the time they left Abramson's office, night had fallen. Olson, Lashbrook, and Ruwet checked in at the Statler, ate dinner, and then went back to their shared room to watch television. At 10 p.m., there was a knock on the door: Dr. Abramson once again, this time carrying a bottle of liquor and some Nembutal pills for Olson.

The next day—the day before Thanksgiving—Olson returned to Abramson's office for another therapy session. The doctor invited Olson to spend Thanksgiving with Beaver Abramson and his children at their house on Long Island. Olson said he wanted to go back to his family.

That evening, Ruwet and Lashbrook took Olson to the theater. The show was a Rodgers and Hammerstein production, *Me and Juliet*, a meta-play about a behind-the-scenes love triangle between two electricians and a chorus girl working in a Broadway musical. Just before the intermission, the main character, Jeannie, is shoved onstage without warning. A spotlight operated by her embittered ex follows her as she ruins the fictional musical performance. She is nearly killed when a sandbag falls from the light bridge above the stage and narrowly misses her. Then the curtain falls.

When the lights came up for intermission, Olson seemed worse than ever. He got up, paced the lobby, and asked to leave. Back to the Hotel Statler they went.

Before dawn on Thanksgiving morning, Lashbrook awoke to find Olson missing. Eventually, he discovered Olson in the lobby, disoriented. Olson told them he had thrown his wallet "down a chute someplace" while walking the streets of New York at 4 a.m. Despite this, Ruwet boarded a plane with Olson to take the chemist home to his family outside Washington, DC, for Thanksgiving. Lashbrook stayed behind in New York City. Within a few hours, however, Lashbrook received a call from Ruwet saying that the plan to return Olson to his family had been called off: Olson was doing even worse and should be hospitalized.

Rather bizarrely, Lashbrook headed back to Washington, flew with Olson back to New York, and *again* took Olson to an emergency 4 p.m. therapy session with Abramson—who presumably had been forced to abruptly bow out of his own family's Thanksgiving celebration. Olson and Lashbrook shared a sad Thanksgiving dinner together at a restaurant in Cold Spring Harbor, spent the night at a nearby guesthouse, and hitched a ride with Abramson to Manhattan in the morning. By this point, Abramson had spent much of the past four days with Olson and Lashbrook. He recommended that Olson be hospitalized at Chestnut Lodge, a private psychiatric institution that was the favored facility for Agency operatives in danger of "cracking up."

That evening at the Statler, for the first time since his ordeal began, Olson telephoned his wife. He seemed to be in a better mood. Lashbrook and Olson had two martinis each at the Statler's fabled cocktail lounge, where glamorous big band orchestras had once serenaded wartime GIs and their dates, though now the crimson velvet interior was showing signs of age.

And then they slept—until, by Lashbrook's account, at 2:30 in the morning "I was awakened by a loud noise. Dr. Olson had crashed through the closed window blind and the closed window and he fell to his death from the window of our room on the tenth floor [*sic*] of the Statler Hotel."

At 9:15 p.m. the following evening, Lashbrook arrived once again at Abramson's office. This time he was accompanied by a CIA security agent, who took the opportunity to eavesdrop as the two men spoke privately behind a closed door. The security agent overheard Abramson saying that he "wanted to be kept out of the thing completely." In a short time, however, he "changed his mind and agreed to assist Lashbrook." The two men "listened to portions of a conversation which had been recorded"; the security agent couldn't be sure, but it seemed to him that it was a therapy session between Olson and Abramson. Then they both had a much-needed drink. He was "worried as to whether or not the deal was in jeopardy," Abramson told Lashbrook.

But the deal remained intact. Abramson continued with his CIA-funded work in psychedelic therapy. In fact, in the months that followed, he expanded it.

After she returned from Manus, Mead had cocktails with Abramson. It was April 1954, some five months after Olson's death. Abramson asked her to sign on as a consultant for his ongoing LSD experiments, and Margaret Mead said yes.

Later in their collaboration that summer, Abramson invited Mead over for dinner, then gave her a tour of his laboratory at Cold Spring Harbor on Long Island Sound. He showed her his acid-fried snails and his beloved Siamese fighting fish, their iridescent crimson tails twisting languidly amid the aquarium bubbles. Abramson explained to Mead that he had been dosing successive generations of the fish with LSD for nearly three years.

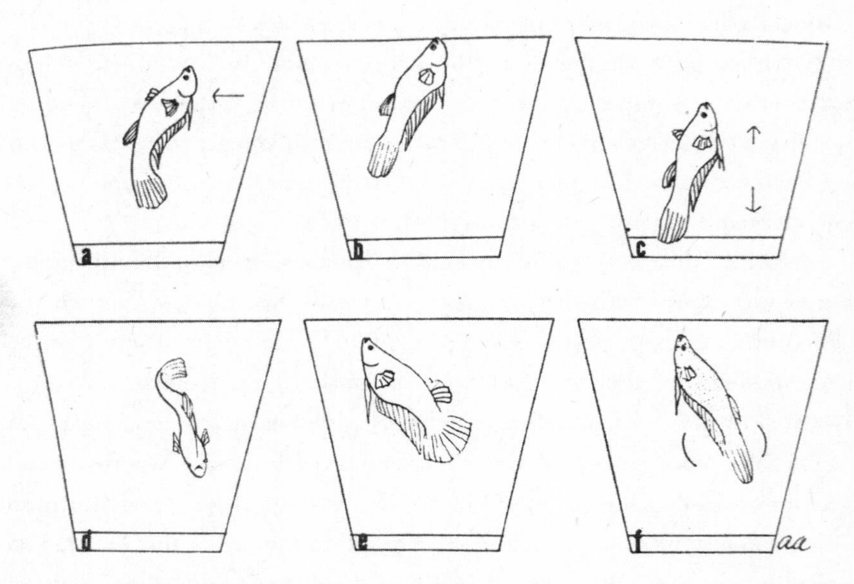


**Fig. 1. The six diagrams illustrate the nine criterions or effects of LSD 25 upon the fighting fish, *Betta splendens* Such effects are essentially exaggerated postures or caricatures of the normal fish.**

Harold Abramson's Siamese fighting fish, as portrayed in a 1954 article detailing his LSD experiments. (Abramson et al., "Psychobiological Effects on the Siamese Fighting Fish," *Science* 120, no. 3128 [December 10, 1954]: 991.) *Public domain.*

"Learned its non-addictive character," Mead wrote in her notes of that evening. "Explored truth aspects and decided it is not a 'truth drug.' "

Afterward, Abramson played her a recording of an LSD patient—"a chemist working with Abramson." The man "felt he was a bump on a log," Mead wrote. It cannot be proven either way, but the possibility exists that the voice on the recording belonged to Frank Olson.

--------

Who *was* Harold Abramson, really? His books—sensitive, thoughtful, learned books, filled with apparently genuine empathy for his patients—reflected his fascination with the psychosomatic causes of childhood asthma. One physician peer remembered him as a "profound psychoanalyst and man of letters" who spent a quiet but productive life "probing into the mysteries of human existence, aspirations and suffering."

Other stories emerge from those closer to him. Abramson's son, a retired judge, offered up the following anecdote, and nothing else: "He told me (after an official secrecy period had expired) he and his colleagues (they were all MDs, I believe) proposed a plan whereby they would volunteer to parachute into Germany and put LSD into the civilian water supplies as a means to end the war. FDR would not approve the plan because he felt it would be condemned by the international community. Small wonder that after the war we thought the Soviets planned to do that to us."

It is an impossible scenario—Abramson began working with LSD in 1949 or 1950. But the fact that he told some version of this tall tale to his son says something about the man. Murray Jarvik, a research assistant of Abramson's during the 1953–55 LSD trials, recalled him this way in a published reminiscence:

> Dr. Harold Abramson . . . was very smart but somewhat peculiar. He was a physical chemist, a psychiatrist and an allergist, all rolled up into one. I remember that he had interesting connections. Margaret Mead used to visit him. He would invite us to his rather palatial house out in Huntington, Long Island. I spent one summer at Cold Spring Harbor Laboratories looking at the effects of LSD on snails and fish and other things.

Jarvik's son was more blunt. "My father thought he was crazy," he said in an interview.

For her part, Margaret Mead thought he was brilliant. In a published review of his book *The Patient Speaks*, she praised him as "a physician who combines a thorough knowledge of natural science methods with a genuine respect for the intangible and often incommunicable processes of psychotherapy." Abramson's use of recordings in his therapeutic practice was particularly innovative: he recorded hundreds of hours of conversations, then replayed snippets of them for his patients. It was a cinema verité approach to therapy that resonated with Mead's experimental fieldwork style, with its combination of documentary photography and direct interventions in the culture she was observing.

Perhaps it was this shared interest that led Abramson to bring Mead on as a consultant for his LSD experiments. Or perhaps it was her fame—for a man conducting covert research for a top-secret government agency, Abramson was oddly attention-seeking, granting interviews about his LSD research with *Time* magazine, among other media outlets.

Whatever the reason, Mead's role was soon enshrined in the CIA's bureaucracy. On June 9, the day after Mead began her research into LSD, Sidney Gottlieb granted Abramson an additional year of funding. Abramson had explained that now the goal was not just to understand the effects of the drug, but to develop new methods for studying its role in society using personality tests such as the Thematic Aperception Test, or TAT. Mead regarded herself as an expert administrator of precisely those personality tests.

The question, then, is not why Abramson wanted to work with Margaret Mead. The question is, why did she want to work with him?

CHAPTER 15

# The LSD Sessions (1954)

> The graveyard...the point of extreme transition, standing between old and new, past and future.
>
> —*Margaret Mead, "Preliminary LSD Memo," September 25, 1954*

WITH HER PAPER-CLUTTERED DESK spotlit underneath a pendant lamp, Mead's office in the western turret in the American Museum of Natural History had begun to take on the look of a Broadway theater set. All around her in the shadows hung masks and carved figures, as if the museum dioramas from the floors below had begun to creep into her workspace.

Back in her spiritual home after spending the last six months of 1953 in Manus, she felt refreshed, confident. She was already thinking through her book about her experience; *New Lives for Old* would be the title, and it would describe the Noise in relatively benign terms. True, it really *was* an apocalyptic cult, she wrote, complete with mystical "prophetic dreams" and the promise of "a utopia to be immediately established on earth." But who said utopian dreams were entirely bad?

For Mead, this was not just a matter of survival in one village on one island. All human societies, everywhere, needed to learn to change in similar ways and with similar speed, avoiding the excesses of what she called "the apocalyptic cult aspect" but also recognizing that, as she later put it, "conscious intervention in the process of cultural evolution and human survival" was now the most urgent task that humanity faced. This was the goal of her science. And the things she saw and heard in the spring and summer of 1954 made her think that psychedelics might be part of it.

After all, Abramson hadn't just shown her Siamese fighting fish. He had shared the results of an LSD trial that resulted in what he called "loss of fear of homosexuality": a woman racked by anxiety over her suspicion that she might be a lesbian had, under the drug, resolved her fears. Mead, who was in a committed, years-long relationship with Rhoda Métraux by this time, absorbed this information with great interest. Perhaps LSD could speed up the rate by which global culture adapted to rapid change. Perhaps, as she wrote to herself in June 1954, the drug could be a "short cut" for pulling society "away from mysticism and escape."

It is worth pausing on this. If Mead had publicly backed psychedelics in the mid-1950s, history would have changed. A case could be made that with Einstein's death in 1955, Mead became the world's best-known living scientist. In that year (by one measure) her name appeared in print more frequently than that of any other scientist alive at the time. Among dead ones, remarkably, she was within shooting distance of Charles Darwin and Isaac Newton. Even science-fiction novels were not free of her. In Robert Heinlein's *Citizen of the Galaxy* (1957), the book's orphaned protagonist is mentored by a space anthropologist named "Doctor Margaret Mader" who teaches him how to navigate the cultural shifts he faces as he skips from ship to ship, planet to planet.

What drove Mead's popularity was her ability to wed an urgent call to action—humanity needed to expand its "awareness," its collective consciousness, in order to survive—with an implicit optimism. A young Carl Sagan was among Mead's avid readers. As a student at the University of Chicago, he became fascinated with the ways her work "gave you a view of the arbitrariness of cultural mores, cultural systems." Sagan was drawn to the "tremendous optimism" of "the idea that you weren't jostled about by the winds of the world. That you could do something" to change the future. When an eighteen-year-old Sagan wrote his first piece of popular science writing, a radio script called "Ad Astra," he dreamed of scoring an interview with Mead.

If a figure such as this had embraced psychedelics before they acquired their social stigma and legal restrictions—before the advent of tranquilizer drugs and antidepressants, before the triumph of a hypercapitalist global pharmaceutical industry—it is possible to imagine a very different history of drugs in the twentieth century. In fact, it may have even led to a different world.

--------

Only with hindsight did Harold Abramson's research assistants, Murray Jarvik and Conan Kornetsky, realize that their work with Abramson was CIA-funded. At the time, they were idealistic young scientists from working-class backgrounds who were trying to solve big problems. Jarvik (like so many other drug researchers of the era, a second-generation Jewish American with relatives who had died in the Holocaust) had a very personal reason for his research into the causes of schizophrenia: his own mother suffered from the illness. He was a natural tinkerer with an orientation toward real-world problem solving. In high school, he had won first prize at a science fair for building a functional, inexpensive "iron" lung out of wood.

By the summer of 1954, Jarvik and Kornetsky were leading clinical trials of LSD on volunteers, including themselves, under Abramson's supervision. One day, their boss told them that they would have a chance to work with the celebrated Dr. Mead.

In the humid summer of 1954, in fact, Mead made studying LSD something close to her full-time job. On Wednesday, June 16, she had lunch with her friend Gotthard Booth and mentioned the new project. Booth reminded her, she wrote, of "my earlier refusal to take Mescal[ine], said I'd said it while [we] were going somewhere in a taxi, must have been early 40's," after which he explained "he HAD taken mescal[ine] in experiments in Germany, and thought he owed much of his insight to it." With Booth, Mead thought more about LSD and "began to [d]evelop idea of who should and shouldn't take it."

And then came a friend of her daughter's named Ralph Blum, who had participated in one of the LSD studies at Boston Psychopathic. A Harvard undergrad, Blum had been in a control group for comparison with "psychotics." But what happened when the control group experienced life-changing results?

"It made a difference in my life. It began to move the log jam of my old consciousness," Blum recalled of his first psychedelic experience. "Although

I wouldn't use them until much later, it gave me a new set of optics." For Blum, participating in the Boston trial had been transformative. He became fascinated by the anthropology of the abnormal—from trance states to traditional magical practices to the possibility of ESP. Mead mentored him, and in return he became what anthropologists call an "informant." Mead had many contacts among psychedelic researchers at this time, but the twenty-two-year-old Blum (who would later collaborate with Timothy Leary in the Harvard psilocybin experiments of the early 1960s) became her most important contact among the new generation of psychedelic *users*.

--------

When she was interviewed in 2022, Marianne Weltmann, a ninety-one-year-old opera singer based in Seattle, was among the world's last surviving MKULTRA test subjects. Weltmann worked as a secretary for Dr. Jarvik at Mt. Sinai Hospital—a day job while she studied to be a singer at Juilliard. She found the work boring. Little wonder, then, that she and another secretary jumped at the opportunity to try the experimental drug that their bosses were studying and sampling themselves: LSD.

The experiment took place in the evening. Weltmann was an adaptable and outgoing young woman, so she was not particularly fazed by the arrival of Margaret Mead and another man named Dr. Caligor, an expert in psychometric tests. Mead also brought along a powerful funder, John Eberhardt, the director of NIMH's extramural research, who hovered in the background as an "interested observer." Weltmann remembered it as an experiment organized by Mead, whom she found impressive but "too serious."

As for Mead, her notes described Weltmann's trip this way:

> Graceful dark little German Jewish girl...studying to be an opera singer...wonders if it is wrong to try to be a dead and gone culture... likes her job but feels that the people given shock aren't really psychotic, talks to them when acting as assistant and feels contact with them...is concerned with keeping control, will keep it, brought up rigid German, her parents now know better, look what happened.

Abramson was not present (Mead had dinner with him beforehand, along with his secretary Naomi, "in a state of giggles from another drug"). At dinner, she had gone over the assumptions she wanted to test. Ralph Blum had spoken of "extraordinary speed of communication" supposedly allowed by psychedelics. Mead wanted to test the secretary's reaction times. More esoteric was what Weltmann remembered as Mead's interest in psychic powers. Mead had apparently heard rumors that psychedelics might play a role in such phenomena as clairvoyance, and she tested them by measuring Weltmann's ability to guess the symbols on cards.

Mead left that night feeling flush with potential—the state she liked best. The following day, she began typing the document entitled "Preliminary LSD Memo," now in her archive at the Library of Congress. It is a strange artifact, filled with typos and hasty asides, and imbued throughout with a suggestion of looming discovery. "I am tending to label any new or good ideas as associated with the project," she noted, "because I am watching my dreams and my dreams are being theoretically fertile at the moment." In it, she speaks of open and closed systems, the French Revolution, "liquid" dreams, and graveyards. There are glimmers of the manic state she experienced when she developed the Squares with Bateson twenty years earlier.

"Things are now under way," she wrote to Ralph Blum a day later, apologizing for delaying her letter until "I got something started on LSD." She and Abramson's team, she wrote, were applying tests of "continuity of personality" to psychedelic patients, "and it looks very promising. I am going to take it—the drug I mean—myself early in November and we are setting it up experimentally with various people making predictions on how it will affect me. So people who have given me projective tests etc are being asked to make sealed predictions. Would you like to make one?" Mead ended by thanking him for being "the living demonstration of what is proving to be a very fertile research possibility."

--------

When it came time to write about her strange months in Manus, Mead found a way to put a positive spin on both the cultural changes on the island and the potential of psychedelics. "Drugs, such as mescal or LSD," could, she speculated at the end of *New Lives for Old*, which she wrote during the period of her involvement in Abramson's psychedelic research, lead to the acceptance of "new patterns." The Noise, Mead wrote, was just one part of a continuum of "conversion experiences": others included the altered consciousness of the Balinese trance dancer, the alcoholic's moment of clarity, and the experience "evoked by drugs." All involved "the complete destruction of the past." These moments of rupture would be the defining experiences of the twentieth and twenty-first centuries. We could not survive if we rejected them. Instead, we needed to understand them—even embrace them.

But a strong note of ambivalence also crept into Mead's account. Technological innovation might allow for faster cultural evolution, she wrote. But when, in the same passage, she compared LSD to brainwashing in the Korean War, she was speaking from experience. She knew enough by then to see that any drug capable of transforming one's sense of self could also be repurposed as a technology of control. Mead believed in psychedelics' ability, as she put it in a letter to Blum on September 24, 1954, to be "integrative and insight giving"—but only if pursued "in a responsible experimental spirit."

--------

One example of how damaging this work could be had come a few months earlier, though Mead was not aware of it.

On April 15, 1954, George Hunter White traveled to Atlantic City for an MKULTRA conference of sorts—one of the few occasions when the various branches of the CIA's drug program came together. The participants included Harold Abramson, Robert Hyde, Sidney Gottlieb, and Paul Hoch. All of Hoch's test subjects for his experiments with LSD were institutionalized; a large proportion were people of color, unaware that their physician was feeding their private medical data to the CIA. Hyde, too, had continued testing LSD on his patients at Boston Psychopathic, without informed consent.

This was their *public* work. What was discussed at Atlantic City is unknown, but the presence of a CIA physician named Dr. Pelikan offers a strong hint. At the time, so-called Pelican teams were assigned for drug-assisted forced interrogations of suspected Communist agents.

Mead was almost certainly not privy to meetings like the Atlantic City conclave, which represents the still-mysterious inner core of Cold War drug research. But she strayed surprisingly close to it. George White's diary shows that he met with Abramson at his lab in the same week that Mead sat in on Weltmann's LSD trip. She maintained her friendship with James Hamilton during the many years in which the genial San Francisco psychiatrist headed two long-running MKULTRA subprojects. And the "LSD" file in Mead's archive contains a record of a telephone call with Paul Hoch, along with the words "will send list." There is also the question of her friendship with Harold Wolff, the neurologist whom Allen Dulles and his wife had tasked with healing their son, and who had in the process become the beneficiary of considerable CIA funding for research relating to altered states of consciousness. Throughout 1954 and early 1955, Mead's longtime romantic and professional partner Rhoda Métraux worked closely with Wolff as an employee of his CIA-funded research organization, the Society for the Investigation of Human Ecology (SIHE). Mead played a role, too, working on an informal basis as a researcher. Métraux later wrote that Mead's research team at Columbia was the "prototype" for Wolff's organization, which she described as a "medically-oriented" spin-off from it. Métraux—no stranger to the spy business, having led an OSS research team—wrote the following cryptic account of this period of their lives: "Margaret Mead was a realist who mastered the dark side of her vision."

These are scraps and hints, fragments of a larger story that cannot be fully integrated at the distance of seventy years. But they are enough to conclude that, despite her genuine good intentions, Margaret Mead likely contributed in some way to a project that has gone down in history as among the twentieth century's most infamous abuses of medical research ethics.

--------

In the end, Mead appears not to have taken LSD. On November 8, she announced that her schedule had gotten too busy. "I think I'll have to put off LSD possibly until the first of the year," she wrote Booth, whom she'd asked to supervise her trip. On the same day, she told Harold Abramson the same thing. "I've not forgotten," she added.

Then, in January 1955, Mead decided to permanently postpone her LSD

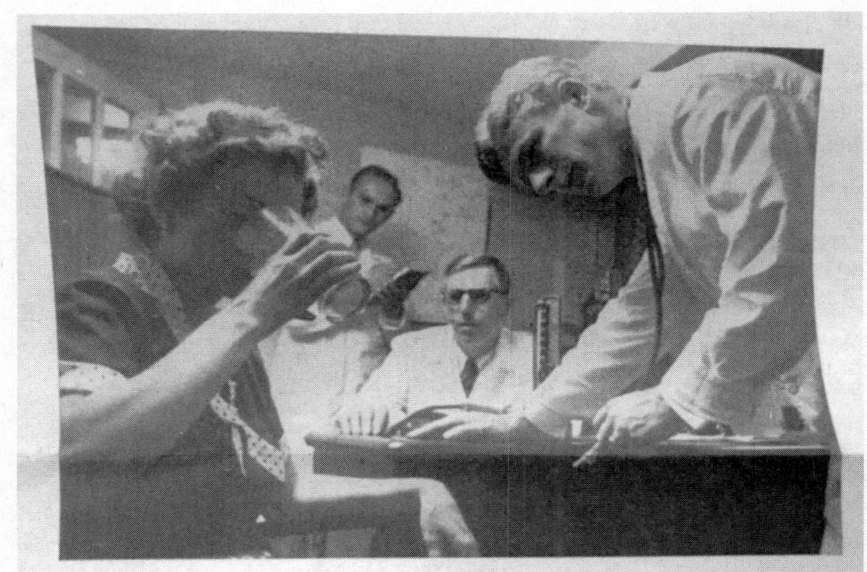


A photograph from Mead's personal clipping of Robert M. Goldenson and Leo Rosten's article about LSD experimentation, "Step into the World of the Insane," *Look* magazine, September 21, 1954. *Courtesy Library of Congress.*

trip. There was more behind her decision than a crowded schedule. She had been reading studies and articles that showed psychedelic patients revealing truths they wanted to keep under wraps. In the same period, she was being investigated, once again, by the FBI—and was coming to the conclusion that she would move in with Rhoda Métraux.

As John D'Emilio writes in his history of gay rights in the 1950s, the years of 1953 and 1954 marked a "retreat to respectability" among LGBT activists who feared that the McCarthy-era prosecution of leftists would spill over into scrutiny of their own membership rolls. The retreat to respectability was, above all, a push to defend the principle that supposedly deviant sexual identities were in fact normal—by *behaving* normally in other domains of life. In June 1953, for instance, representatives of the Mattachine Society, an early gay rights organization, contacted Karl Bowman at Langley Porter, seeking his aid in legitimizing LGBT identity among physicians.

Mead's bisexuality had already nearly destroyed her career once before. In the summer of 1939, Benedict was living among the Blackfeet people in Montana, and Mead was pregnant in New York City. Throughout June and July, letters flew back and forth between the two women about a disastrous love triangle in their group of friends. It came down to a test of loyalties, with Mead favoring one friend (the wife conducting an affair) and Benedict favoring the other (the angry husband, seeking a divorce for infidelity).

Then one day the scorned husband turned up outside Benedict's remote cabin in Browning, Montana. He asked her to testify in his divorce trial. The wife responded with a chilling threat: if Benedict testified, she said, she would reveal Benedict's relationship with Mead to newspapers. Mead panicked, placing a long-distance call to Benedict that appears to have threatened a break between them. In the end, Benedict stepped back from testifying, and Mead worked to smooth out the conflict in her own quietly effective way. But the fear remained.

In a letter to her loved ones from 1955, Mead directly referenced her life's sphinxlike quality. "I have become increasingly conscious of the extent to which my life is becoming segmented, each piece shared with a separate person," she wrote. "Since Ruth Benedict died there is no one alive who had read everything that I have written... Since the break up of my marriage, far less of my life has been shared with one person... It has not been by my choice of concealment that anyone of you have been left in ignorance of some part of my life which would seem, I know, of great importance." Instead, this concealment had been "only from the exigencies of the mid-twentieth century when each one of us—at least those of us who are my age—seems fated for a life which is no longer sharable." Her daughter, Mary Catherine, wrote that her mother kept a great many secrets, and "she clearly believed that the keeping of these secrets was correct and responsible behavior, a precondition to her availability to do the work that she felt was important."

Margaret Mead took LSD seriously as a potential pathway for cultural evolution. But when she decided to move in with Métraux, her risk calculation changed. LSD, after all, was being used at that very moment in interrogations of suspected Soviet agents. She could not guarantee that she would not let something deeply personal, something secret, slip out during her trip.

She could not guarantee that she would maintain control. And if, like Frank Olson, she did slip in that fraught moment, the whole edifice of her selfhood, her life's grand project, might well collapse.

Margaret Mead moved in with Rhoda Métraux in the spring of 1955. They lived together for another twenty years.

CHAPTER 16

# Gadabout (1954–55)

> You have forgotten yellow lights of San Francisco coming on,
> The bridges choked with cars, and islands in the fog.
> Or have forgotten why you left or why you came to where you are,
> Or by what roads and passages,
> Or what it was, if anything, that you were hoping for.
>
> —*Weldon Kees, "Travels in North America," written in March 1954 while he was working with Gregory Bateson*

AT THE BEGINNING OF 1955, George Hunter White's official employer, the Bureau of Narcotics, transferred him to San Francisco. Here, the rotund, mystically inclined former OSS officer would continue the real-world, covert testing of LSD and related compounds that he had begun in New York City. The move had been prearranged by Sidney Gottlieb, the young head of the CIA's newly formed chemical division. White opened a $1,500 bank account for his alias, Morgan Hall, and met with Bateson's old friend from his OSS service in Burma, the psychiatrist James Alexander Hamilton, along with two of Gottlieb's deputies, to plan a new LSD experiment in San Francisco. At the end of March, Sidney Gottlieb flew in to confer with White and Hamilton about LSD experimentation in the city by the Bay. He stuck around afterward to help White and his wife, Albertina, unpack.

Afterward, Hamilton and White set to work. They rented a beautiful apartment at 225 Chestnut Street in the city's Marina District, with large bay windows that offered peaceful views of sailboats scudding past Alcatraz Island

and the Golden Gate Bridge. White's taste for interior furnishings was becoming a passion, and this time around he spared no expense. He hired Maurice Sands, proprietor of the city's most prestigious design firm, to personally oversee the decoration. They aimed for a bohemian look, outfitting the apartment with a $44 telescope, red curtains, an African textile, a standing ashtray, a high-quality radio-phonograph combination set, two portrait busts, an easel with an "unfinished painting," one high-quality Waring brand blender, a well-stocked bar, and—to relieve the surveillance team stationed in the hidden part of the apartment—"1 Handi-dandi portable toilet." The surveillance expert White hired to install the concealed microphones was also the best of the best; he would later consult on the Francis Ford Coppola film *The Conversation.*

The first experiments with LSD began in May. In his "day" job as a Bureau of Narcotics officer, White regularly interacted with sex workers and drug dealers in San Francisco's red-light district, which was then known as the Barbary Coast. He recruited at least one sex worker to help him lure "johns" to the 225 Chestnut Street apartment, where they were surreptitiously dosed with LSD through spiked drinks or cigarettes, then recorded. Not for nothing did White give the operation the cryptonym "Midnight Climax." Throughout June 1955, Hamilton and White could be found touring the bars of the Barbary Coast with this goal in mind.

They had strayed into the world of Gregory Bateson's close friend, the brilliant, deeply depressed poet Weldon Kees, in what was—most likely—the final month of Kees's short life.

--------

Kees had spent 1954 bouncing between an increasingly dysfunctional personal life and the Bateson group, which had by this time moved from the Langley Porter Clinic to Stanford. In his poetry, Kees wrote of the horrors of "the University of California's atom bomb," scientists who "doped retarded monkeys with cocaine," and, of course, the darkness underneath the bright fluorescent lights at Langley Porter. During the height of the McCarthy era, the themes in his poetry seemed to take over his life.

The Army-McCarthy hearings, which ran through the spring of 1954, riveted millions (including George Hunter White, who watched them avidly).

But for Kees's wife, Ann, the high drama unspooling on her television screen triggered a psychotic break. She was convinced the FBI was tapping their phones. By the Fourth of July, Ann was drinking liquor from morning to evening and claiming that FBI agents were stationed outside their home. Weldon managed to convince his neighbor, a young doctor, to give Ann a shot of sodium amytal. After it wore off, Kees called up Dr. Saxton Pope, a psychiatrist friend from the Langley Porter Clinic. Pope had some experience with delusional states. At the time, in fact, he happened to be overseeing a research project—led by a young Berkeley psychology PhD named Timothy Leary—that measured "fantasy and imaginative expression." He prescribed Ann another sedative, phenobarbital, and checked her into Langley Porter.

Kees's poem "The Clinic," which he dedicated to Bateson, had referred to World War II–era narcosynthesis drugs. Now he watched helplessly as Ann sank deeper into an alcoholism-induced psychosis that these same drugs failed to tame. Meanwhile, Kees's dependence on amphetamines was beginning to worry friends. "He's the answer to all my ills," went one of his songs from this period. "Lives on air and those little pills."

Kees filed for divorce. The day before the paperwork went through, he took the borrowed camera he'd been using to make films with Bateson and headed over the Golden Gate Bridge, driving aimlessly amid the rolling golden hills and crashing surf of Point Reyes. He began planning a short film called *The Bridge.*

--------

A newly single Kees moved out of the dreary bungalow he and Ann had shared in the industrial city of Richmond and into a charming one-story gingerbread Victorian cottage in San Francisco, some fifteen blocks west of George Hunter White's pad. This put it in easy walking distance of the city's fabled Barbary Coast red-light district, a tangle of cabarets, nightclubs, and bordellos bordering Chinatown.

Kees had always been a heavy drinker. Now he began to spend more and more time in bars, alone or with new friends. His favorite was the Purple Onion, a basement tavern where a young Maya Angelou had just begun performing. Here he got to know Phyllis Diller—on her way to being one of the first successful female stand-up comedians—and cast her in a variety show and poetry reading he organized, the Poets' Follies. For the same show, in January 1955, Kees recruited Lily Ayers, one of the Bay Area's best-known strippers, to read T. S. Eliot poems.

But what occupied his attention most in this period was a screenplay. It was for a film called *Gadabout*, a Cold War espionage thriller set in San Francisco.

The central character in the film was an eccentric scientist who came "from the world of big government-funded research institutes." On the first page of the script for *Gadabout*, Kees scribbled three words: "Lysergic acid" and "CIA." This is a remarkable fact. If his screenplay had been produced in 1955, the film would have been the very first fictional account of LSD to appear in any format—in any language.

Kees's biographer has suggested that this cryptic note and its timing were "more than coincidental." The note, he argues, may be evidence that Kees had somehow become involved in George Hunter White's Midnight Climax operation. This possibility cannot be ruled out. Kees, after all, moved in the same underworld—a late-night San Francisco of strippers, amphetamines, and Barbary Coast bars—that White and Hamilton trawled for unwitting experimental subjects.

An alternative (or additional) source for Kees's knowledge, however, was Gregory Bateson himself. Bateson, after all, was in direct contact with Hamilton, Abramson, *and* White in these years. It seems likely that part of his fear of militarized science derived from hints of MKULTRA projects dropped by these men.

Another clue comes from the screenplay of *Gadabout* itself: the film's antihero, Helwig Ennis, bears a more than passing resemblance to Bateson. Both Bateson and the character of Ennis were scientists who had gone through messy divorces, then became involved with coworkers. The actor whom Kees imagined playing Ennis, Robert Ryan, was the same unusual height as Bateson (six foot four) and had the same guarded look. In the film, Ennis is introduced to LSD by way of a Harold Abramson–like psychiatrist who is secretly working for an intelligence service. But most telling of all is the yearning of Kees's character to flee his increasingly dangerous life in California for a place that Bateson knew well: Port Moresby, New Guinea. Port Moresby was a refuge during Bateson and Mead's earliest fieldwork together in the 1930s, their port of call for obtaining new supplies and for recuperating after bouts of malaria. And it was a place that Mead had just visited the year before, but which Bateson had not seen since the aftermath of the feverish spring of 1933, the time and place when he had first stretched out his hand toward scientific greatness.

Is it pure coincidence that, of all the places in the world, Kees's scientist character, caught between spies, jealous colleagues, and experimental drugs, yearns for this one place that held such rich emotional resonance for Gregory Bateson?

--------

Weldon Kees's final piece of published writing appeared in the *New Republic* under the title "How to be Happy #1053." In it, he bitterly skewered a book of pop psychology. Self-knowledge, he wrote, won't save us from a world filled with "violence and irrationality, with so many human beings murdering themselves, either literally or symbolically."

In retrospect, it reads something like a suicide note. The day the piece appeared in print, July 18, 1955, was the day that Weldon Kees disappeared. The police found his 1954 Plymouth Savoy on the north side of the Golden Gate Bridge. The keys were still in the ignition. Kees had carefully folded his Langley Porter Clinic lab coat and left it on the passenger seat. A search of his apartment revealed that he'd left behind his cat, Lonesome, but taken his sleeping bag, savings account book, and wallet.

Kees's disappearance closely mirrored the plots of the noir films he loved. On February 7, 1954, for instance, he watched *The Big Heat*, a crime drama about the suicide of a police officer involved in a corruption scandal. On February 18, it was *The Glass Wall*, about a refugee from behind the Iron Curtain who is hunted by spies through New York and, at the film's end, contemplates jumping off the United Nations Building.

In the aftermath, Kees's friends began to piece together a narrative that was no less cinematic. It was one that had two possible endings: Mexico or death. At about 4 p.m. the day before, Kees had called his friend Janet Richards.

"Things are pretty bad," she recalled Kees saying to her. "I may go to Mexico. To stay." She never saw him again. The same afternoon, Kees called another friend, the film critic Pauline Kael, to ask, "What keeps *you* going?" Kael remembered the Kees of this period rewatching the film of schizophrenic children he had made with Gregory Bateson and becoming "terribly distressed."

Medicine, he said, had not healed them. Medicine had made them *worse.*

--------

Weldon Kees had just missed the beginning of what has become known as the San Francisco Renaissance, a flowering of countercultural art and poetry.

Allen Ginsberg made his way to San Francisco in June 1954, where he moved in with a girlfriend named Sheila—he was still closeted—and her child. The trio found a quaint Victorian apartment boasting a picturesque view of the Sir Francis Drake Hotel. Ginsberg treated himself to a new tweed suit from Brooks Brothers, cut his hair, and landed a job at a market research firm. And he began seeing a psychiatrist at the Langley Porter Clinic, Philip Hicks, "to overcome a block in his writing," as Hicks wrote in his notes. But it was also to work toward accepting his homosexuality, which Hicks, unlike Ginsberg's past doctors, did not condemn.

That October, Ginsberg experimented with peyote alongside Sheila and Neal Cassady in their apartment overlooking the Sir Francis Drake. The foggy San Francisco night cast an ancient, haunting aura over the 1920s-era building: "a Golgotha-robot—eternal—smoking machine crowned visage," as the poet put it in a letter to Jack Kerouac. In the same letter, Ginsberg mentioned his new friendship with poet Kenneth Rexroth, Weldon Kees's longtime rival. Ginsberg's experience with peyote would inspire what became one of the most celebrated and controversial poems of the twentieth century: "Howl." He first read his poem while standing alongside Rexroth at a reading at the Six Gallery on October 7, 1955.

By the time this happened, Weldon Kees may have been three months into his new life in Mexico.

More likely, his bones were being pulled out into the Pacific by three full moons' worth of estuary tides.

## Summer 1955

Gregory Bateson rarely spoke about the death of Weldon Kees. There is no entry in his archival finding aid for "Kees, Weldon." No surviving letters. In Bateson's semi-official biography, Kees is mentioned only once.

But Kees was among his closest friends, and the loss shook him. It may have also helped shape the theory for which he is best known. Bateson premiered his "double bind" theory of schizophrenia at a small conference held at Sea Island, Georgia. Like Ginsberg's reading of "Howl," it took place in October 1955. He described the double bind as a situation in which a paradoxical communication from an authority figure demands contradictory responses: "I cannot survive if I do not obey and to obey would be to die." This, he thought, was at least a partial cause of schizophrenia.

In *Hidden Valley Road*, his bestselling history of a family whose six schizophrenic children became a textbook case among researchers at NIMH, Robert Kolker portrays Bateson's double bind theory as a significant wrong turn in the study of mental illness. Having, Kolker writes, "invented this theory without so much as ten minutes of clinical psychiatric experience," the anthropologist supposedly "helped to turn mother-blaming into the industry standard for psychiatry." In truth, Bateson saw his double bind as a characteristic of *families* and *societies*, not just mothers. He spoke of a "constellation . . . of trauma" that forced individuals diagnosed with mental illness into paranoia and withdrawal because their social circumstances made it impossible to communicate effectively with those around them.

But the bigger problem with this depiction is that it assumes Bateson was standing outside of his theory, gazing at it from behind the safety of a two-way mirror. The truth was very different.

The person suffering from a double bind, Bateson wrote, found it impossible "to judge accurately what a person really means." This person might "then be excessively concerned with hidden meanings," paralyzed by the notion that those around them are hiding covert activities. Such a person, Bateson wrote, might "become suspicious," detach from the world, and "concentrate on his own internal processes." This was an apt description not just of a schizophrenic patient, but of Bateson himself. Following the trauma of his experience in World War II, Bateson wrote a letter to Maya Deren describing himself as a "statue." She had the "key" that would "unlock" him, he said, but this possibility terrified him: Bateson said he was so paralyzed by fear that he doubted whether it would ever be possible for him to truly "come alive." After the death of the twins carried by his wife, Betty, at the end of 1952, Bateson again fell into a kind of social paralysis, refusing to engage with his family even as he obsessively studied how otters played with one another at the San Francisco Zoo. And throughout this period, he watched as people who had been—who *remained*—close friends, people like James Hamilton, Harold Abramson, and to an extent even Margaret Mead herself, moved into a world of covert activities and psychological manipulation that he looked upon with horror precisely because he had helped inspire it.

In short, Gregory Bateson didn't just diagnose the double bind. He *lived it*.

--------

Weldon Kees had lived two lives, as an artist and as a scientific researcher. He missed his shot at glory in both. After his disappearance, those glories came surprisingly fast.

On the one hand came the San Francisco Renaissance of art and poetry, a movement that Kees deeply influenced. On the other came the Bateson group. The double bind theory went public when, in 1956, Bateson and his acolytes published their findings in a widely cited journal article. Finally, Bateson's meandering path through science seemed to be leading toward the cathedral of his father. Finally, he seemed to be making world-changing discoveries. And he was doing so on his own terms. His emphasis on schizophrenia as a product of a *communal* breakdown rather than an individual failing was an eye-catching departure from the norm. It pushed back both on Freudian psychoanalysis and on the theory that mental diseases could be reducible to chemical imbalances in the brain. And it also turned away from tranquilizers, the newly faddish cure for the Age of Anxiety.

Margaret Mead was part of the cultural consensus that created the tranquilizer era. Writing for the *New York Times* in 1956, in an article titled "One Vote for This Age of Anxiety," Mead argued that if a society's most pressing problems were anxiety and stress, it meant that it had escaped humanity's traditional enemies of warfare, famine, and disease. The rising significance of psychiatry and psychiatric drugs in modern life was, she argued, not a sign of growing mental illness, but a sign of progress. It led to "a world in which no individual feels that he need be hopelessly brokenhearted, a failure, a menace to others or a traitor to himself... a society where there will be freedom from want and freedom from fear."

Alert and attentive, the patient participates in a P.T.A. meeting, following her second capsule of Meprospan-400 taken with the evening meal. Meprospan-400 does not decrease her mental efficiency or interfere with her normal activities or behavior.

Details of 1961 advertisements for the new anxiety medications Meprosan and Trancopal, which were both introduced in the mid-1950s (*New York State Journal of Medicine*, January 1, 1961). *Public domain.*

Mead's article spoke to the new pharmaceutical reality of 1956, for it now appeared that drug researchers really *had* developed a cure for fear. A new category of psychiatric medication, called "minor tranquilizers," was revolutionizing both the business and the science of drugs. The first approved drug in this category, Miltown, had already been sampled by one in twenty Americans by the time Mead's article appeared in 1956, just one year after its release. Competing anti-anxiety drugs like Librium and Valium followed close on its heels. By 1965, 25 percent of American adults polled in a Gallup survey had used a minor tranquilizer. This meteoric rise was due in part to the labors of a New York psychiatrist turned medical advertiser named Arthur Sackler.

Along with his two brothers, Raymond and Mortimer—who would later helm Purdue Pharma, the developer of Oxycontin—Sackler developed an innovative approach to selling anti-anxiety drugs: direct-to-physician marketing. Miltown, Librium, and Valium, not LSD, became the first breakthrough psychiatric drugs directly targeted at "normal" people. Meanwhile, a so-called major tranquilizer, chlorpromazine, was being hailed by many—including Bateson's colleagues at the Palo Alto VA—as a chemical cure for schizophrenia. Bateson took an interest, taking notes about his coworker Leo Hollister's investigations. In one notepad from 1956, he scribbled that the drug "cuts the fear." But, he wrote, it also "diminishes the joy effect."

No single drug could truly heal the psychic ills of life in the 1950s, Bateson believed. But his skepticism toward tranquilizers was not a rejection of drug-based interventions as a whole. Bateson and the group of young researchers he had assembled were about to become the pioneers of a new kind of psychedelic therapy.

CHAPTER 17

# The Learys and the Batesons (1955–57)

> O, Child of Buddha Nature, listen. It is due to your own past actions that you are now suffering in this way. No one else is responsible... Recognize, now, that this is the intermediate state!
>
> —The Tibetan Book of the Dead, *part 2, as translated by Gyurme Dorje*

## Summer 1955 to Fall 1956

TIMOTHY LEARY WAS DEAF in one ear. It was a memento from Army basic training in 1942, where a bad case of bronchitis left him with a lifelong impairment. He didn't talk about it much, but anyone who got close to him saw that it shaped his life. "The deafness isolated him," said one man who knew him in his early twenties. "It removed him from close contact... there was some change in the way you approached Tim because of the deafness."

Afterward, Leary was surprised to find that his newly acquired disability seemed to be leading him toward his life's work—and to love. Leary became a specialist in administering personality tests to soldiers with hearing loss. In 1944, when he reported for duty at a clinic for the deaf in the Pennsylvania coal country, a young woman named Marianne was the first person he met. "I fell in love on the spot," Leary recalled. Intelligent, ambitious, and with a beautiful soprano singing voice, Marianne Busch came from a strict Catholic background just like Leary. She shared with him a half-articulated desire to burst loose from it. Leary encouraged her to pursue a career as a speech pathologist while she helped him focus on his newly discovered passion for psychometric tests. By August 1947, Tim and Marianne were living in a small apartment in Berkeley, California. Marianne was eight months pregnant with their daughter. Her husband was preparing for the first day of UC Berkeley's PhD program in psychology.

Photographs of Marianne Leary from the early 1950s, preserved in Timothy Leary's archive at the New York Public Library. *Courtesy New York Public Library (photographer unknown).*

Throughout his early life, Leary was determined to avoid the fate of his absentee father, a dentist named Timothy Leary Sr. who suffered from a debilitating addiction to alcohol. But his father's legacy was strong. When young Timothy Jr. was accepted to West Point as a teenager, his tight-knit Irish Catholic family foresaw a glorious future in the military—until he was expelled after getting blackout drunk on cheap whiskey at the Army-Navy football game. As the 1950s began and the Learys welcomed their second child, the old specter emerged once again. Like the Keeses and the Batesons, the Learys felt caught between their creative ambitions and the containment model of domestic life in the age of Eisenhower. Marianne went into occasional flights of panic about the danger of atomic warfare. Leary grew an unflattering mustache and found that the waistbands of his pants no longer fit. They sloshed their way through twice-weekly cocktail parties with Leary's grad school friends, fancying themselves intellectuals as they listened to James Joyce short stories while pouring strong martinis or debated existentialism over smoky glasses of mescal at a Oaxacan restaurant in Oakland.

Leary enthusiastically and unabashedly chased women. By 1954, he was conducting an affair with a secretary from his office: a thirty-two-year-old married woman named Mary Della Cioppa whom he called "an Audrey Hepburn twin." By this time, he had finished his PhD and was working at the Kaiser Foundation Psychiatric Clinic in Oakland. His closest friend was an equally hard-drinking and even more libidinous UC Berkeley psychiatry resident named Frank Barron.

A ladies' man from a Pennsylvania coal-mining town with a taste for cathartic literary fiction—Truman Capote's *Other Voices, Other Rooms* was a favorite—Francis Xavier Barron was a natural foil to Leary. They had been working together since 1948, when Barron was only twenty-four and Leary twenty-six. Both had served in the Army, but whereas Leary's service in World War II consisted of office jobs, Barron had seen war up close as a field psychologist in Europe. He became the chief ringleader for what Leary thought of as his "bardic" tendencies: drunkenness, mysticism, and machismo. But he was also an important intellectual force in the Bay Area psychological community, helping originate the scientific study of creativity. Barron's Rolodex was stuffed with interesting names, and he enjoyed nothing more than smashing them together like the high-energy particles that careened into one another in the cyclotron—the particle accelerator used in atomic energy experiments—that lay hidden behind a security fence in the Berkeley Hills.

It was Frank Barron who first introduced Tim and Mary. The affair became an open secret among their circle of friends; "everyone was aware of it," said one. By the summer of 1955, Marianne was actively planning for her husband to leave her. Perhaps she would finally go back to school and become a speech pathologist like she had always hoped, she told friends. Or maybe she would move to Switzerland with the kids and learn how to ski. She drank more and more.

--------

The last note in Leary's archive before Marianne's death is scribbled on the back of a photograph of their son, John. He and Marianne went to "a terrible party" on October 15, 1955, Leary wrote. Six days later, on the day before Leary's thirty-fifth birthday, he awoke at dawn to find Marianne's place beside him in bed empty. There was a note on her pillow. "My darling, I cannot live without your love," it read. "I have loved life but have lived through you. The children will grow up wondering about their mother. I love them so much and please tell them that. Please be good to them." He went downstairs. The heavy redwood door of their garage was closed. Behind it was the hum of an engine. Marianne was in her nightgown, sitting behind the wheel of their car, asphyxiated. Their two children, Jack and Susan, were still asleep.

Leary later recalled the last thing he ever said to Marianne. It had been 2 o'clock that morning, just before they went to bed. They were both drunk on martinis. Marianne told Timothy she knew about his mistress and was "incredibly unhappy."

"That's your problem," Leary replied.

The psychiatrist who extolled the virtues of connecting mental anguish to a wider social web, seeing the individual as part of a fabric of relationships, had, in the end, reduced his own misdeeds to a problem for someone else to solve.

Newspapers throughout California ran the story. "Psychologist's Wife Ends Life After Conference," went the headline of an article in the *Long Beach Independent*. After his wife died, Leary put an ad in his own local paper: "Housekeeper wanted to live in motherless home. Two children 6 and 8. Must understand and love children."

--------

By 1956, Leary was reassembling himself with some success. A nanny had been found for the children, he had married Mary Della Cioppa, and his psychology research was progressing. And then another disaster threw his life out of joint. Leary's doctoral adviser at Berkeley, Hubert S. "Hugh" Coffey, was among his closest friends. The Leary house at 1230 Queens Road in Berkeley was less than half a mile from Coffey's residence in the Berkeley Hills, and the two men saw one another frequently. In September 1956, Hugh was arrested for "public lewdness" in a men's restroom in Oakland.

Soon after the arrest, Leary received a letter from Berkeley's dean referencing the "tragic situation." The dean thanked Leary for his support of Coffey.

Then the story hit the newspapers. Word of Coffey's arrest for "making an improper suggestion to a plainclothes officer in Oakland's Lakeside Park" appeared in the *San Francisco Chronicle.* A hastily torn-out copy of the article is preserved in Leary's archive.

Amid the investigation of his mentor, Leary's drinking escalated. "Written when blind drunk," reads one document in Leary's archive from this time. Another records his first day of work at the Kaiser Foundation Psychiatric Clinic in Oakland. "Kids are late to school; get to clinic at 9:30 and can hardly walk," Leary writes. "Leave then come back—should have cancelled... I'm drunk etc." This behavior was not unusual—even though Leary's patients at the hospital included chronic alcoholics. "Tim drank all the time," remembered one of his colleagues. He favored California white wine sold in half-gallon bottles.

Meanwhile, Leary raced to finish a book based on research he'd conducted with his friends Merv Freedman and Abel Osorio. According to Freedman, their project was "responding to a question posed by Jurgen Ruesch and Gregory Bateson: '...can the processes of daily living become a focus of scientific endeavors?'" But Freedman and Osorio did not remain Leary's friends for much longer. Leary alienated both men by portraying their collective work as a project he had led.

Leary worked as a research associate at the Langley Porter Clinic in 1955, and likely met Bateson at that time. He certainly resembled Bateson in his disarmingly informal approach to psychotherapy, which often involved befriending patients rather than simply treating them. "He wanted a model where, instead of detachment, psychologists practiced involvement," said Richard Alpert, a.k.a. Ram Dass. "When he and our grad students answered patient requests for help, instead of meeting at an office, they went to the person's house. They'd sit at the kitchen table, drink coffee, and ask 'Ok, what is the situation, and how can we help?"

--------

Murray Jarvik, the young doctor who assisted Harold Abramson during the LSD trials of the summer of 1954, would later claim that he had inspired Timothy Leary's interest in psychedelics by telling Frank Barron about Abramson's LSD experiments. Toward the end of his life, another distinguished scientist made much the same claim. Jack Block, a longtime psychology professor at UC Berkeley, believed that it was *he* who had first "turned on" Barron and Leary to psychedelics.

Block counted Gregory Bateson as one of his most significant mentors. During the early 1950s, he was Bateson's research assistant in his study of "paradoxes of communication." One day, he might drive across the Golden Gate Bridge with Weldon Kees to film guide dogs; on another, he would visit the curmudgeonly poet Kenneth Rexroth to observe how he interacted with his newborn baby. The work was at times mystifying, but Block began building his life's work out of it. If he could invent reliable tests for assessing emotional states over time, he reasoned, then psychology might finally become a truly quantitative science. The underlying question was fascinatingly fundamental: are we truly the same person from moment to moment? Almost all of us *feel*, intuitively, that there is an inner core, a sense of self, that persists across the days, months, and years. But could this be proven? And what about drugs that promised to dramatically alter that sense of self?

These questions—along with his reading of Aldous Huxley's *The Doors of Perception*—led Block to find some mescaline of his own. In 1956, sitting in an interview room at the Langley Porter Clinic, Block tried the drug for the first time, "observed by Gregory Bateson." Block loved it. Mescaline "seemed to enrich and remarkably broaden perception, whereas tranquilizers dulled and narrowed perception." The experience was "profound . . . something cosmic and religious in a Zen-like way."

Block became a committed psychedelic user, and Bateson continued to take an interest in his trips. When Bateson and Aldous Huxley met at Stanford in April 1957, Bateson followed up to tell him about the mescaline trips he had been observing at Langley Porter. "I wish we could have had time to talk quietly about some of these things," he wrote to Huxley afterward.

For his part, Block was speaking more and more loudly about his own experiences. He remembered thinking that if only the leaders of the United States and USSR could experience psychedelics together, the Cold War would end. Before long, he was conducting clinical trials of mescaline's remarkable effects on what he called "ego structure" and creativity among patients in the Bay Area and telling his friends and colleagues about it. Two of them were Frank Barron and Timothy Leary.

## 1957

Following Marianne's death, Timothy Leary began calling himself an "existential psychiatrist." The scientist should be a participant in real-world struggles, he argued, a figure who didn't just design and observe laboratory studies but who moved in the world, finding new ways to reach patients in their homes, in community centers, and in the streets of cities.

He was also finding new ways to position himself as a visionary scientist. "I thought I knew how humans could direct their personal evolution," Leary remembered. In notes for an unpublished memoir, Frank Barron scribbled his impressions of Leary at this time: "A phony persona...acting...Tim's life was centered on the lost father." Elsewhere, Barron wrote of his friend, "One of Tim's attributes was his strong desire to get his name in the paper." Timothy Leary Sr. had not only abandoned his family, but entirely disappeared: a friend recalled Leary seeing an old man passed out on the sidewalk, and speculating, only half-jokingly, that it might well be his father lying there.

By 1957, Leary seemed to have decided to do exactly the opposite of his father. He would be everywhere.

Nineteen fifty-seven was the year that Timothy Leary made his name as a scientist. His breakthrough was an idea for visually mapping the components of an individual personality on a wheel-like pattern. Though it had been developed by half a dozen Berkeley researchers, this soon became known as the "Leary Circle." It was strikingly similar to the theory of "the Squares" that had emerged out of Mead and Bateson's feverish collaboration on the Sepik River early in 1933. Leary mapped personality on two poles: a north-south

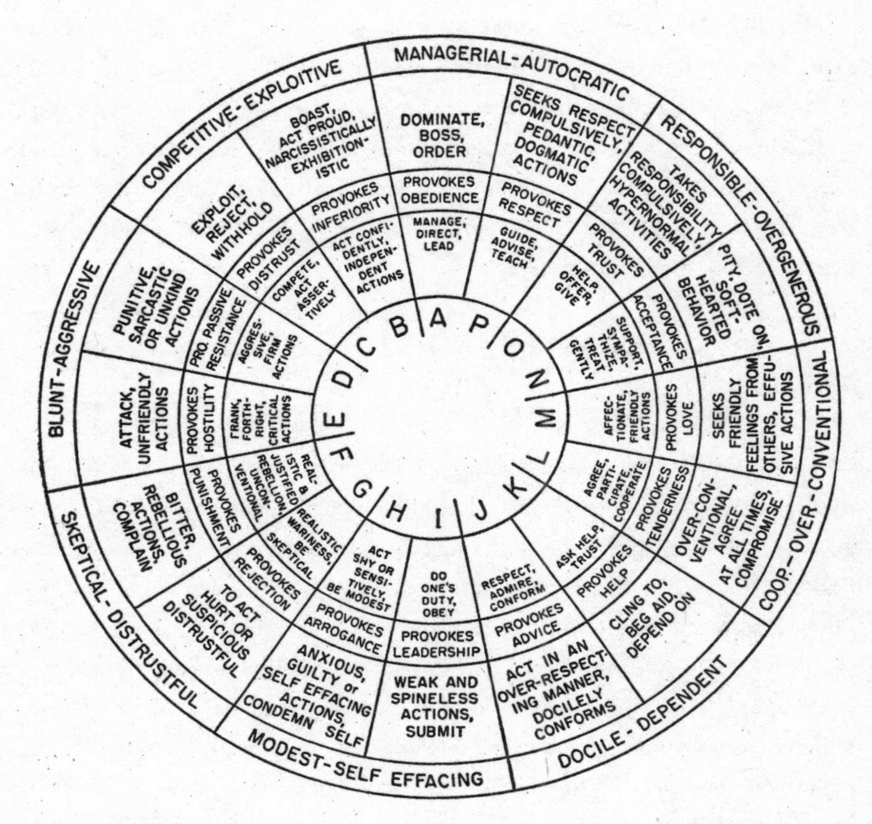


An early depiction of the “Leary Circle” personality mapping system in Timothy Leary, “The Theory and Measurement Methodology of Interpersonal Communication,” *Psychiatry* 18, no. 2 (1955): 147–61. *Courtesy Taylor and Francis.*

line running from submissive to dominant, and a west-east line running from aggressive to cooperative. Franz Boas had warned Mead that the idea of precisely plotting human personality on a chart was a dangerously bad one. It encouraged an illusion of certainty, he said, offering new ways for prejudice and bias to be scientifically “proven.” The Margaret Mead of 1933 abandoned the idea. The Timothy Leary of 1957 was becoming famous for it.

Or at least famous among the relatively small community of personality psychology. He soon found that this was not enough for him. Because Leary, like Mead in 1933, didn’t just want to publish books and articles and contribute to the patient labors of science, to add his anonymous labor to the building of a multigenerational cathedral. He wanted to save the world.

--------

In the late 1950s, psychedelics were not yet illegal, not yet stigmatized, and still perceived to be latent with untapped potential energy for changing society. And Jack Block, Bateson's younger protégé, was a true believer.

"He was *extremely* positive about his experiences with LSD and mescaline during that time," Block's daughter recalled. By 1957, Block had left the Langley Porter Clinic to join UC Berkeley as a professor of psychology. He spent much of his time in the hallways and conference rooms of Berkeley's Institute for Personality Assessment and Research (IPAR), the center where Frank Barron now worked. The two men disliked each other from the start—they were both deeply competitive, and both working in the same field of personality science. They also both felt a certain strain from the Cold War ambience at IPAR, which was primarily funded by military grants. Block called IPAR "a derivative of the derring-do wartime OSS," led by Donald MacKinnon, who had worked on psychological assessment with Bateson and Hamilton in the Office of Strategic Services. "What we had from the OSS was the legacy of the undercover agency," Barron remembered. "I even wondered in dark moments whether I was unwittingly working for the CIA."

Block, too, began to harbor doubts. "There was a feeling of military funding, undercover. Some FBI agents came to Tolman Hall, the Psychology Department at Berkeley, and were investigating my father," Block's daughter said. "It created conflict for him to take the military funding, which was part of the mescaline studies, which were funded by the Department of Defense or something like that... there was some part of this that made him uncomfortable."

Nevertheless, Jack Block remained deeply committed to the utopian potential of psychedelic therapy. "He considered them to be among the most important and transcendent experiences of his life," said his daughter. "He felt that it was an experience that everyone should have." Block saw the experience as a "dissolving of self" that unlocked deep mysteries about the human mind and "made him a better person."

Frank Barron was aware of Block's mescaline experiences and did his best to get his colleague and rival to share the drug with him, but to no avail. "He was intrigued by word of my mescaline adventure," Block recalled, but "I wasn't talking about it"—at least not with him. Barron began looking elsewhere for a connection to the world of psychedelic therapy. And he began discussing the concept with his friend Timothy Leary.

--------

In 1957, while Block was going deeper into mescaline, Leary was trying to reshape himself as a visionary scientist, welcoming the attention he was beginning to receive as the inventor of the "Leary Circle." This alienated many of his closest friends and colleagues, who believed he had unfairly claimed individual credit for work done as a group.

Leary responded to their mild criticisms by diving headfirst into a midlife crisis. He quit his job at the Kaiser Foundation Psychiatric Clinic in Oakland, packed up the family home, and bought one-way tickets to Spain for himself, his son, Jack, and his daughter, Susan. Here, he tried and failed to write his follow-up academic work, which would be called *The Existential Transaction*. It was agonizing, as he remembered it. He spent his mornings, Leary recalled, "sweating in a small room in a Spanish house adding and subtracting long columns of figures" relating to quantitative measures of personality. By the early afternoon, frustrated, he would end work and walk to the nearest bar, where he would "drink and talk detached-zombie-fashion with the expatriates and leave abruptly and run back to the house and continue the paralyzing calculations, sweating in panic." He would later claim that this experience triggered a mental and physical collapse so total that it resulted in a kind of rebirth.

"I died. I let go. Surrendered," Leary wrote. "I slowly let every tie to my old life slip away. My career, my ambitions, my home. My identity. The guilts. The wants. With a sudden snap, all the ropes of my social self were gone."

--------

Margaret Mead, as it happened, had already coined a phrase to describe this state: Leary was having a "transformation experience," a "violent" and "complete" break with the past after which the individual feels "reborn." This had been the subject of her research on the island of Manus in 1953, when she was studying the aftermath of the cultlike religious movement there known as the Noise. And she had come to see these transformations as a source of chaos, pain, and confusion—*unless* they were accompanied by a set of techniques that allowed an individual or a society to "revise and integrate" their old ways of life rather than reject them. For a time, she hoped that psychedelics could play a role in precisely this. They had a part to play, in short, in the goal that continued to define her life: leveraging the power of science to help create a new culture capable of adapting to rapid change, a culture that used science to ensure collective human flourishing and expand the limits of conscious experience rather than as a way of asserting power.

"What will have to happen before we have constructed a world which takes into account that instead of near-starvation we can hope for food for all?" Mead had asked at the outset of her 1955 book about her experience on Manus. What was needed, she answered, were tools for integrating the past wisdom of human cultures into a new form—"twentieth century housing for twentieth century people," as she put it.

Margaret Mead believed at that time, in 1955, that mescaline or LSD might allow the user to "reorganize" themselves in such a way that they could better adapt to a rapidly changing social order, heading off the mystical "fanaticism" that so often accompanied transformation experiences. This is what she had meant when, typing her thoughts about LSD late at night in the summer of 1954, she wrote of the "new possibility" that LSD might offer a path "away from mysticism and escape." And for a brief period of a few years in the late 1950s, this dream still seemed achievable.

CHAPTER 18

# Out of the Lab (1957–59)

> If the doors of perception were cleansed every thing would appear to man as it is, Infinite. For man has closed himself up, till he sees all things thro' narrow chinks of his cavern.
>
> —*William Blake,* The Marriage of Heaven and Hell

## Spring 1957 to Summer 1958

*BEEP... BEEP... BEEP.*

On October 4, 1957, a new noise appeared in the sky. The American satellites that chased Sputnik into space tended to emit frightening squeals that could be picked up via radio receivers back on Earth—the uncanny train whistle of Explorer 1, for instance, or Vanguard 1's haunted bird call. But there was something oddly approachable about the chirping Soviet satellite. Something human. It sounded... *lonely.*

The Soviet triumph caught many by surprise, but not Mead. On the day before Sputnik launched, she was already preparing a conference about the future of humanity in space. The Space Race had begun, and Mead was part of it from the beginning.

Just as mescaline users after World War I described their experience in terms of warfare—exploding shells of light, colors like trench lines on maps—psychedelics in the year of Sputnik became noticeably cosmic. "Psychoanalysis resembles Galileo's telescope," wrote the British psychedelic psychiatrist Humphry Osmond in 1957. Osmond, a soft-spoken and reserved man with a deep sensitivity toward the needs of his patients, had been experimenting with the use of mescaline and LSD in psychotherapy since the early 1950s; it was he who had introduced Aldous Huxley to the mescaline that inspired him to write *The Doors of Perception*, and it was in their conversations that they coined the word "psychedelic." In the aftermath of Sputnik, however, Osmond had a new metaphor to contextualize both his own experiences with psychedelics and what he was coming to think of as their role in the ever-present question of how humans could survive the twentieth century. Psychedelics, he said, are "more like the radar telescopes now being built to scan the deeps of outer, invisible space. They are not convenient. One cannot go bird watching with them. They explore a tiny portion of an enormous void." But it was a void that humans would have to learn to inhabit in order to survive. "I believe that these agents have a part to play in our survival as a species," Osmond concluded.

As Sputnik's lonely chirp echoed through the skies, it became emblematic of a broader shift in the application of Cold War technology. No longer the sole domain of covert military operations, rockets and atomic power were now being harnessed for peaceful pursuits. This transition was mirrored in the realm of psychedelics, which were also experiencing a metamorphosis. Early media coverage in 1954–55 had piqued public interest in LSD and mescaline, paving the way for their emergence from the confines of the laboratory. By the late 1950s, thousands of people around the world were participating in various forms of psychedelic therapy (not all of which included professional guidance). Psychedelics seemed to be on the cusp of mainstream adoption. Sputnik had heralded a new era of exploration in the deep space of the cosmos. Now the emergence of psychedelic therapy promised to reveal the depths of the human mind.

--------

In the weeks just before Sputnik, a young woman took LSD on a CBS television broadcast that appeared across the United States. It was for a series called *Focus on Sanity*.

"I wish I could talk in Technicolor," the woman murmurs at one point, staring off-camera. Her eyes are practically glowing, her expression a mixture

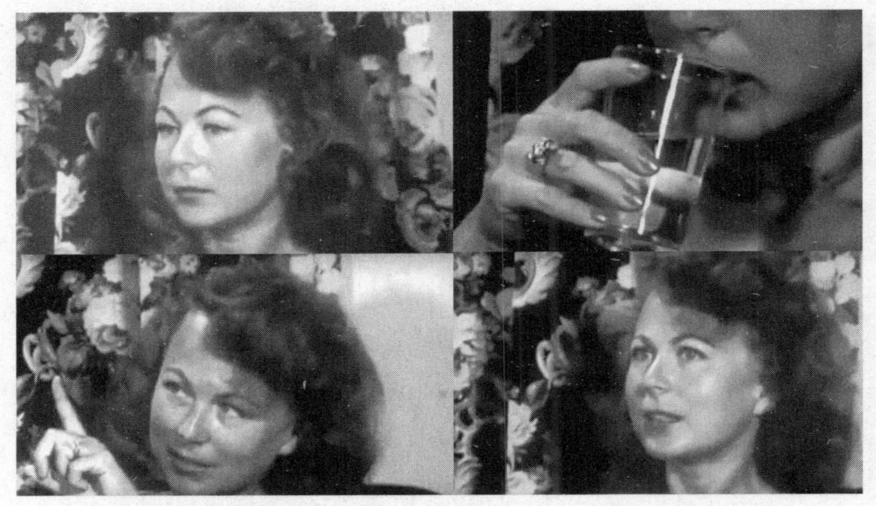

Scenes from the September 1957 CBS broadcast of *Focus on Sanity. Public domain.*

of trancelike intensity and excitement. "Everything is so beautiful and lovely and *alive*."

Then she turns to the doctor who gave her the LSD. He can't see it, she realizes, this newly noticed aliveness inside her and in everything around them. "*It passed right through me!*" she says at one point. "I can see all the molecules. I... I'm *part* of it. Can't you see it?"

The doctor, approaching middle age, is handsome in a nondescript way, like a drawing in a menswear ad. He is wearing a neatly pressed suit. He stares at her calmly. His name is Dr. Sidney Cohen.

"I'm trying," Dr. Cohen says.

They go back and forth like this for some time. Then a look of innocent, sympathetic sadness passes over the woman's face.

"If you can't see it, then you'll just never know it," she concludes. "I feel sorry for you."

--------

The woman in the footage was, and remains, anonymous. But another like her is known.

Betty Eisner was a married mother of two, on the cusp of her fortieth birthday, when she spotted something interesting on the notice board of the psychology department at UCLA. Eisner was finishing up her PhD—getting up before dawn to write before her young son and daughter woke. But something about this notice grabbed her attention and kept hold of it. It asked for volunteers to test a "new and unusual drug." Eisner, who had been fascinated by a magazine article about psychedelics earlier that year, correctly guessed it was LSD. On October 10, 1955, she became the first research subject for the LSD trials of a young UCLA professor: Sidney Cohen. She was entranced by the experience but annoyed by Cohen's attempts to administer personality tests during her trip.

"It was as though I was in the middle of a wide wonderful pasture—free and green and full of sunlight, and something was going on back at the fence that they wanted me to do," Eisner remembered. The fence was the personality tests, the questioning, the world of everyday science that, in her normal life, she desperately wanted to enter. The green pasture, Eisner felt, was a space where something profound but deeply hidden could be found, if only you spent enough time there. This, she realized, was the great paradox of life. "Leaving the pasture is the motion of trying to go—trying to get some place is the difficulty—it is the cause of the descent from Eden. Because the minute that one tries to go someplace or to 'be' someone or something, then one is not content to let things be."

Afterward, Eisner returned to the world of ordinary science. She received her PhD, which was delayed a year because "both kids got the mumps." And then she signed up to work in Cohen's lab, this time as a scientist rather than a test subject.

"During this period, the fall of 1956 and early 1957, there was a boiling activity," Eisner recalled. "We read report after report—dozens—of people who had taken LSD and/or mescaline. And we discussed them, Sid and I." William Griffith Wilson, the legendary founder of Alcoholics Anonymous (known to the group only as "Bill W."), paid them a visit one day to try LSD himself; he seriously contemplated making the drug the "thirteenth step" of his famed twelve-step program. ("I am certain that the LSD experience has helped me very much," Wilson wrote to a writer and amateur psychedelic researcher named Gerald Heard. "I find myself with a heightened color

Betty Eisner (shown here at left in 1970) and Sidney Cohen (shown at right in 1965, in a pamphlet advertising a lecture he gave on "Sensory and Extrasensory Communication") had by the late 1950s emerged as the two leading figures of psychedelic science in Southern California. *Courtesy Hachette Book Group (Eisner); and Clare Boothe Luce papers, Library of Congress (Cohen).*

perception and an appreciation of beauty almost destroyed by my years of depression.") With Cohen, Eisner dreamed of large-scale studies of the effects of psychedelics on creativity, alcoholism, depression, cognition. She began seeing herself as part of a movement.

One of Eisner's closest colleagues described her as "an extremely serious person" who became wholly devoted to the idea that psychedelics could expand human potential. She was not alone. More than anywhere else, it was in California in the 1950s, in clusters of researchers centered on UCLA and Stanford, that modern psychedelic therapy emerged. As with much else in midcentury California, this variant of psychedelic science arose not just in obscure academic conferences and locked laboratories, but under the glare of television lights and in the columns of mass-market newspapers and magazines; from the start, it was a field that generated substantial media attention, interweaving applied scientists with enthusiastic amateurs.

Aldous Huxley, the longtime Los Angeles resident and author of the dystopian science-fiction novel *Brave New World*, was an especially important early contact for both Eisner and Sidney Cohen. Huxley's *The Doors of Perception*, a book that recounted his experiences with mescaline, had helped spark the initial wave of popular interest in the therapeutic potential of psychedelics, and by 1955 he was an enthusiastic user of LSD as well. That same year, a banker named R. Gordon Wasson and his Russian-born wife, Valentina, were seeking out Indigenous *curanderas*, healers, to sample psilocybin mushrooms in Mexico. Wasson's 1957 account of these experiences, a *Life* article entitled "Seeking the Magic Mushroom," introduced psilocybin mushrooms to a wide audience and helped to legitimize their use in psychedelic therapy.

Gordon Wasson got the credit, but it was Valentina Wasson—a physician and expert mycologist—who did much of the legwork. This was typical of the era. To an underappreciated extent, many of the pioneers of psychedelic therapy in the 1950s were women. In Buenos Aires starting in 1956, for instance, a group of Argentine psychoanalysts led by Luisa G. de Alvarez de Toledo began developing an innovative psychedelic therapy technique that combined LSD with talk and group therapy in comfortable, nonclinical settings that integrated food and calming music. Their goal, as with Eisner's group in California and the emerging psychedelic therapy research taking place in Palo Alto, was not to replicate psychosis but to trigger new insights in the patient that "facilitated emotional unblocking" and the "surfacing" of unconscious conflicts—not to mention what Toledo and her team called "oceanic experiences of identity with the cosmos." Female scientists like Luisa de Toledo in Argentina and Betty Eisner in California, working behind the scenes as they struggled against ever-present misogyny, have not made their way into the mainstream history of psychedelics. Yet they pushed the field forward during the short-lived but remarkable period in the late 1950s when psychedelic therapy was legal and widely used around the world.

One of the more important, and least studied, chapters in the history of psychedelics at this time took place in a cluster of cabins amid the redwoods of the Santa Cruz Mountains, just south of Stanford. The Sequoia Seminars had been established in the late 1940s by Emilia Rathbun, a Mexican American activist and spiritual seeker, along with her husband, Henry, a distinguished

Stanford Law School professor. The group they led was an odd mixture of an intentional Christian community, a summer camp for Silicon Valley notables, and an intellectual retreat that emphasized experimental psychotherapy. Betty Eisner attended regularly. So did several influential members of the early Silicon Valley tech community, most notably Willis Harman, a Stanford electrical engineering professor who would later become a psychedelic researcher. By the summer of 1957, when the "magic mushroom" was making headlines, Betty Eisner was already leading LSD workshops for the core membership of the Sequoia Seminars. Eisner's work proved to be an important catalyst, helping to inspire Joe K. Adams and several other Palo Alto psychiatrists to begin experimenting with the drug in the same year. More than this, though, it introduced a social and recreational element to what had formerly been treated primarily as a laboratory experiment. The groups of people in the redwoods of 1957 using psychedelics improvisationally, seeing them as spurs to creativity or tools of personal insight, were small in number and demographically quite limited—overwhelmingly white, middle aged, and upper middle class. But they were also developing a template for a psychedelic experience mingling science with mysticism, therapy with recreation, that would explode into public consciousness with the hippies of the 1960s.

With all these changes, however, Harold Abramson remained the grand old man of psychedelic science. He had not given up his habit of conducting weekly LSD-laced dinner parties at his palatial home on Long Island. Remembering this period of his life years later, Abramson said, "It was all I could do to prevent all of Brookhaven, people in the school system, friends, and so on, to come to dinner with us on Friday evenings to take LSD." On a Friday evening toward the end of May in 1957, Gregory Bateson was one of those who knocked on Abramson's door. Bateson took only 35 milligrams, a small dose. Even then, he recalled, "I got myself into an interesting bind." One of the other attendees, tripping on a higher dose, asked him how he felt.

"Not so bad," said Bateson. "I think I could still lecture a class."

"Why don't you?" the other man replied. "What are you doing in California?"

Lacking a PhD or a medical degree, Bateson was by this time an adjunct lecturer in anthropology at Stanford University and a full-time "ethnologist" at the Palo Alto Veterans Administration hospital. He apparently perceived the man's comment as a condescending jab. True, Palo Alto was no longer the sleepy farm town it had been before World War II. Alongside neighboring Mountain View and Menlo Park (where Bateson lived), the region now hosted fast-growing tech start-ups such as the Ampex Corporation, Shockley Transistor Corporation, and Fairchild Semiconductor. Nevertheless, when viewed from the New York City metro area, Bateson's life in the foothills an hour south of San Francisco may have been hard to fathom. *What are you doing in California, working at a mental hospital in the suburbs?* the man may as well have asked.

But things were changing quickly. To a striking degree, the world of psychedelic science in the 1950s was not just a field but a social circle. It was based largely on who you knew—and few outside the field's inner circle had an inkling of how influential Bateson's once-sleepy corner of California was becoming. By the end of the 1950s, it is likely that no region on earth harbored more psychedelic researchers per square mile than the area of orchards, ranch houses, and research labs around Stanford University. As Margaret O'Mara documents in her book *The Code*, Silicon Valley's origins lay in military funding for advanced communication technologies during the Korean War. A parallel infusion of funding was reshaping psychiatry and pharmacology at the same time and place. Cross-pollinations between these two worlds were frequent. Gregory Bateson—whose home sat less than a thousand feet from the garage that would become the first headquarters of Google—thus found himself not just witnessing but shaping the place that would become known as Silicon Valley in the years of its birth.

Entanglements between Stanford-area technologists and psychedelic science were already forming in the redwoods of the Santa Cruz Mountains via the psychedelic workshops led by Betty Eisner. But the spring of 1958 marked a key inflection point for the rise of psychedelic science in Silicon Valley. This was the time when Gregory Bateson and his protégé Donald deAvila Jackson, a young psychiatrist with a nascent interest in LSD-assisted therapy whom Bateson, four years earlier, had hired to work alongside Jack Block and Weldon Kees in his Macy Foundation–funded study of communication, began planning an organization called the Mental Research Institute (MRI).

The MRI was an independent organization, but it was also a direct outgrowth of the Bateson group, drawing its leadership from the same team that had coauthored Bateson's early papers on his "double bind" theory, and with an advisory board that included Jack Block. Supported by an initial seed grant from the Macy Foundation, it was administered by Stanford University and run out of a suite of offices in downtown Palo Alto, just one block from the famous garage where the original tech company in the area, Hewlett-Packard, had been founded.

The MRI was of critical importance in the history of psychedelic science not just because of the studies conducted there, but because of the social circle of researchers it brought together. One of the institute's first employees was Dr. Joe K. Adams, a soft-spoken young Texan who had left a cozy position as chair of the psychology department at Bryn Mawr College to begin working as a research associate at Stanford. Here, he met Gregory Bateson, who he soon decided was "one of the cleverest men alive."

Another new arrival at Stanford in 1958 was Frank Barron, the creativity researcher who remained Timothy Leary's closest friend. Barron had won a fellowship from Stanford's Center for Advanced Study in the Behavioral Sciences, hoping to study ways of fixing the communication breakdowns and conflict that he believed were contributing to the nuclear arms race between the Soviets and the Americans. Based on what he had heard of mescaline, psilocybin, and LSD (including from his former colleague and rival, Jack Block), Barron wondered whether these substances might have some role to play in ending the Cold War. In July 1958, he began hunting for a contact in Mexico who could arrange for him to use psilocybin mushrooms. The person he found, the Mexican American anthropologist Anita Brenner, was a longtime friend of Margaret Mead's who had studied alongside her and Ruth Benedict under Franz Boas in the 1920s. With her help, Barron began planning a trip to the Mexico City lab of an eminent professor of neurology, Efrén C. del Pozo, to observe his experiments with "creative individuals while under the influence of hallucinogenic substances."

The following January, in 1959, Barron flew to Mexico City to participate as a test subject in del Pozo's experimental trials with psilocybin mushrooms. Not long after, he met Joe Adams and enrolled as a volunteer for the Mental Research Institute's experiments with psychedelic-assisted therapy.

These two psychedelic experiences, one in Mexico City, the other in Palo Alto, planted a seed that stayed with Barron the rest of his life. He believed that psychedelic drugs could save the world, *in a literal sense.* He came to see them as the trigger for a mystical transition in human consciousness, a moment when history would turn on a hinge and everything would be different, ending wars and unlocking a new renaissance of creative expression. Barron's attraction to psychedelics, in other words, was at least as much about utopian yearning as it was about his subfield of academic psychology. And it was one that had been, in ways both visible and invisible, shaped by the ideas and social patterns of the Macy circle.

## Summer 1958 to early 1959

Where was Margaret Mead amid all this? Since 1954, when she had suddenly changed her mind about using LSD during yet another FBI investigation into her personal life, she had been taking on increasingly significant leadership roles as a scientific administrator, including a one-year term (1957–58) as president of the World Federation for Mental Health. The organization she represented was at this time beginning to sponsor psychedelic research. But Mead was moving into loftier territory. In 1958, the first year of the Space Race, she repositioned herself as an expert in the anthropology of human spaceflight. On a return trip to Bali later that year, for instance, she studied how Balinese children perceived Sputnik.

Even as the eyes of the world turned to the cosmos, however, the seeds Mead had planted in helping to originate and plan the Macy "Problems of Consciousness" conference series—the first interdisciplinary scientific forum that featured extended discussion of psychedelic substances—were taking root. Not just Harold Abramson, but half a dozen other participants in that series were, by 1958, deeply engaged with psychedelic science. And the Macy Foundation was planning its first conference devoted entirely to LSD.

--------

Direct-to-physician marketing—the extraordinarily lucrative advertising strategy that the Sackler brothers were perfecting—did not go unnoticed by Sandoz Pharmaceuticals. In 1958, a vice president at Sandoz named Carlo Henze wrote to Harold Abramson to suggest that Sandoz could sponsor a Macy Foundation conference devoted entirely to the potential of psychedelics for use in psychotherapy. Henze was, like the Sacklers, not just a PR man, but a distinguished physician in his own right. He had been part of the Alsos Mission, a secret team dispatched to Europe in 1943 to document Nazi Germany's progress in biological and chemical weapons. Now, however, he was concentrating on building Sandoz's market share, and LSD and psilocybin (which Albert Hofmann was working on synthesizing at that very moment) numbered among their most promising pharmaceutical offerings.

The resulting Macy conference, on "the use of LSD in psychotherapy," met in April 1959. It was the first international gathering of the world's leading psychedelic scientists. Bateson was there, along with Frank Fremont-Smith and Harold Abramson, as well as Sidney Cohen and Betty Eisner. There was also Ronald Sandison from England, one of the most optimistic psychedelic therapists, who had been using the drug with his patients at Powick Hospital, a progressive new institution that sought to reform the older model of mental institutions as spaces of collaboration between patient and physician rather than as prisonlike fortresses. They went around in a circle explaining how they had got there. Bateson said his LSD interest began with Abramson but was "revived" by Mental Research Institute employees Charles Savage, Joe K. Adams, and Don Jackson; he was most impressed, he said, by the way it broke down the barriers between doctor and patient, "perceiver and the thing perceived."

Harold Abramson, who at fifty-eight was among the older participants in the conference, was by this time looking backward at his own past, not forward to new beginnings. At one point, he alluded to the Frank Olson case—the only public occasion when he even hinted at being involved. Abramson spoke of "rumors . . . of suicides caused by LSD" that in fact circled back to "the same case." In the aftermath, he complained, "it seemed that I had become a sort of psychiatric Dracula." Yet the drug *was* safe, he insisted, if administered in places with access to reliable medical care.

--------

A few weeks after the conference, back in the Bay Area, Bateson ran into an acquaintance: the poet Allen Ginsberg, who shared Bateson's passion for William Blake. In mid-May 1959, at Bateson's invitation, Ginsberg drove to the Mental Research Institute in Palo Alto to serve as a test subject in an LSD therapy session overseen by Joe K. Adams. The setting was an innovation on the sterile laboratory environments of the past. The MRI's research facility was in fact a quaint Victorian with cozy oak paneling and a welcoming porch; the chairs were comfortable, the lights dim. Ginsberg was invited to bring along his own special totems to customize the room: a colorful Tibetan "ghost trap" and a recording of Gertrude Stein.

Afterward, Ginsberg contacted his father, Louis. "I went down to Stanford Univ. the other day to be subjected to a research experiment with a new drug—LSD-25 (Lysergic Acid)," he wrote. "It was astounding." The experience inspired a new poem, one seemingly influenced by Bateson's fascination with cybernetic machines and feedback loops. Ginsberg's "Lysergic Acid" described the drug as "electricity connected to itself" and a "conscious sentient part of the interrelated machine." It was perhaps the earliest poem about AI in history. And it had been inspired by Gregory Bateson and LSD.

After his trip at the MRI, Ginsberg spent the night at Joe K. Adams's house. Adams spoke freely, telling Ginsberg—who he knew was gay—about his own sexual experiences with men. This impressed the poet, who valued frankness above all other traits. Ginsberg, whose life had given him good reasons to fear and dislike psychiatrists, began to see *this* psychiatrist as a kindred spirit.

Adams and Bateson both believed that the value of psychedelics lay in their ability to break down the barrier between the perceiver and the thing perceived—and between doctor and patient, professional and amateur. But this breaking down of barriers had its downsides. The symbols of medical authority, the trappings of degrees and licensing boards and Hippocratic oaths, were not well regarded by Bateson and the young men he mentored. They were seen as vestiges of an older time. But they were also the hard-won legacy of reformers and radicals of earlier generations, the product of centuries of bitter lessons learned. In the medical clinics where psychedelic therapy was emerging, those lessons would soon be learned again.

--------

But before the fall to earth, there was a brief and dazzling period of progress. Throughout Silicon Valley and beyond in 1959, the belief was spreading that psychedelics could change the world. The most optimistic, and perhaps the strangest, proponent of this view was a mirthful middle-aged man who claimed to be a millionaire uranium miner. He called himself "Captain" Al Hubbard. Though many around him assumed he had a military background, Hubbard's self-appointed title actually came from a brief period when he piloted a private yacht off the coast of Los Angeles in 1940. Likewise, Hubbard has often been described as the "Johnny Appleseed of LSD" and even as the inventor of psychedelic therapy. However, no evidence indicates he used the drug prior to 1955.

Al Hubbard was in fact a confidence man of the old school. In the 1920s, the Kentucky-born Hubbard bilked a priest by selling him a "perpetual motion machine" that supposedly generated energy from changes in atmospheric pressure. By the mid-1950s, having obtained a fake PhD from a diploma mill, he was presenting himself as a consultant who advised scientists on the nature of the psychedelic experience. It was in this persona that the staff of the Mental Research Institute first encountered him in 1958.

It was in this time and place that one of the most influential ideas in psychedelic science was developing: "set and setting." Early psychedelic therapy sessions typically took place in forbidding, institutional atmospheres featuring buzzing electric lights and white-coated doctors. By contrast, the work at the MRI and the parallel psychedelic therapy being conducted by researchers like Betty Eisner and Humphry Osmond introduced innovations that still define the therapeutic modality of psychedelic therapy today. These included attention to the *setting* of a trip, such as warm ambient lighting, soft cushions, and customized music, and to the *mindset* of the patient. And though it is an innovation that has in the past been credited to Eisner and Osmond—or to Al Hubbard, who worked with both—it was also being pioneered at the MRI

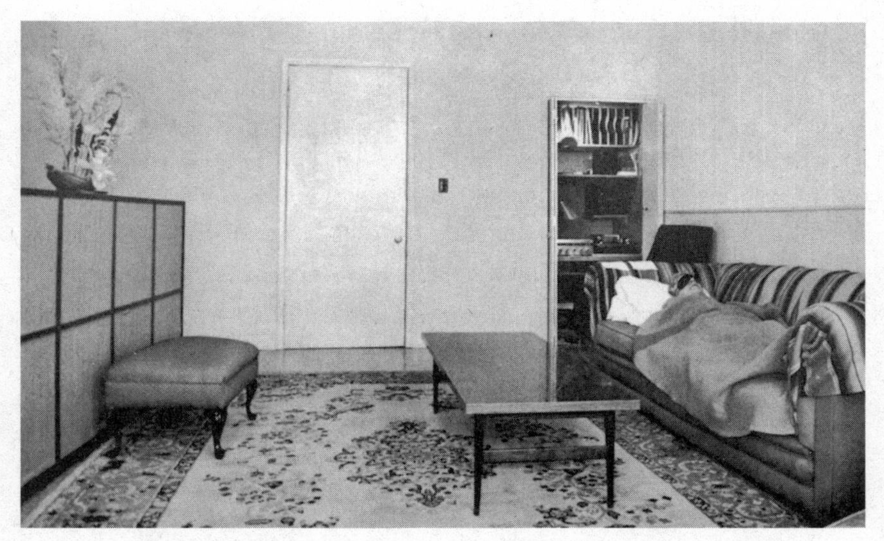

The psychedelic therapy room used by one of Al Hubbard's clients, the psychiatrist J. Ross MacClean, who operated a small mental health facility, Hollywood Hospital, in British Columbia, Canada. This image appeared in the appendix of McClean's 1963 article "LSD-25 and Mescaline as Therapeutic Adjuvants," but it depicts an approach to set and setting that had been developed in the late 1950s. *Courtesy of the New Westminster Archives, British Columbia, Canada.*

by Joe Adams, an outgrowth of his long conversations with Gregory Bateson about the anthropology and science of consciousness.

Adams's awareness of the aesthetic and cultural context of a psychedelic therapy session had helped to overcome Ginsberg's fear of mental health institutions. And Ginsberg was not the only participant who came away from his trip at the MRI feeling as if something profound was in the process of being discovered. Yet, as we'll see, the flowering of psychedelic therapy at the MRI would not last for long.

--------

According to recent studies using advanced brain imaging techniques, psychedelics are capable of integrating, or "hyperconnecting," disparate regions of the brain. At the cellular level, psychedelics may even encourage new communication pathways between neurons (neuroplasticity). But the psychedelic state can also *inhibit* communication. LSD, for instance, has been shown to increase communication between brain areas involved in sensation, but to *decrease* communication between areas related to planning for the future. In other words, psychedelics can be both disruptive and integrative.

In the 1960s, psychedelics would have this paradoxical effect on society at large. Increasing communication could also increase noise, creating not useful signals but misleading and confusing static.

Ted Schwartz, Mead's young protégé during her time studying the Noise in Manus, came to know this firsthand.

After the failure of his marriage to Lenora, Schwartz had married another cultural anthropologist and student of Mead's named Lola Romanucci. For two years at the end of the 1950s, as Bateson was delving into psychedelic therapy, Theodore and Lola were living in a small village in Mexico. They were part of a study of Mexican rural life headed by Erich Fromm, who was based in the resort town of Cuernavaca in the volcanic mountains south of Mexico City. On an August day in 1960, fresh off checking in with their boss in Cuernavaca, the Schwartzes visited a group of psychologists vacationing at a villa nearby. One of them was a man approaching his fortieth birthday. He had laugh lines around his eyes, a square jaw, and a badly healed broken nose—the relic of a drunken fight with his now-deceased first wife. He introduced himself as a lecturer in Harvard's psychology department. Then he boasted that he had found the "sacred mushroom" of Aztec lore and asked if they would like to try some.

Theodore and Lola were sitting across from Timothy Leary on the evening of his second ever psychedelic trip.

The night went badly. Leary remembered it as ruined by the earnest concern of the two anthropologists, who tried to take the experience seriously and worried about its effects. He found it tedious when they sang Mexican folk songs they had learned over two years of fieldwork. Leary was more curious about the Berkeley undergrad who was swimming in his pool.

Years later, Theodore Schwartz attempted to write what he called "the Great Anthropological Novel" based on his experiences with the Noise and what followed in Mexico and the United States. He spent a day in February

1971 with Lola and a tape recorder working out the book's structure. ("Three wives, affairs...," he says dreamily at one point of the book's Theodore Schwartz–like hero—then catches himself. "*You're* the last wife," he reassures Lola, who would divorce him soon afterward.) The narrator of the book, in fact, was to be a fictional Lola, who would discover the male anthropologist's archive and piece together a story to explain his violent death.

The first part of Schwartz's novel would follow the anthropologist and his mentor, a character based on Margaret Mead, as they studied the emergence of a Noise-like cult on an island off the coast of New Guinea. For the second part of the book, covering the 1960s, Schwartz imagined the same characters watching in horror as the phenomena they observed on the island become "mirrored" back home.

"At first the university seems so detached" from the cult, Schwartz says. "It is *studying* the field. At the end it *becomes* the field...There's an increasing noise and turmoil."

The main character, Theodore concludes, will die in a student riot, after "seeing the cult-like aspects, the paranoia, of my own society—mounting and mounting."

"Well, I disagree with this," Lola says upon hearing her husband's plans for fictional martyrdom.

But Theodore won't budge. "No," he says. "This is the way it ends. You can't have an ending otherwise." After all, Schwartz explains, "in the end, you know, there's you looking through all these files...this tremendous mass of material and knowledge."

At this, Lola gives a tiny laugh.

"I will burn them," she replies.

## Part III

# THE NOISE

*(1959–80)*

We began to have feelings which I believe are best described by the word "weirdness." The feeling was that we were up against the edge of a vast uncharted region... The feeling of weirdness came on us as the sounds of this small whale seemed more and more to be forming words in our own language.

*—John C. Lilly, 1962*

If we ever did get efficient enough at planning a Utopia it might be the end of us. The real end.

*—Allen Ginsberg at the Dialectics of Liberation conference in London, July 1967, responding to Gregory Bateson*

Stills from the 1959 television special *The Fine Line*, in which Bateson (top row and at bottom left) appeared alongside Harold Abramson (bottom row, middle) and a colleague who took LSD on camera. *San Francisco State Special Collections and Archives.*

CHAPTER 19

# The Fine Line (1959–60)

THROUGHOUT 1959, PSYCHEDELIC THERAPY captured the public's attention in a way it would not again until the 2020s. In April of that year, the newspaper columnist Joe Hyams published the first in a series of articles which revealed that Cary Grant, one of the world's most bankable movie stars, was an enthusiastic user of LSD. This was no mere recreational drug for the Hollywood icon. Cary Grant credited psychedelic therapy with a personal and professional breakthrough. "All the sadness and vanities were torn away," he told Hyams. "I was pleased with the hard core of the strength I found inside of me. I think I've always been a pretty fair actor. Now I know I'm going to be the best actor there is."

In the days after this article appeared, Hyams received a deluge of telephone calls and nearly eight hundred letters from readers. "Psychiatrists called, complaining that their patients were now begging them for LSD," he remembered. "Every actor in town under analysis wanted it." As rival reporters jumped on the story, Grant's friends confirmed that a transformation really had taken place. "The changes in him as a result of the treatment have been extraordinary," the playwright Clifford Odets told one journalist. "He's bloomed...why, he's almost like a kid." In September 1959, a long profile entitled "The Curious Story Behind the New Cary Grant" appeared in *Look* magazine, reaching well over five million readers. Those readers learned from Grant's doctor, Mortimer Hartman, that LSD was a "psychic energizer" that "intensifies memory and emotion a hundred times." Hartman was a radiologist by profession. But after being introduced to psychedelics by Betty Eisner and Sidney Cohen earlier in the decade, he decided to partner with a psychiatrist, Arthur Chandler, to open one of the world's first psychiatric offices devoted entirely to psychedelic therapy. Unsurprisingly, given what *Look* described as

Grant's "state of euphoria about his self-discoveries" (which included a psychedelic vision in which "his mind seemed to leave his skull and visit outer space"), those clients now began flocking to Hartman's Beverly Hills office.

Even after Dr. Hartman's license was suspended by the California Board of Medical Examiners in October 1961, Cary Grant continued his psychedelic therapy sessions. The actor began taking LSD under the supervision of Dr. Oscar Janiger of Beverly Hills, a close friend and colleague of Betty Eisner and Sidney Cohen who shared their belief that psychedelics could have a potentially revolutionary impact on society. Janiger's files contain a remarkable transcript of Cary Grant speaking into a dictation machine about the events of one LSD trip in Janiger's office—the actor's seventy-second—which took place as the astronaut John Glenn was being recovered from his orbital rocket flight. "He, in a way, is the product of the united thought-patterns and endeavors of so many scientifically-minded men," Grant said into the tape recorder as the acid was kicking in that afternoon. "Perhaps of the united knowledge of *all* men up to its present evolution today." Then he began pondering Hegelian dialectics ("everything creates its opposite and therefore, cyclicly, itself").

Another less public yet equally influential figure advocating for the benefits of psychedelic therapy was Clare Boothe Luce, former congresswoman, ambassador, and wife of publishing magnate Henry Luce, the founder of *Time* and *Life* magazines. A glamorous Broadway playwright turned Republican member of Congress, Luce became the first woman to serve as U.S. ambassador to a major nation when President Eisenhower appointed her ambassador to Italy in 1953. But throughout her adult life, Luce was dogged by a deep-seated depression that had been worsened by the 1944 death of her only child, Anne, in a freak car accident in Palo Alto. Tranquilizers and alcohol merely numbed her pain.

Then, in 1958, Clare Boothe Luce met Sidney Cohen. Under his guidance, as well as that of the English writer Gerald Heard, she began LSD therapy. "For almost 58 years—all my conscious life," she wrote in 1959, "I felt certain that I would be deserted, rejected, 'let down,' denied by everyone or anyone I came to love." But her psychedelic experiences in the spring of 1959 had changed that, Luce believed. (Remarkably, during her very first acid trip, Luce received a telephone call from then vice president Richard Nixon, who was beginning to plan a presidential run against John F. Kennedy and counted Luce as one of his close friends. She decided it would be better to call him back.)

Luce was the psychedelic patient from the 1950s whom we know the most about. Her trips were transcribed and annotated, and her activities were recorded almost daily by photographers, newspapers, and in the archives of fellow VIPs like Nixon, who kept a voluminous file on her. But though she was an exceptional figure, in other ways she was typical of the midcentury psychedelic patient: female, middle aged, white, and haunted by grief or trauma but not diagnosable with a psychiatric disorder—a "neurotic," in the terminology of the age. And she was also typical in another way: she bought into the techno-utopian belief that she was living through an era of wonder drugs. One of Alfred Kinsey's informants in his studies of sexuality, the transgender pioneer Christine Jorgensen, once wrote that the hormone pills she began taking at the end of the 1940s made her think of the atomic bomb. But the atoms contained in these tiny pills promised, she wrote, not death and destruction but "another kind of explosion." A wrenching but peaceful transcendence.

A possible future was contained in this remark: the dream that, like the atom bomb, transformative new drugs could be repurposed from wartime use. From the 1940s through to the end of the 1950s, millions still believed in the dream of atomic power so cheap it was free, of nuclear-powered airplanes and even cars. And in the same era, many of the same people believed that the synthetic drugs which had emerged in the same troubled era as the atom bomb could reshape society toward equally utopian goals. Cohen and Eisner and Luce were among those people.

--------

The summer of 1959 was the final one before the tensions within psychedelic science cleaved it in two. Among other things, that August, Gregory Bateson and Harold Abramson followed Sidney Cohen's lead. They, too, decided to give LSD to a volunteer on TV.

The setting was a television program that aired on the San Francisco Bay Area's CBS station. *The Fine Line* was billed as an investigation of mental illness, but in fact it was primarily a discussion of psychedelics and the double bind. The program featured a dramatic vignette performed by three local

Gregory Bateson discussing "Margaret" and Harold Abramson holding an ampule of Delysid (Sandoz brand LSD) in *The Fine Line* (1959). *San Francisco State Special Collections and Archives.*

theater actors: a teenage girl character (named Margaret) fails repeatedly to communicate with her mother and father while they sit at the breakfast table. There is a fade to black, and then the camera pulls out to reveal Gregory Bateson. He is wearing an extremely rumpled suit and he clutches a cigarette between the knuckles of his right hand. He looks on at the domestic scene as if he has been watching it through a two-way mirror. Margaret, he explains, is stuck in a double bind. "There's a very serious problem there," he tells the program's host. She has "considerable doubt about her identity." Her doubts could lead to madness.

This scene serves as a segue to a deeply strange piece of live television. The viewer learns that a real-life psychodrama is taking place alongside the reenacted one of Margaret at the breakfast table. As the camera swings away from Bateson, the affable host checks his watch and explains that Dr. Harold Abramson of New York has for the past two hours been sitting in a "guarded room above this studio" with one of Bateson's colleagues at the Mental Research Institute, a psychiatrist named James Terrill. And Dr. Terrill has been tripping on LSD. "This drug, lysergic acid, produces a state of induced psychosis: temporary insanity," the host excitedly announces.

The camera then cuts to the room above. Sweating visibly in a tie and short-sleeved white shirt, Terrill sits across from Abramson. Between them is an ashtray filled with ten stubbed-out cigarettes. Terrill determinedly lights an eleventh while staring downward and blinking. "One minute I have the feeling that everything's clear, and then it sort of fades off," he mutters.

Abramson suggests he feels "boxed in."

"It could be," Terrill replies. "It depends so much on what the environment is when you take the drug."

Afterward, with Terrill backstage, having taken a sedative provided by Harold Abramson, the host sits down with Gregory Bateson and Abramson to discuss what has just taken place. "I remember quite vividly when Dr. Abramson gave me a dose of LSD," Bateson says, explaining that parts of his trip had made him feel conflicted and unhappy. Some of Bateson's reticence may have been because he suspected that drugs like LSD had already been weaponized by the intelligence community for nearly a decade. Despite these personal reservations, however, the Gregory Bateson of 1959 allowed himself to be carried along by the enthusiasm of the younger psychiatrists he mentored at the MRI. Although the host of *The Fine Line* repeatedly described LSD as a drug that induced psychosis, other press accounts of the MRI's psychedelic therapy disagreed. Earlier in 1959, for instance, a newspaper article about the LSD trials being conducted there portrayed it as a healing substance that pushed those who used it "out of their limited worlds."

For the younger researchers at the MRI (and above all for Don Jackson and Joe K. Adams, both of whom Bateson had mentored for years by this point), psychedelic therapy promised not just an advance in psychiatric methods, but a transformation of society—a chemical solution for dissolving old structures of elitism, nationalism, and conflict. Joe Adams later wrote that when he heard Harold Abramson describe LSD as producing an "experimental psychosis" on the program, he "privately congratulated myself on being more informed and up-to-date" than the "old-fashioned" Abramson.

Harold Abramson (*left*) and James Terrill (*right*) in *The Fine Line* (1959). *San Francisco State Special Collections and Archives.*

Adams kept the more utopian aspect of his work to himself. As a junior researcher, he did not dare disagree with Abramson publicly. But in *The Fine Line*, there is another man at the table beside Abramson and Bateson: the MRI's director, Don deAvila Jackson. And it is Jackson who comes closest to publicly articulating this view. As if seeking to reassure the television viewers at home who had just witnessed a sweating and downcast James Terrill struggle to articulate anything at all about the psychedelic state he was experiencing, Jackson jumps in to say that his own LSD experiences had been "terrifically pleasant from end to end." Jackson, like Adams, believed that psychedelics could chemically induce what he called "transcendence and a new beginning," providing "new meaning to life." Jackson and Bateson's relationship was by this time splintering into mutual dislike, but for a time earlier in the 1950s, when the two men had published a series of papers on the concept of the double bind and schizophrenia, it had been electric. They believed then that they were moving toward a profound scientific discovery: not just the double bind itself, but the new kind of therapy that would be built on it.

For Jackson, this new therapy—which he believed could transcend older concepts of mental illness by reimagining psychiatric disorders as products of a sick society, not a sick mind—was integrally bound up with the consciousness-expanding and empathy-inducing properties of psychedelics. In *The Fine Line*, Jackson exudes the confidence of a true believer. "I think it's important to point out that the early days of LSD were taken up with a lot of these kinds of experiments," he says, referring to tests conducted in the stressful setting of clinical white rooms with whirring recording equipment and locked doors. "There were a lot of erroneous conclusions drawn." Jackson seems sure that the newer model of psychedelic therapy pioneered by the MRI—friendly faces, welcoming porches, poets in Kublai Khan reveries, warm lighting, music playing—will yield different results.

By this point, the viewer would be forgiven for being confused. Hadn't the host said LSD *induced* insanity? Hadn't Harold Abramson, just a few minutes earlier, described it as capable of inducing over a dozen different symptoms of neurosis? *The Fine Line* captured a period when the polarity of psychedelics in American culture was changing fast. In psychedelic therapy's first decade, nearly every expert had agreed that substances like mescaline, peyote, and LSD caused "experimental psychosis." But as those same researchers began using psychedelics themselves, many concluded that the experience conferred lasting insights. Stuck between an outmoded medical definition and their own reluctance to forsake the veneer of scientific objectivity that would be stripped away if they admitted to their self-experimentation, their contradictory public utterances about psychedelics betrayed an inner epistemological tension.

One of those who best embodied this tension was James Terrill himself. Though the casual viewer might have assumed he was using LSD for the first time on TV, this was not the case. In Abramson's hands was a folder detailing Terrill's extensive experiences with psychedelics. Moreover, Terrill, along with Joe K. Adams, had administered LSD to some sixty patients of his own by this time. In a speech written later in 1959, Terrill cautioned that the expert should be "at home" with the drug's ability to increase creativity, lest he "overlook creative aspects of the patient's thinking and label it all as merely confused or psychotic." In fact, he said, "very few reactions" in his patients could be seen as psychosis. Instead, he noted an increased valuation of creativity, aesthetics, philosophy, and "religious interests."

This ambiguity was an integral part of the first wave of psychedelic science—part, in other words, of how these researchers believed the drug *worked*. Yes, Terrill said, LSD could be "terrifying." But he added, "Several of our professional [volunteers] have remarked that they believed that much of the beneficial effect of LSD was due to a person's having faced a stressful and ambiguous situation and worked it through satisfactorily."

--------

In many ways, this era of psychedelic therapy was narcosynthesis redux. During World War II, psychiatrists had used hypnotic drugs to heal the trauma they envisioned as trapped in the unconscious mind. They believed that a wrenching, difficult drug experience could surface these hidden memories. They spoke of "clearing" soldiers of trauma. In the early 1950s, L. Ron Hubbard had adapted much of this language (and, at least at first, the same drugs) as he sought to turn narcosynthesis into a mass movement. But he had done so outside the scientific establishment. The psychedelic therapy of the late 1950s seemed, for a time, to be entering the scientific mainstream while also offering a revolutionary new improvement on the older drugs (such as sodium amytal), which had dangerous side effects and none of the creativity-inducing, transcendent, or mystical qualities of psychedelics.

Betty Eisner was among the most ardent backers of this view. By the spring of 1958, she had decided that psychedelics were genuinely revolutionary substances that "speed up the process of evolution" by helping their users achieve a state of "clear"—a mind that had been expanded via the integration of repressed unconsciousness traumas into conscious experience. Throughout the last two years of the 1950s, she shared her thoughts at academic conferences, in workshops like those of the Sequoia Seminars, on television and radio appearances, and in newspaper articles. Psychedelics were a pharmacological spur for the rapid cultural change Mead and Bateson had long believed was necessary to heal a sick society.

But even as this belief began to enter the mainstream, it faced a powerful challenge in the form of another new category of drugs: tranquilizers. Where psychedelics demanded that the user "work through" deep-seated memories and beliefs, tranquilizer drugs promised to *remove* stress and ambiguity. Throughout 1959, as both psychedelics and tranquilizers began reaching wider audiences of patients, researchers started choosing sides.

Alongside this public scientific debate on what psychedelics were *for*, revolutionary changes were afoot in the science of consciousness. Through the work of Macy-funded researchers like Walter Pitts, Jerome Lettvin, Mary Brazier, and Warren McCulloch, new frameworks emerged that persist to this day: terms like "artificial intelligence" (coined in 1956), "neurotransmitter" (coined in 1959), and the theory that mental illness is a result of a "chemical imbalance in the brain." Meanwhile, history-changing discoveries of new mind-altering compounds were coming so fast that it was difficult to adapt. In many cases, in fact, the world is *still* struggling to adapt to the new drugs developed in these years of rapid change. In a twelve-month period between 1958 and 1959, for instance, the drug research team at Jannsen laboratories in Belgium discovered both haloperidol—which remains among the world's most commonly used antipsychotic medications—and the fentanyl family of compounds, an enormously potent category of opioid drugs that has been responsible for hundreds of thousands of overdose deaths in recent decades.

Psychedelics were at the center of this activity. Two of the era's most eminent neuroscientists, D. W. Woolley and E. N. Shaw, were, at the time *The Fine Line* aired, entering the sixth year of an epic research project to find the connection between serotonin—which was at the time still thought of as a hormone—and LSD. Evidence was mounting that LSD did *something*, though no one was quite sure *what*, to the serotonin receptors of the brain. They also discovered that the brain's "serotonin system" seemed to be a crucial part of consciousness itself. They spoke of a "serotonin key" that fits into brain receptors "as a key fits into a lock." Could similar substances unlock the same receptors in subtly different ways? Was this, in fact, what LSD and related psychedelics were doing?

This idea of a "chemical key" pushed into the mystical for psychedelic researchers like Betty Eisner. "I think of it like a very rusty door which has never been opened before consciously," she wrote. "There are many keys to the door: stress, solitude, meditation, limited sensory environment, drugs, etc. But these don't seem to work until the rusty door has been forced open once. That is what LSD or mescaline can do."

Bateson, in his quiet way, encouraged the view that LSD could unlock psychic doors. In addition to Ginsberg, for example, there was the Stanford anthropology student who thanked Bateson for his guidance as a mentor, then acknowledged the help of "two LSD sessions, under the direction and with the indispensable help of Dr. Joe Adams of the staff of the Mental Research Institute." Bateson might have shown some ambivalence on *The Fine Line*, but the fact remained that he had publicly revealed his own use of LSD and his friendship with Harold Abramson, perhaps the most prominent psychedelic researcher in the world. As psychedelics reached the peak of their public acceptance via the coverage in *Time* and *Life* magazines and in television appearances, Bateson remained an important gatekeeper in the emerging field of psychedelic therapy.

--------

And then Joe Adams had a catastrophically bad trip.

Negative experiences—"bad trips"—are not unusual in psychedelic therapy. They are understood to be relatively common reactions that can be anticipated and mitigated. Adams's trip was something different, so *legendarily* bad that it lived on as a kind of warning bell in the collective memory of psychedelic scientists for decades afterward.

The events of that day are still hazy, but the following account can be pieced together from surviving records: One afternoon in December 1959, Adams and James Terrill—the young doctor whom Harold Abramson had dosed with LSD on television earlier that year—arranged for Al Hubbard to demonstrate his psychedelic therapy technique to them. But the "Captain" failed to show. Undeterred, Adams decided to take a large dose of the drug on his own, without a trip-sitter. Adams's MRI colleague Charles Savage wrote that the unfortunate young psychiatrist was then "immediately plunged into hellfire," imagining that "he was the infant Jesus, about to be sacrificed." And though this dysphoric part of the trip mercifully ended, Adams's overall psychedelic state did not.

Adams himself described what followed as an ecstatic experience that felt physically "somewhat like an orgasm in every part of my body." For months thereafter, he felt "much more open, aggressive, spontaneous, and unguarded than I had been before." The only person in his life who responded well to the change was, he claimed, his two-year-old son. Others, such as his wife, who ultimately divorced him, "were appalled and considered me 'disturbed.'" Adams underwent two long-term psychiatric hospitalizations, then authored a rueful academic article praising psychedelic therapy while admitting it had triggered his own psychotic break. He left the MRI soon afterward, though he remained close friends with Gregory Bateson.

The end of 1959 marked an inflection point for what could have become a very different history of drugs in the twentieth century. Stories like that of Joe K. Adams, coming on the heels of the 1959 Macy conference on LSD and the appearance of his colleagues on *The Fine Line*, offered a warning that the utopian promise of psychedelics was counterbalanced by an ability to cause harm.

There had been many other warnings before this, to be sure. Psychedelic research—*all* drug research of the 1940s and 1950s—was built on enormous social disparities between the patient population and the scientists studying them. Racism, sexism, the inherent power imbalances of institutional spaces like psych wards and prisons: these were integral to the set and setting of early psychedelic experiments, and Mead and Bateson and their circle turned a blind eye to them. But such failings were common to all branches of medicine and science in the 1950s. As they confronted stories like that of Joe Adams, the researchers most intimately involved in psychedelics began to realize that there was something *distinctively* destabilizing, and perhaps distinctively dangerous, about the same mind-altering properties that figures like Don Jackson and Betty Eisner credited as pathways to collective transcendence.

## 1960

Around the same time that *The Fine Line* was airing on Bay Area television, Timothy Leary was hosting Frank Barron at a rented apartment in Florence, Italy. The book he had begun writing in Spain, *The Existential Transaction*, was now finally coming together. Though the manuscript is lost, quotes that survive show Leary embracing an anthropological approach to psychiatry, reimagining the therapist as a "participant observer," someone who actively intervenes in the life of the patient rather than analyzing it from outside. For all his rhetoric about taking an active role in society, however, it looked to his peers that Leary had retreated from it. Approaching forty, he had alienated most of his colleagues back in Berkeley, was nearly bankrupt, and had no income despite his extravagant multimonth family vacation.

Frank Barron, by contrast, seemed to be thriving. His experience with psilocybin mushrooms in Mexico City and with LSD at the MRI had encouraged him to further pursue research into the origins of creativity, which he now imbued with a profound significance. As Leary remembered it, by the time he showed up at Leary's doorstep in Florence, "Frank had become passionate in his belief that only psychology, by effecting a fundamental reprogramming of the human mind, could prevent a nuclear holocaust." And he believed that psychedelics were the key. Leary remembered being taken aback by Barron's almost cultlike enthusiasm as he spoke about "William Blake revelations, mystical insights, and transcendental perspectives produced by the strange fungi."

But by this time, rebirth was what Leary thought he needed. Leary and Barron were both planning to spend the summer of 1960 in Mexico. There, the two men decided, they would meet and take psilocybin mushrooms: "the tool for [the] behavior change we were seeking."

--------

As the 1960s began, psychedelic therapy seemed poised to become a significant new treatment on a global scale even as the science of psychedelics remained a relatively small and esoteric field, one that had largely grown out of the ethos of the Macy circle.

In just a few years, however, the field would be torn apart by internal conflicts and external pressures. What had once been conceivable as a unified project would fracture into mutually hostile camps. The hippies versus the spies. The skeptics versus the true believers.

And Timothy Leary would be standing at the center of the fault line.

CHAPTER 20

# Anthropologists from the Twenty-First Century (1960–61)

> Some therapists, even, have been known to make this error, of mistaking themselves for prophets, and their pharmacopoeia for a philosophy of life.
>
> —*Gregory Bateson, "Diet or Pharmacopoeia," September 1949 (unpublished manuscript)*

## Summer 1960

THE SUMMER OF 1960 found Timothy Leary and his ten-year-old son, Jack, crossing the Mexican border in a battered Ford. The air was filled with the scent of blooming jacaranda trees and the sound of chirping birds as they reached their destination: the ancient city of Cuernavaca. Within a few days of their arrival, Leary had rented a villa on the edge of town complete with a sparkling pool and panoramic views of the surrounding volcanic mountains. And he had welcomed two additional guests: Dick Dettering, a semantics professor at San Francisco State; and his wife, Ruth, a former nurse. As Jack wandered the rugged hills, the adults sat poolside sipping daiquiris and Mexican beer from morning to evening.

Leary had two goals for his vacation. One was to finish his book. The other was to find psychedelic mushrooms. The second goal proved easier thanks to the help of a Mexican German anthropologist named Lothar

Knauth, an expert in the Indigenous cultures of Mexico. On the second weekend of Leary's vacation, Knauth arrived at the villa's doorstep with a bag of twisted brown fungi. Apart from Ruth Dettering, who was pregnant, each of the adults present on that day resolved to try the drug: Leary and Dick Dettering, Knauth and his American girlfriend, her college-age daughter, and the daughter's friend. Jack Leary played nearby with an elderly iguana he had adopted as a pet.

It had been over a year since Frank Barron urged Leary to experiment with psilocybin mushrooms. Now they lay in his hand. He hesitated for a moment, taken aback by the unexpected smell, which reminded him of "certain New England basements." Then he gulped them down along with a slug of Carta Blanca beer. Their effects came on slowly, beginning with a sense that everything around him was more alive in some indefinable way. Then they really kicked in. He felt he was witnessing the history of all living things in reverse, experiencing "a trip through evolution . . . to those ancient mid-brain projection rooms: snake-time, fish-time, down-through-giant-jungle-palm-time, green lacy fern leaf-time." It was, Leary wrote afterward, "without question the deepest religious experience of my life."

--------

Left unsaid, however, was what this experience told him *about* his life.

Timothy Leary often evoked the "evolutionary" scale of psychedelic experiences, such as their supposed ability to unlock ancient collective memory. But in his many writings about his own early trips, he offered little insight into one very specific set of memories—those of a widowed psychologist with a drinking problem and an unfinished book. On the day of his first psychedelic trip, Leary's fortieth birthday was a few weeks away. And so, too, was the fifth anniversary of his wife Marianne's suicide. In all of Leary's numerous descriptions of his first psychedelic experience, no mention of her can be found. Nor did he dwell on his daughter, Susan (who had been deposited with family friends in Berkeley earlier that summer), or his son, Jack, who was still whiling away a Cuernavaca afternoon with his iguana.

Timothy Leary's thoughts instead turned to his career. He found himself laughing at what he called "the narrow arrogance of scholars." He then fell into a reverie resembling Arthur Stoll's vision of a pageant of global cultures. But if Stoll's cathedrals, caves, masks, and skyscrapers had evoked the continuity of the human spirit across time and space, Leary's vision was more like a sprint through a kitschy art gallery. "Palace by Nile / temple near Hong Kong," he wrote in a poem afterward. "Babylonian boudoir / Bedouin pleasure tent."

At length, Leary shook himself away from the pleasure tents of the mind. He had suddenly "realize[d] responsibility," as he put it—but not for Jack. Instead, he worried that he was neglecting his "role as host." He headed to the deck to tell his guests that drinks were available in the kitchen. Then he sat alone for a long time, staring up at the clouds moving across a darkening sky.

--------

One of the distinguishing features of the psilocybin mushroom is its ability to communicate. As with all fungi, it transmits signals via a network of delicate tendrils—a mycelium—which carries chemical messages over vast distances. But the psilocybin mushroom also possesses a more mysterious ability, a kind of cross-species neurological mimicry. When it is consumed by humans and processed in the liver, it produces a chemical called psilocin. This compound interfaces with the brain's communication pathways by stimulating serotonin 2A receptors, involved in the retrieval of memories. In a sense, then, psilocybin mushrooms really do "talk" to us, joining some small part of themselves to our brains and altering our conscious experience in unpredictable ways.

Timothy Leary knew little about fungi, and scientists of his time were just beginning to understand neurotransmitters. But he *did* know—knew, somehow, with a deep conviction—that the mushroom had a role to play in his own attempts to communicate. More specifically, Leary believed that it could help him score some serious points with his boss.

When he visited Leary in Florence the previous summer, Frank Barron hadn't just urged his friend to try psychedelics. He had also put him in touch with an eminent Harvard psychology professor, David McClelland, who happened to be vacationing in Florence. Beguiled by Leary's down-and-out charisma and his hints of a looming breakthrough, McClelland offered the younger man a job as a lecturer at the institute he directed, Harvard's Center for Research in Personality, which he accepted. Now, a year later, Leary had dutifully followed his boss to a *second* vacation spot. McClelland was spending the summer in Tepoztlán, which was only ten miles away from Cuernavaca. Leary began making plans to invite McClelland to take mushrooms with him and explore the psychological insights they offered.

And he didn't stop there. For Leary also hoped to connect with another famed expert in the science of consciousness and personality: Erich Fromm, the legendary German psychoanalyst, who was overseeing a research project about Mexican village life at a villa just down the road from Leary's own.

In the days after his first trip, Leary drove to meet Fromm's three research assistants. One was already known to him, a Harvard anthropologist named Michael Maccoby. The other two were a married couple who had been recommended to Fromm by his old friend Margaret Mead. These turned out to be none other than Ted Schwartz and his second wife, Lola (who, much like Schwartz's first wife, Lenora, was a young cultural anthropologist whom Mead had taken under her wing). Seven years earlier, Schwartz had traveled with Mead to the island of Manus to study the aftermath of the cultlike movement known as the Noise. Now he was present at another movement in the making.

With a convert's enthusiasm on that day in Cuernavaca, Timothy Leary urged the Schwartzes and Maccoby to take mushrooms. Maccoby remembered being confused when Leary asked if he wanted to experience "hyperconsciousness." As a Harvard undergraduate, one of his friends had participated in the early LSD trials at Boston Psychopathic. Unlike Ralph Blum (who volunteered for the same experiments and found them enriching and transformative), Maccoby's friend became "delusional" and dropped out of school. Maccoby politely declined the offer of hyperconsciousness. But then one of his colleagues, a Mexico City psychoanalyst, got word of the proposal. "Dr. Beltran" (Maccoby used a pseudonym to protect the man's anonymity) volunteered to try psilocybin along with Leary. It was one of Leary's first experiments with the drug, and it went badly. Maccoby had agreed to drop Leary off at Dr. Beltran's house with the mushrooms, where the two men would trip together, and then return at 4 p.m. to conduct post-experience interviews.

He arrived to find the Mexican psychoanalyst pacing outside, "agitated and almost incoherent."

"He is a vampire," Maccoby remembered the distraught man saying to him. "He wants to suck my blood. I am going to kill him." Leary kept insisting that everything was fine, but it was obvious that something had gone badly wrong. "We really need to split," he remembered urging Leary as he hustled him into his car.

"Was Beltran just delusional," Maccoby wondered afterward, "or did he sense a vampire-like quality in Leary, a need to live off of others?" Maccoby himself observed that "there was a deadness about Leary." It was as if the psychologist retained a deep-seated psychological distance from his own life despite all his protestations of transcendence, interpersonal connection, and spiritual awakening.

Timothy Leary's summer afternoon with Dr. Beltran symbolized the paradox of his life: the tension between his idealism and his disregard for others. Making bold scientific claims came easily to him. Honestly facing up to the risks involved in psychedelic research did not. As a therapist with over a decade of experience, Leary knew enough to realize that the Mexican psychoanalyst's break with reality had revealed the need for well-defined safeguards when it came to psychedelic experimentation. But he never mentioned this early warning sign in any of his published writings, nor does he appear to have discussed it with those he subsequently offered psilocybin to. In the months and years to come, Leary swung between describing psychedelics in clinical and messianic terms, but he always seems to have remained secure in his belief that there was little need for self-reflection or for caution. Armed with this beguiling but toxic combination of traits, Timothy Leary would spark a mass movement—but also trigger a massive backlash.

--------

By this time, one of Leary's new friends from Harvard had joined them in Mexico: an assistant professor at Harvard's Department of Social Relations named Richard Alpert. The son of a Connecticut railroad executive, Alpert would later spend time at an ashram in India, where he became known as Ram Dass. Under this name, he became a popular spiritual teacher and author, spreading a psychedelic take on Hindu and Buddhist thought to millions in the West. But the Richard

Alpert of 1960 was a young Ivy League professor, and he looked the part. His suits were from Brooks Brothers. His passions included psychological assessment forms, Rorschach tests, and electronic gadgets. His face was alert and youthful, almost baby-like were it not for his thinning hair and horn-rimmed glasses.

But his cherubic looks disguised a rare tolerance for risk. A good example was how Alpert (who had spent the first part of the summer working as a consultant near Stanford) reached Mexico. He wanted to fly there, but no one would rent him a plane because it was considered too dangerous given his status as a novice pilot who had never flown solo over a long distance—not to mention his rudimentary Spanish. Undeterred, he purchased a Cessna 172 from his flight instructor in Palo Alto without explaining the reason. And then, entirely alone, he flew it to Mexico City over a period of three days.

Richard Alpert (Ram Dass) with one of his motorcycles in an undated photo from the 1950s. *Estate of Ram Dass.*

Timothy Leary met him at the airport. As with everyone that week, Leary began talking about the mushrooms. Alpert promised to visit, and to bring along their boss, David McClelland. But by the time they got to Leary's villa, Alpert recalled, "Tim had shared them with other visitors, and he wasn't sure he could get more."

--------

These "other visitors" were Lothar Knauth and his entourage, plus Ted and Lola Schwartz. According to Leary, the Schwartzes believed that taking psilocybin mushrooms was a necessary part of their anthropological fieldwork. Ted Schwartz, he wrote, "wanted to take them because it was like a duty." Leary describes the day he offered mushrooms to them—his second experience with psychedelics—in his memoirs. The Schwartzes arrived around two in the afternoon. While Leary's other guests swam in his pool, Leary fixed drinks for Ted and Lola and discussed what he called "the mushroom situation."

Although the thirty-two-year-old Schwartz was nearly eight years younger than Leary, the older man came to see that day as a clash of generations. "The young kids"—Mandy and Betty, the daughter of Knauth's partner and her college friend—were, he decided, "the psychedelic generation... the first people I had talked to who were not automatically, reflexedly frightened at the idea of expanding consciousness," Leary wrote. By contrast, Leary depicted Schwartz as a "pleasant intellectual chap" who utterly failed to embody the correct vibe. The anthropologist sat staring at the mushrooms for a long time, asking questions about their place of origin and how they were used by locals. Leary interpreted this as a delaying tactic. It seemed to him that Schwartz was terrified. "I could see drops of sweat on his brow just below the hair line," Leary wrote. "He didn't look happy." Eventually, Schwartz ate the fungi. Leary headed for the pool. He had no time for earnest anthropological questioning. He had bigger plans.

Leary had promised Professor McClelland that he would produce something truly innovative, some paradigm-shifting new psychiatric technique that moved beyond talk therapy. Something that would justify the risks of hiring a man who had flamed out in his previous jobs. Psychedelics, he now believed, were precisely that.

## Fall 1961

The delicate mycelial structures of psilocybin mushrooms and the gnarled orange roots of peyote cacti extend deep into the ground. But the intellectual roots of the psychedelic research project that Leary, Richard Alpert, and Frank Barron developed at Harvard in the months after Cuernavaca were shallow.

"We saw ourselves as anthropologists from the twenty-first century" dropped into "the dark ages of the 1960s," Leary later explained. Like L. Ron Hubbard a decade earlier, Leary portrayed himself as a lonely prophet surrounded by enemies. The truth was different. Exactly thirty years before Leary's first trip, Margaret Mead was living among the Omaha people in Nebraska, collecting information about peyote and its role as an engine of consciousness expansion and cultural change. This was such a common experience among the students of Franz Boas that it was almost a rite of passage. Allen Ginsberg, too, was surprised to learn that Timothy Leary seemed unaware that, as Ginsberg put it, "every poet in San Francisco had lived with Indians and taken peyote and mescaline long ago." Leary had not stumbled onto a long-hidden mystery. He had simply stumbled onto the history and anthropology of psychedelics, having been guided there by Mexican scientists who had far more knowledge in the field than he.

In their tendency to portray their work as a radical break from the past, the group that developed around Leary differed sharply from other psychedelic researchers of the period, who believed they were contributing to an ancient body of knowledge. In the 1957 article in which he coined the term "psychedelic," Humphry Osmond credited what he thought of as fellow experimenters from Mesoamerica, Amazonia, South Asia, and sub-Saharan Africa. "We inherit their secrets," Osmond wrote. "Let us honor them." The pharmacologist Joel Elkes, likewise, was voicing a widely held belief when he wrote that substances like LSD must be used with "the care and reverence that the Mexicans gave" to peyote and psilocybin. Even at the RAND Corporation in Santa Monica—that citadel of Cold War militarism—researchers studying the effects of psychedelics on creativity supplemented their clinical trials by carefully reading, and citing, Margaret Mead's description of the Omaha peyote users.

Leary and his circle were part of a complex history of cross-cultural knowledge exchanges involving psychedelics. But this was a history they mostly ignored. In its place, they cultivated a new perception of psychedelic science as the product of a visionary group of researchers at Harvard University. It was a deeply polarizing take—not to mention deeply inaccurate. But, at least for a time, it made them the center of the story.

--------

In November 1960, Allen Ginsberg arrived at Leary's family home in Newton, Massachusetts, along with his partner, Peter Orlovsky. They were there to try psilocybin, a new synthetic form of the psychedelic compound in psilocybin mushrooms that LSD's inventor, Albert Hofmann, had developed in his Swiss lab. Leary had been put in touch with the poet by the narcosynthesis pioneer John Spiegel, who was now a Harvard professor and consultant for Leary's nascent psychedelic research project at the university. Leary welcomed them warmly. Cocktails were served with lunch; Frank Barron and Leary helped themselves, while Ginsberg and Orlovsky abstained. No matter. There would be more with dinner, after the psilocybin kicked in.

Leary remembered that day as a decisive moment. Under psilocybin's influence, he, Ginsberg, and Frank Barron developed a utopian plan that involved an end to nuclear war and the beginning of a new era of expanded consciousness and cultural creativity. "First we would initiate and train influential Americans in consciousness expansion," Leary recalled. "They would help us generate a wave of public opinion to support massive research programs, licensing procedures, training centers in the intelligent use of drugs."

From that day forward, Leary and the group gathering around him began working to actively transform global society by providing psilocybin to the world's most influential people. Their list of possible collaborators included everyone from Margaret Mead and Gregory Bateson to the jazz musicians Charlie Mingus and Dizzy Gillespie. Leary even began to speak, half-seriously, about "turning on" John F. Kennedy.

This was an inherently elitist project, the epitome of the top-down approach to psychedelic science. It was an approach that Mead had long disavowed. But in 1960, the timing was particularly bad.

--------

Margaret Mead was one of the most hopeful scientists in history. By 1960, however, anyone who knew her well could see that her celebrated optimism was facing challenges. On December 1, 1959, her sister Priscilla Ann slit her wrists, distraught after her husband of twenty-four years, the humorist and writer Leo Rosten, told her he wanted a divorce.

Mead discovered the body. At the hospital, she identified herself as "Mrs. Bateson." Perhaps she feared a stranger would notice her. Or perhaps she was in such a state of shock that she forgot that she, too, was divorced.

Mead had learned from past experience to fear shortcuts to utopia. Now she learned from the death of her sister, married to the ever-striving Rosten—a man who had spent 1959 in a whirlwind of activity, bouncing from editing *Look* magazine to writing a bestselling novel while making dozens of public appearances—that enormous professional ambition could have unexpected downsides. Soon afterward, Mead developed a new interest in scientific and medical ethics. This included reassessing her own complicity in the misdeeds of scientists during World War II and afterward.

Mead was not alone. For the first generation of psychedelic scientists, the boundless future was narrowing, the field of possibility growing smaller. And the regrets among researchers who had once been driven by utopian energy were piling up.

--------

It was against this backdrop that Mead received a letter from Timothy Leary. The date was March 22, 1961. It invited her to try mushrooms with him.

Ever the networker, Leary tried to ingratiate himself by mentioning their mutual friends. "Ralph Blum has told me of his conversation with you about research on psilocybin and Mexican mushrooms," he wrote. Then he referenced a quip Mead liked to make in her public speeches regarding the most prominent pathways for mind expansion: studying animals, studying babies, anthropological fieldwork, psychoanalysis, a religious conversion, a psychotic break, or having an affair with a Russian. "Dave McClelland has told me of your six methods for expanding awareness," Leary wrote. "We hope that mushrooms (given under collaborative circumstances) can be added to your list."

Blum, still Mead's trusted protégé, was by this time an active collaborator with the Harvard group. In addition to proposing what Leary called "interesting research designs," Blum numbered among a tiny handful of insiders, such as Allen Ginsberg and Aldous Huxley, to whom Leary and Alpert entrusted psilocybin "without personal supervision." Around the same time that Blum spoke to her about Leary's psilocybin project, Margaret Mead began researching a series of articles that revealed her to be grappling with her legacy of collaboration with the U.S. military and intelligence services. Even a future without armed conflicts, she wrote, would be threatened by the "subtler warfare of control of the mind" that modern science had made possible. And she described her fear that "new drugs" could "provide new temptations to exploit other men."

The Korean War was over, but the memory of supposedly brainwashed soldiers lingered in popular consciousness. As atomic tensions escalated in 1961, the possibility that mind-altering drugs could be used to precipitate global nuclear war cast a haunting shadow on the hopeful promise of mind expansion. Leary's letter to Mead in the spring of 1961 came, in fact, at a time when this fear of drugs as tools of psychological manipulation was reaching its cultural zenith. In the month that Leary wrote Mead, Ken Kesey was working as an orderly at the Palo Alto VA hospital, the same institution where Gregory Bateson worked as a medical anthropologist. Kesey had already begun writing *One Flew Over the Cuckoo's Nest*, a celebrated novel in which new drugs like Miltown and Thorazine appear as pharmacological weapons. In one scene in the novel, a character imagines a psychiatric nurse firing a gun filled with the drugs to "tranquilize all of us completely out of existence." Meanwhile, throughout March 1961, Hollywood producers vied to sign Frank Sinatra to star in a film adaptation of the bestselling novel *The Manchurian Candidate*, a sinister tale of Communist brainwashing and presidential assassination.

Even Aldous Huxley, among the most outspoken and utopian proponents of psychedelic therapy, could not escape the ambient dread surrounding the concept of brainwashing. In January 1961, Huxley spoke at a widely publicized conference called "Control of the Mind." His speech linked Margaret Mead's anthropological research to his hopes for a future in which psychedelic drugs might become integrated into the education system to increase appreciation for beauty. Other speakers at the event were more dystopian. The *Oakland Tribune* placed its coverage of Huxley's speech alongside another report of a talk at the same conference. "New Mind Control Pill Can Make Man Believe Anything," this headline declared. The next month, the *New York Times* ran an article on the future of psychiatry that featured a boast from Amedeo Marrazzi, a military-funded drug researcher who had worked with Harold Abramson on coordinating and analyzing covert psychedelic experiments in the 1950s. Dr. Marrazzi, the paper reported, envisioned psychiatry of the future as "a kind of 'therapeutic brainwashing'" using mind-altering drugs.

It was also around this time that Mead, who had again fractured the ankle she had broken in New Guinea decades earlier, began carrying a curious forked wooden staff called a "thumb stick." When it was pointed out that her stick made her look like a shaman or "mythical sage," she admitted that this was intentional. "In Israel I look like a prophet," she said. "In Greece, like Tiresias from *Oedipus Rex*. I can fit into their myths." But she always kept these roles at a sardonic distance. She joked that the staff should not be taken as a sign that she was starting a cult of her own.

Perhaps a hint in Leary's letters—or in her conversations with Blum—warned her that there was something more seriously cultlike afoot at Harvard. Whatever the reason, Timothy Leary's letter to Margaret Mead received no reply.

--------

Leary had more success—at first, at least—with another prominent figure he contacted during this time: William S. Burroughs. The writer and drug enthusiast arrived on Leary's doorstep in September 1961 and offered to participate in and observe the Harvard group's psychedelic experiments. But Burroughs soon became disillusioned. "Situation here worse than I can tell you," he wrote a friend after spending several weeks with Leary, whose heavy drinking and inappropriate behavior disconcerted even the legendarily druggy Beat Generation icon. "DO NOT TAKE ANY HALLUCINOGEN UNDER ANY CIRCUMSTANCES." Burroughs had apparently learned something during his visit which convinced him that the "money and power" of Clare Boothe Luce was secretly backing the Harvard psilocybin project. "The money comes from Madame Luce and other dubious quarters," he warned Ginsberg, without providing further explanation.

Burroughs's suspicion, which remains unproven, revealed the growing political divide among those who experimented with psychedelics in the 1950s and 1960s. Luce and her husband, Henry, publisher of *Time* and *Life*, two of the most influential magazines in the world at the time, were staunch Republicans and anti-Communists who supported the CIA and its covert operations. They were also ardent proponents of psychedelic therapy, especially Clare, who had used LSD to cope with the death of her daughter and had introduced it to many of her friends and associates. Their personal beliefs influenced a series of largely positive articles on psychedelics that appeared in *Time* and *Life* throughout the late 1950s and into the early 1960s. For Burroughs, who was a radical critic of American society and culture, the Luces represented the intrusion of big money and Cold War politics into Leary's world. He feared that they would try to co-opt or control psychedelic research for their own agenda. It is unclear how much this was a paranoid speculation on the part of Burroughs—who was certainly no stranger to that mode of thought—but there are indeed indications that Luce may have been an anonymous donor to the project.

Much later, a researcher thought to ask Ginsberg himself for his *own* recollection of the first trip with Timothy Leary and Frank Barron, the event that Leary described as a utopian moment on a beautiful sunlit day in the late fall of 1960. Ginsberg's memory diverged sharply from Leary's. "Leary had this big beautiful house," the poet recalled,

> and everybody there was wandering around like it was some happy cocktail party, which was a little shocking to me at first because I still thought of myself as a big, serious religious meditator. And they were all so cheerful and optimistic and convinced that their kind of environment would be welcomed as a polite, scholarly, socially acceptable, perfectly reasonable pursuit.

The fragments of the Bateson group could have pointed out some of the flaws in this reasoning. And so could Ginsberg. On that day in December, Ginsberg did not remember giving his new friends a benediction.

Instead, he remembered giving them a warning: "You have no idea what you're up against."

CHAPTER 21

# Planes of Fracture (1961–62)

> I have told him repeatedly that the only attitude for a researcher in this ticklish field is that of an anthropologist living in the midst of a tribe of potentially dangerous savages. Go about your business quietly, don't break the taboos.
>
> —*Aldous Huxley on Timothy Leary, December 1962*

## Fall 1961 to Fall 1962

TOWARD THE END OF 1961, a series of harrowing reports began surfacing in the international press: newborns whose mothers had taken the sedative drug thalidomide were suffering from severe birth defects. It turned out that the German and Swiss firms responsible for testing thalidomide had not investigated its effects on pregnant women—or even pregnant lab rats. This was no isolated case of carelessness. Just as the earliest airbags and crash test dummies were designed for "average" passengers who were always assumed to be male, sexism and racism profoundly shaped midcentury drug research.

The thalidomide scandal, which had in part been uncovered by a female FDA employee named Frances Oldham Kelsey, signaled a new awareness of that bias. Influenced by the civil rights movement and the first stirrings of second-wave feminism, medical and drug researchers began reflecting on their own complicity in systems of power and institutionalized abuse. The field of bioethics was coming into being.

Psychedelic therapy grew up alongside the civil rights movement, bioethics, and the nascent sexual revolution, and its proponents shared many of the same stated goals. But in the opening years of the 1960s, psychedelic science remained primarily a man's club, overwhelmingly white, and largely ignorant of the changes happening beyond the walls of research universities and psychiatric clinics. Many psychedelic researchers in 1961 still thought they were unlocking "a new beginning," as the leaders of the Mental Research Institute had put it a year earlier. But though they were mostly unaware of it, a door was also closing on them—a vanishing opportunity to join psychedelic science to a larger transformation of racial, gender, and class consciousness in the making.

--------

The sole, short-lived exception to the Leary circle's unofficial "men only" rule was a young anthropologist named Jean Houston. A recent Barnard graduate with an intense interest in both mind-expanding drugs and human sexuality, Houston would later become so close to Margaret Mead that she described herself as Mead's "second daughter." Timothy Leary remembered her with something like awe, writing that she "stalked regally through our house" as she "urged us to go respectably public." Houston herself remembered it differently. She recalled telling the group that "male scholars taking drugs and sitting in the lotus position" was unlikely to lead toward the widespread adoption of psychedelics as a transformative tool. Houston was one of the few women (arguably the *only* woman) in Leary's orbit whom he treated as a peer rather than as a sex object. But in the end, Leary wrote, "Houston never joined our project. We were too raffish."

"Raffish" was a polite way of putting it. Leary and his circle were blatantly misogynistic even by the standards of the time, and at least one member of his research team, a therapist, later faced credible charges of sexual assault from former patients. Moreover, despite their alliance with the openly gay Allen Ginsberg (and Richard Alpert's own uneasy status as a closeted gay man), the Harvard psychedelic researchers portrayed homosexuality as a form of "sexual pathology" that psychedelics might *cure*—not as something that psychedelics could help the user *accept*.

The contrast with Margaret Mead and her world could hardly have been starker. On September 11, 1961, she appeared on a groundbreaking documentary called *The Rejected*, which aired on KQED in San Francisco and was syndicated across the country. It was the first program on American television to address homosexuality openly and sympathetically. Surrounded by artifacts from New Guinea and armed with charts showing human sexuality as a spectrum rather than a binary, Mead challenged the notion that homosexuality and transgender identities were "unnatural." She argued that they were part of the rich diversity of human potential, and that embracing this truth must be seen as part of a larger collective project of expanding beyond a single cultural frame.

In the realms of both altered states and sexuality, Mead and Bateson saw themselves as part of a multigenerational scientific endeavor that spanned cultures, reaching back into a distant past and influencing untold generations to come. Mead was always comfortable with publicity, but she had been sincere when she wrote to Ted and Lenora Schwartz, "Science is a cooperative task; individual geniuses advance IT, not themselves." After all, who remembers the names of the cathedral builders?

Alfred Kinsey had betrayed this vision when he made himself the center of his sex research. He had lost sight of science as a method of collective consciousness expansion, of species-level uplift. Mead never forgave him for it. Now, armed with good intentions but a corrosive egotism, Leary was about to do the same thing with psychedelic drugs.

--------

But first, the apocalypse. Or at least the closest approach to it in human history up to that point.

The Cuban Missile Crisis began on October 16, 1962, when American spy planes flying over Cuba spotted evidence of Soviet nuclear installations. The following evening, John F. Kennedy mulled over the question of whether he should launch a preemptive strike on the island. It was an action that would bring humanity closer to the brink than ever before. His advisers began speaking in term of tens of millions or hundreds of millions of casualties. For now, though, they kept the news quiet.

That same evening, Frank Barron was in a San Francisco lecture hall.

Standing before a large audience, he was introducing a weekly lecture series that he called "The Silent Revolution." It was a deeply optimistic exploration of the expanding powers of human consciousness and creativity.

In the weeks that followed, as the world hurtled toward the nuclear precipice—and millions of people experienced nightmares and panic attacks and unprecedented existential dread—weekly speakers stood at a podium at the First Unitarian Church in San Francisco and gave earnest lectures on such topics as "imagination as an instrument for human survival." The series had eight speakers in all. But in Barron's memory, two stood out: Margaret Mead and Timothy Leary. Mead's talk concerned the "revolution in women's consciousness." Leary spoke on the "extension of human consciousness."

The themes were similar—but the speakers were on very different trajectories. By this time, Leary and Alpert were moving fast into mysticism. Throughout 1962, Leary began describing the psychedelic state in terms of a "drug-induced satori," a kind of chemical enlightenment by which, as he put it in one speech, "the cortex can be cleared." Of what, exactly? Simply *cleared*. Talk of unconscious trauma or latent psychosis was for the old-fashioned psychiatrists. Leary was now operating in a more abstract and mystical domain. Some began publicly questioning whether what they were doing even *was* science.

"Going clear" is a phrase redolent of Dianetics. But if the Harvard group seemed to be following L. Ron Hubbard's playbook, they were even more influenced by Alan Watts, an English writer, onetime Anglican minister, and spiritual seeker whose gnomic statements about Buddhist renunciation and the meaning of life contrasted jarringly with his womanizing, hard-drinking personal life. Starting in 1961, Watts had begun working closely with Leary on fund-raising for the Harvard psychedelic project, and by 1962 he was helping to push the group in a distinctly mystical direction. Psychedelics, Watts wrote, "make the spotlight of consciousness a floodlight." In that brighter light, the illusory nature of the world, of science, and of all earthly desires—including the desire to build a better world using applied science, or to avoid being destroyed by nuclear war—was supposedly revealed. Watts later described "a terrible argument" he got into with Mead around this time that concerned precisely this. When he expounded on the futility of trying to prevent the accumulation of nuclear weapons, Watts claimed, Mead became enraged, calling him "a phony swami who believed in retreating from facts."

In the decade that followed their divorce, Mead and Bateson had grown apart in many ways. But there was one issue on which they remained in harmony. The scientist must *never* retreat from facts. From their first conversations on the lotus-strewn waters of Lake Chambri in 1932 to their trancelike years in Bali, from Manhattan during the darkest months of World War II to the espionage and anxiety of the Cold War, Mead and Bateson pursued goals—like curing schizophrenia, legalizing homosexuality, and expanding human potential through the study of altered states of consciousness—that many considered outlandish, unrealistic, even utopian. But they never strayed from pursuing these goals *as scientists*. They stayed within the limits of the facts as they saw them. Mead argued that scientists were the true utopians *precisely because* they faced and accepted these limitations. She rejected the "curiously tasteless . . . blank white spaces" of those who promise "heaven in this world and the next." For her, the most "vivid utopia" was the one conjured, in the real world, by scientists like herself.

As the utopian scientific ethos of the 1930s was submerged under the oceanic dread of the atomic age—and as young people increasingly sought a life raft in the form of what Mead, in her LSD memo, had called "mysticism and escape"—Mead and Bateson faced a point they could not cross. Or rather, a beachhead they refused to retreat from. Over the course of 1962, it became clear that the most visible group of psychedelic researchers in the United States, and probably the world, was sprinting headlong into the frothy white static of a false utopia.

--------

Bateson, meanwhile, was heading toward a new career.

One sunny morning in Palo Alto, the Scottish psychiatrist R. D. Laing followed Bateson to a meeting at the Mental Research Institute. Bateson shuffled into the room first. He was punctual; the others were not. They trickled in one by one, without greeting or apology. Jay Haley, Bateson's faithful disciple, turned his chair around and put his feet up on the windowsill, facing away from everyone else. Don Jackson came last. He was the founder of the institute, but no one seemed to care. Laing, who had flown there to learn their methods, realized in that moment that he was too late: "The group had sort of fallen apart." He came away from his visit thinking that never in his life had he "seen anyone look with more contempt or scorn before" than Bateson did when he conversed with Don Jackson.

Laing knew something about scorn. He had spent years working with people in extreme emotional distress, people who felt rejected by society and by themselves. He admired Bateson as a brilliant and sensitive man who cared deeply about human suffering. But something had clearly gone wrong within the very group of psychiatrists who spoke confidently of psychedelic drugs as a source of interpersonal "transcendence." Jackson had been one of the original followers of Bateson in Palo Alto, working alongside Weldon Kees and Jack Block. They had appeared on TV together, talking about the promise of psychedelic drugs. They had shared meals and stories at Bateson's home. And then their relationship had fallen apart.

The timing of their conflict suggests one explanation. It was in the months following Joe Adams's disastrous psychotic breakdown at the beginning of 1960—months when Jackson had insisted on pushing further with psychedelic research rather than stepping back from it—that Bateson began to slowly detach himself from the circle of psychedelic researchers and unconventional psychiatrists he had helped create in early Silicon Valley.

--------

Gregory Bateson was moving on in his personal life, too. In 1961, he married for the third time, and this new relationship carried him not to the psychiatric clinic, but to the icy tide pools of the Pacific coast.

Lois Cammack, age thirty-four, was a social worker at the Palo Alto Veterans Affairs hospital (where Bateson worked as a medical anthropologist). As a grad student in the mid-1950s, Cammack had taken LSD as a volunteer in an experimental trial of the drug. She had found the experience entrancingly beautiful and meaningful. But though they shared similar professional interests, their courtship played out around an unlikely passion that took them far away from the psych wards and counseling rooms: octopus collecting. She and Bateson would hop in his old-fashioned military-issue jeep, Cammack remembered, "and go down to La Jolla camping along the way. We would collect octopuses and bring them up to the hotel room, where we had all these little bubblers and plastic bags... then [we would] marathon back to Menlo Park, so that the octopuses wouldn't die, and we could get them in the tanks." The Batesons filled their home with dozens of burbling fish tanks sheltering octopuses with names like "Valentino," "Freddie," and "Virgil."

They also visited Joe K. Adams, the former leader of the psychedelic experimentation at the Mental Research Institute. Adams was now living in the same remote stretch of Pacific coast where Gregory and Lois collected their octopuses. For a confused period in the spring and summer of 1962, Adams continued to see himself as a professional psychiatrist. But though he still submitted grant applications and wrote essays, they now bore titles like "Sexual Aspects of Class Warfare" (published in a 1962 issue of the *Mattachine Review*) and "A Gymnasium for the Production of Dionysian Rites and Other Health-Giving Rituals." He also served as an adviser for the newly founded Esalen Institute in Big Sur, a haven for yogis, alternative psychoanalysts, Buddhists, and other spiritual seekers that helped crystallize what came to be known as the Human Potential Movement. In the fall of 1962, Bateson visited Esalen to teach a class there with Adams—and to commiserate about their shared decision to quit the world of professional psychiatry.

In their darker moments, Bateson and Adams believed that it was manipulation and power that had motivated their former colleagues all along. The two men decided to take LSD together one last time, sitting in a seafront garden at Esalen rather than in the cozy but still institutional quarters of the MRI. Bateson stared for a long time at the Pacific waves crashing far below them. "This stuff is all very well," Bateson recalled saying to Adams, speaking about the psychedelic experience. "It's very pretty but it's trivial... It's like the patterns of breaking waves or glass. What I see is only the planes of fracture, not the stuff itself."

Bateson kept thinking of those waves in the months that followed—months when the planes of fracture in his life became more and more evident. And as he detached himself from the world of psychedelic science in early Silicon Valley, a new path to a kind of scientific utopia began to present itself.

It was there in the waves that greeted him every time he drove down the coast on octopus hunts with Lois.

Perhaps the stuff itself was in the *ocean* itself. Perhaps it was down there with the octopuses and the dolphins. Without telling his colleagues, Bateson quietly began making plans to leave California for good.

CHAPTER 22

# The Deep End of the Pool (1963)

> His research involved putting electrodes in monkeys' brains and hooking them up to a big display board. Lights coming on and off would show what the brain was doing. And then he got into this business of dolphins being able to communicate. That was after he left NIMH. And I don't know that that ever made any sense.
>
> —*Virgil Carlson, in an NIMH oral history, recalling John Lilly*

## Spring 1963

IN MAY 1963, AS the Bateson family was packing up their Menlo Park bungalow, Timothy Leary and Richard Alpert were clearing out their offices at Harvard. Their downfall came at the hands of a young *Harvard Crimson* reporter named Andrew Weil. In a series of articles that spring, Weil had charged the two men with using hallucinogens to manipulate and seduce Harvard undergraduates. He pointed to Ronnie Winston, a Harvard student whom Alpert had befriended in the fall of 1962. Though Alpert denied a sexual relationship with Winston, he conceded that he had given Winston psilocybin on demand, without safeguards. At one point, Alpert admitted, he had even decided to show off his piloting skills by flying Winston in his personal plane while they were both on LSD, barely managing to land. Harvard's administrators summarily fired both Leary and Alpert.

Weil's reporting would have far-reaching consequences for the burgeoning psychedelic movement. But it was not the only sign that the winds were changing. In April 1963, undercover narcotics agents arrested two men who had approached psychedelic researchers at the International Foundation for Advanced Study in Menlo Park, a competitor and peer organization of the Mental Research Institute. The men asked if "anyone wanted to purchase a lifetime supply" of LSD. And then they revealed their possession of what authorities described as "three pints" of the drug. If true, this was a colossal amount, sufficient to provide a standard 100-microgram dose to roughly fourteen million people. More busts would follow.

This was a time of transition for psychedelics as they moved from the realm of scientific and medical research to the domain of recreational use. By 1963, as the researchers in Menlo Park now realized, nonscientists had begun to recognize that it was possible to purchase or steal supplies of legal psychedelics like Sandoz-manufactured LSD or psilocybin. Some opportunists began acquiring psychedelic compounds to resell on the burgeoning black market; others simply gave them away to friends.

That summer, Timothy Leary and Richard Alpert set off for the Mexican beach town of Zihuatanejo with a new plan. They would host a psychedelic retreat under the banner of their newly formed nonprofit, the International Federation for Internal Freedom (IFIF), a pilot project that had shown promising results the year before. This time, they were determined to make it bigger and better. Leary dreamed of creating a permanent psychedelic paradise on the beach—"Hotel Nirvana."

Leary spent part of his time that summer writing a speech he planned to deliver at the upcoming American Psychological Association conference, the leading professional forum in his field. "A profound transcendent experience should leave in its wake a changed man and a changed life," he wrote, recalling the afternoon in Cuernavaca when he had first experienced psychedelics. "Since my illumination in August, 1960, I have devoted most of my energies to try to understand the revelatory potential of the human nervous system and to make these insights available to others." In this speech, he directly responded to Bateson and his most famous idea, the theory of the double bind. Leary believed he was no longer subject to the "tribal concepts" that anthropologists used to explain human development. Now the stakes were higher, the goals loftier: "No spiritual game can be lost. The choice is not double-bind, but double-win."

Hotel Nirvana was not to be. Mexican authorities were tipped off by an anonymous psychoanalyst in Mexico City—possibly the same one who had called Leary a "vampire" in the summer of 1960. Reports of drug use and reckless behavior circulated in the media, and the police promptly shut down the retreat. Leary, Alpert, and their group attempted to relocate to Dominica, a tiny island in the Caribbean. Alpert even sold his beloved motorcycle to raise money to grease the wheels with government officials there. But within days, Leary was expelled again, this time accused of heroin trafficking. The charges were false, but the damage was done. Alpert was horrified to learn that the money from the sale of his motorcycle had been spent on a speedboat. The two men reunited on a beach in Dominica, where they proceeded to fight on the sand. Though they would continue as leaders of the IFIF for three more years, Leary and Alpert's relationship was never the same.

Lisa Bieberman, a math and philosophy major at Radcliffe, was among those who devoted herself to Leary and his group, which she called "the Movement," during these years in the wilderness. The Harvard circle, she believed for a time, was "an indissoluble family, destined to go forward, hand in hand, to win souls and bring in the Kingdom." And yet just a few years later she had concluded that "the LSD story up to now has been a tragedy." Leary and his circle had allowed "a tool of tremendous potential value for science, medicine, and personal life enrichment... to become the plaything of unscrupulous cultists."

## 1963

Why did the Harvard project fail? And, as was so often charged afterward, were Leary and his associates responsible for the global campaign to criminalize psychedelics that followed?

The answers to these questions are complex. Leary and Alpert argued that the project had turned on too many important people, making it dangerous. But the Harvard psychedelic project was not particularly large or impactful relative to others of the era. Nor was it particularly scientifically innovative. From their research into psychedelics as spurs to creativity or mysticism to their investigations of psychedelic therapy's effects on recidivism rates among prisoners, the Harvard group was not asking new questions or exploring them in novel ways. They were not a model of successful behavior change either, beset by frequent arguments, substance abuse, and troubling evidence of a pattern of sexual exploitation and violence.

But this was not the only cause of the discord in the psychedelic research community in these years. The forces arrayed against the psilocybin project at Harvard were not Nixonian reactionaries—they were members of the Macy circle who had been experimenting with psychedelic therapy and related techniques for decades. Many of the most ardent opponents of the "second wave" of psychedelic research, in other words, turned out to have been key participants in the *first* wave. Their criticisms carried weight, especially when set against the cartoonish accounts of Leary and his circle who spoke vaguely of opposition from "the establishment."

One of their leading opponents was Roy Grinker, an attendee of the Macy conferences on consciousness and one of the founders of narcosynthesis. Grinker's former colleague, John Spiegel, was one of Leary and Alpert's mentors at the early stage of the Harvard psychedelic project, introducing them to Allen Ginsberg and encouraging their experiments. But soon Grinker (who was no stranger to altered states *or* self-experimentation) argued that those who "administered the drug to themselves" were no longer "competent investigators" because they were "enamored of the mystical hallucinatory state." Likewise, Harold Abramson acknowledged Leary's "earnest desire to support the public use of these 'mind manifesting' drugs" but wrote of the "chasm" between Leary's claims of "a short cut to Nirvana for the millions" and the more complex reality.

Mead joined her old allies and friends. Scientists, she argued, had failed to integrate these substances into everyday life. In Balinese culture, hallucinatory trance states were not the domain of a sect of countercultural radicals—they were a traditional expression of *mass* culture. Psychedelic researchers had sought a shortcut to true cultural change, and it had backfired. Jean Houston, with her warning that it was hard to imagine a social revolution being led by bead-wearing middle-aged men meditating in locked rooms near Harvard, had been proven right. Rather than helping to integrate society around common goals, they had sped up that society's fracturing. Opposition to consciousness-expanding drugs stemmed from "our dislike of being alone," Mead came to believe. The substances she once saw as potential tools of cultural transformation had instead become engines of isolation.

This perception of psychedelics as a source of isolation was not unfounded, despite Leary's claims that LSD and psilocybin could unite the world. By the time Harvard fired him, Leary was living in a communal house with Richard Alpert and several others. Led by Leary, the residents sealed off one room in the house, concealed its outlines with paisley wallpaper, and cut a hole in the floor so that the space could be accessed only via a ladder from the cellar. It was here, amid red pillows, candles, and a smiling bronze Buddha statuette, that Leary spent much of his time.

"It was easy to forget, on drugs or straight, where you were in the house or indeed on the planet," Leary recalled. "It was an early isolation tank." In the fall of 1963, Leary and his followers began creating a countercultural enclave at the Hitchcock Estate, a sixty-four-room mansion in upstate New York owned by one of Leary's wealthy admirers. One of their first experiments involved randomly selecting two members of their group and confining them for a week in a former bowling alley on the mansion's grounds. Sustained by deliveries of food and LSD, the task of this isolated couple was to focus on nothing besides staying "as high in the consciousness as possible," as one member of Leary's circle put it.

--------

As all of this was happening, Gregory Bateson partnered with a scientist who believed that studying isolation was his life's work. By the summer of 1963, Bateson and his family had moved to the island of St. Thomas in the Virgin Islands, where Bateson assumed the title of associate director of the Communication Research Institute. The man who hired him was the CRI's director, John C. Lilly.

Though not as famous as Margaret Mead, Lilly had a bestselling book under his belt, as well as a string of radio and television appearances.

His professional résumé was impeccable: degrees from Caltech and Dartmouth, a stint teaching at Penn, then head of a government research lab at the NIH, and now running his own lavishly funded institute. Moreover, for reasons that remained somewhat mysterious, he had managed to convince the administrators of NASA to fund his efforts to teach dolphins how to speak English.

Born to a bank executive and a stockyard heiress in Minnesota, Lilly had a lifelong fascination with the mystery of consciousness—and for him, to be conscious was to be alone. His job titles over the years included neurologist, inventor, pharmacologist, dolphin researcher, and freelance writer. But Lilly's passion for what he once called "experiments in solitude" ran through it all, from his work in World War II developing tools for high-altitude pilots to his innovative early research involving electrical stimulation of the brain, from his invention of the sensory deprivation tank in 1956 to his later obsession with the journeys into "innerspace" brought on by high doses of LSD and ketamine.

With his odd affect and penchant for unusual headwear (he once conducted a lengthy television interview while wearing a red velour jumpsuit and a raccoon-fur hat), anyone who came to know Lilly well realized that he was profoundly eccentric. Though in public he and his wife, Elisabeth, a retired fashion model, seemed the embodiment of a well-to-do midcentury family—two kids, suburban house—there was a fundamental weirdness at the core. It was not out of character for Lilly to spend eight hours at a time submerged in his homemade sensory deprivation tank. Inside this soundproof, water-filled pod, the user floated as if in the womb, entirely isolated from the world and free to experience the mind-bending effects of prolonged sensory deprivation on consciousness and perception.

But this weirdness was not immediately obvious when Bateson joined Lilly in June 1963. It was easy to be impressed by what Lilly had accomplished in just a few years. The St. Thomas lab of Lilly's Communications Research Institute directly fronted a rocky stretch of pristine coast on the island's south side; cruise ships and millionaires' yachts plied the electric-blue waters just beyond. The lab itself was a modernist concrete building painted bright white. It housed a library; an electronics room; a series of observation decks, walkways, and peepholes; and, of course, a specialized cetacean lab outfitted with extra-large furnishings, lifts, and pulleys to accommodate the eight-hundred-pound bulk of adult bottlenose dolphins. Lilly regularly jetted between the Florida headquarters, the St. Thomas research lab, and fund-raising events on the East Coast and in California.

Bateson had no prior experience with dolphins. But to return to animals was to return to his roots, to the natural history he and his two brothers had done as boys in England. And behind the ghosts of those two long-dead boys was their father's shadow. Here was a last chance at scientific greatness—one last chance to complete the circuit that had been broken by the deaths of Martin and John. One last chance to help the great edifice of science mount skyward.

Or not.

In the months that followed the Bateson family's arrival at the St. Thomas lab, Gregory Bateson and John C. Lilly would take part in a series of experiments that have, in the years since, passed into the realm of urban legend. They have been retold in everything from scholarly studies like D. Graham Burnett's *The Sounding of the Whale* to an episode of *Drunk History*, from a BBC documentary to a *Saturday Night Live* parody sketch called "The Dolphin Who Learned to Speak." The events at the St. Thomas lab also inspired two Hollywood feature films: *The Day of the Dolphin* (1971), a thriller about cetaceans trained to become government killing machines; and *Altered States* (1980), an eerie tale of a scientist who mutates his own body via a combination of psychedelic drugs and isolation tank experiments.

There were more than a few grains of truth in all these depictions, even the most outlandish. What Gregory Bateson did not know when he began working at Lilly's lab was that, within a little over a year, his new boss would begin injecting their beloved dolphins with LSD.

CHAPTER 23

# Carl Sagan at the Dolphin Lab (1963–65)

**Margaret Howe:** Peter was insanely mad . . . His tail went flipping over his head and I think he hit John Lilly with it . . . it was definitely the LSD.
**John Lilly:** Oh, I agree.

*—from an audiotape discussing the dolphins Pamela and Peter, January 15, 1965 (John C. Lilly papers, Stanford University)*

## Summer 1963 to early 1964

ELEVEN-YEAR-OLD ERIC BATESON, LOIS'S son from a previous marriage, was delighted by his family's move from suburban Silicon Valley to a mysterious laboratory on a Caribbean island. "The water was absolutely crystal, and the beaches were gorgeous," Eric remembered. Best of all were the mongooses. Introduced in the nineteenth century to control rodents in the island's sugarcane plantations, the tiny Indian mongoose had become St. Thomas's apex predator—on land, at least. Eric remembered walking through mangrove forests while the small, secretive creatures scurried and leapt in the shade of vine-laden trees. He summed it up with a single word: "exotic."

Eric's stepfather, Gregory Bateson, was more than a bit awestruck himself. There was a profound optimism and self-confidence to John C. Lilly and his employees in these years—as if the Right Stuff of the early space program had sloshed over from Cape Canaveral and into the murky waters of cetacean science. And, of course, there was also the Jacques Cousteau factor. Lilly carefully read the works of the French oceanographer and emulated aspects of his debonair, marine adventurer lifestyle. Lilly's lab was stuffed with gadgets like his portable LINC computer and his handmade electrode caps. He encouraged an esprit de corps among his intrepid "crew." And, like Cousteau, Lilly had a genius for self-promotion. His Hollywood contacts included the actress and scuba diver Zale Parry, Ivan Tors (the producer of *Flipper*), and the actor Lloyd Bridges. Lloyd's son, Jeff Bridges, later remembered Lilly as "a fascinating person for so many reasons" and "a pioneer in the psychedelic realm" who was "really onto something" in his insight that dolphins were akin to an intelligent extraterrestrial life-form inhabiting our own planet. Lilly was part of a global vogue for the mysteries of the deep sea in the early 1960s, a region that was now being explored by "aquanauts" to match the cosmonauts and astronauts of the Space Race. It was a quirky, mysterious, and at times comical world: *The Life Aquatic with John C. Lilly.*

After years of frustrated ambition, Bateson now began once again to think big. Margaret Mead noticed. After visiting the CRI, she praised Lilly and her ex-husband for creating a "new field of investigation" in her 1964 book *Continuities in Cultural Evolution*. Mead may have had them in mind when she described the "massive evolutionary change" that would be necessary for ensuring human survival, change that would come from "clusters of human scientists . . . communicating with one another simultaneously at many levels and within different sensory modalities."

This was a vague thing to hang human survival on. But the science of communication was the fundamental principle not just of the work being done at the CRI, but of the nuclear war scenarios being gamed out by scientists on both sides of the Cold War. New technologies of communication—even, or *especially*, radically strange forms of it—really could save us all, many of these scientists thought. Lilly advocated for appointing a dolphin representative of "the Cetacean Nation" to the UN. He hoped he might live to see contact with alien beings (and play a part in achieving it). He believed that his sensory deprivation tank was a revolutionary tool for expanding human consciousness.

Yet despite Lilly's utopian dreams, his research represented a disastrous

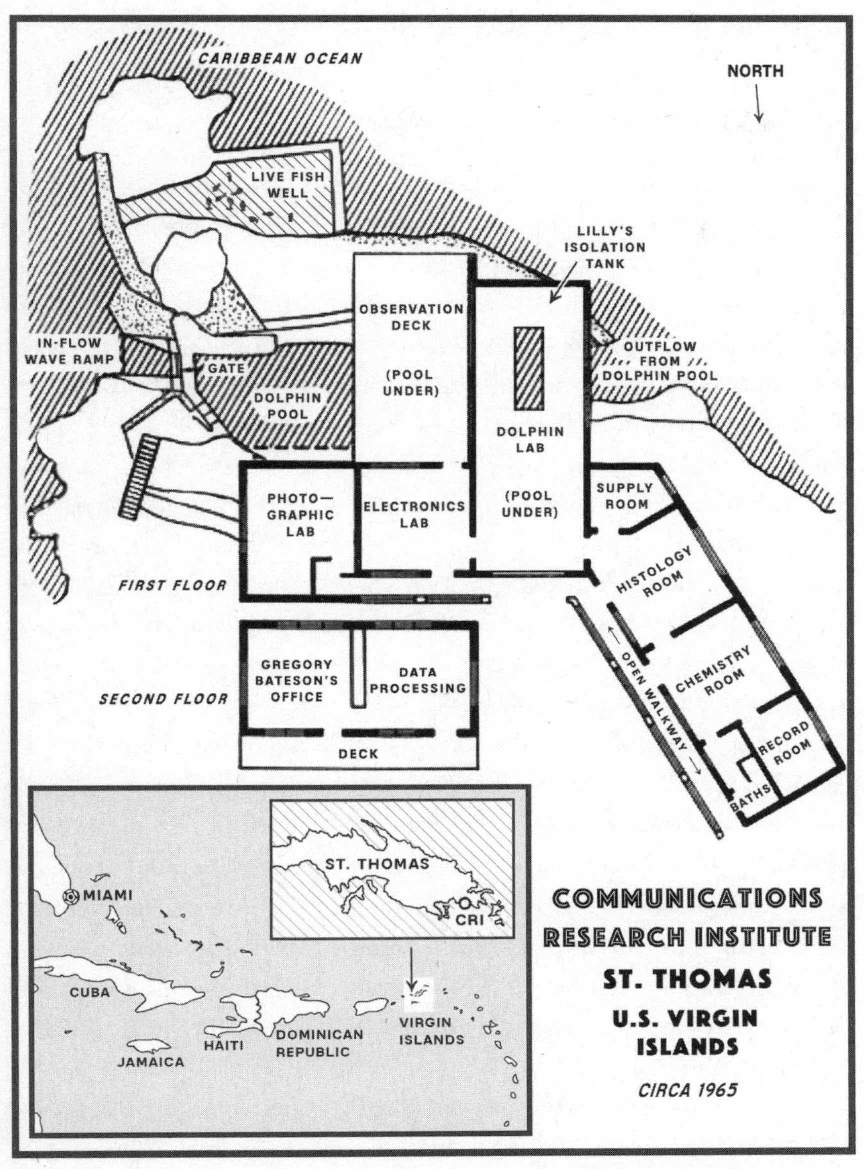


The layout of the Communication Research Institute circa 1965. By the author, based on the map in Lilly, *The Mind of the Dolphin* (New York: Doubleday, 1967), 227. The location of Lilly's isolation tank is approximate.

wrong turn not just for Bateson, but for the entire project of psychedelic science.

--------

In the 1930s, Mead and Bateson had seen themselves as scientists seeking to expand the accepted limits of "normal" human behavior, communication, and consciousness. In the 1940s, they came to believe that this project could help ensure the survival of humanity itself, threatened first by the spread of Fascism and then by the advent of atomic warfare. They regarded both as a product of "pathogenic" cultural patterns—ideas that drive you insane. Now, in 1963, Bateson continued to see himself as pushing the boundaries of science to prevent feedback loops of conflict. Though Bateson was never truly on board with Lilly's project of teaching dolphins how to speak, he spoke vaguely but fervently of the CRI's work as somehow connected to his larger goal of healing a sick society through interdisciplinary science. "I hope from the dolphins we may learn a new analysis of the sorts of information which we need—and all mammals need—if we are to retain our sanity," Bateson pronounced grandly to reporters at a CRI fund-raising gala.

Anyone observing his daily work at the lab might have been surprised by such claims. The truth was, Bateson spent most of his time struggling to learn the basics of dolphin behavior. Each day, he carefully documented such activities as "genital stimulation" and "beak propulsion." Diagrams of dolphin movements covered his desk; photographs of their underwater acrobatics hung in the darkroom. Bateson's notebooks from this period resemble the work of a deranged football coach, with thousands of colorful loops, arrows, and dashes indicating movements in a pool that was perpetually clouding due to inadequate filtration.

And then there were the tapes—hundreds of reels of magnetic audio and Super-8 film. One surviving clip shows Lois Bateson with a dolphin named Sissy. Her auburn hair tied up in a bun, Lois looks apprehensive, as if suddenly aware of the enormous strength of the cetacean she is swimming beside. In another recording, a young CRI staffer named Margaret Howe can be heard struggling to teach a young male dolphin named Peter how to speak in human-sounding phonemes. Progress is slow. "ONE . . . TWO . . . THREE . . .

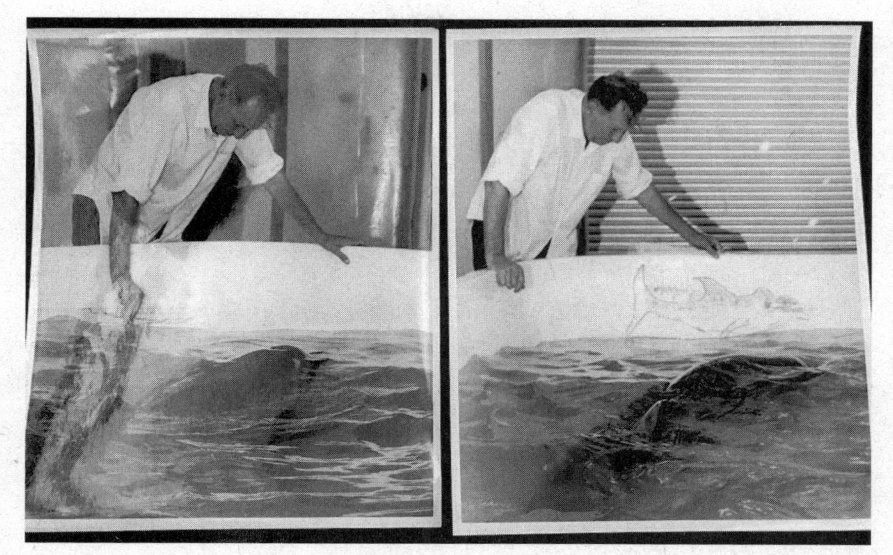

Two of Bateson's personal photographs of his work with the CRI dolphins, circa 1964. *Courtesy Bateson Idea Group. Photographer: Frank Essapian.*

FOUR…FIVE…SIX…," Howe counts, only to be interrupted by a shriek from Peter that sounds more like a satellite signal than a human voice. "Now, now, Peter," she chides gently.

The recordings were grist for Lilly's mill. By 1963, John C. Lilly was absolutely convinced that his dolphins were uttering coherent words, in English—but at a speed so rapid and a pitch so high that only computerized manipulation could make them understandable. The payoff was vague, but in his mind immense. After all, if he could bridge the barrier in communication between these two radically different species, how could Americans and Soviets continue to claim that communication between their own camps was impossible?

--------

Six months into Bateson's time at St. Thomas, a young Harvard assistant professor named Carl Sagan arrived for a tour. Not yet famous, but already a brilliant writer, the twenty-nine-year-old astronomer saw the visit as a tropical diversion from his research into the question of why Venus's atmosphere was so blisteringly hot. But Sagan was also visiting in a semi-official capacity: he and Lilly belonged to the same secret society.

In 1961, Sagan had helped plan one of the first conferences on extraterrestrial life. The organizers had sought not just scientists interested in first-contact scenarios but also someone who *already* spoke to "aliens"—or at least the closest parallel that Earth afforded. John C. Lilly was the obvious choice, impressing Sagan with his utopian aspirations and Hollywood glamour. Afterward, Sagan founded a whimsical scientific fraternal organization, the Order of the Dolphin, that was partially inspired by Lilly's own work. Now, three years later, Sagan was finally visiting Lilly's "aliens" in person.

Sagan's first encounter was with a young male dolphin named Elvar. "We entered a medium-sized room at the far corner of which was a large polyethylene tank," Sagan wrote. A dolphin was floating inside it, observing Sagan "with his head thrown back out of the water."

"Carl, this is Elvar; Elvar, this is Carl," John C. Lilly said by way of introduction. Then, as Sagan recalled, "Elvar promptly slapped his head forward, down onto the water, producing a needle-beam spray of water that hit me directly on the forehead."

"Well, I see you two are getting to know each other," Lilly said, then left the room.

The dolphin rolled over to present his belly. Sagan rolled up his shirtsleeves and began scratching Elvar like a dog. The dolphin demanded more, then more again, until Sagan tired of the game. Elvar responded by leaping so high that "only high tail flukes were in contact with the water." As he did so, he emitted a single shrill syllable. It sounded to Sagan as if he had shrieked, "More!"

Sagan "bounded out of the room" and excitedly announced the news to Lilly, who seemed entirely unsurprised. "Good, that's one of the words he knows," was all Lilly said in reply.

Sagan was delighted by the trip. Afterward, however, he began to wonder about the quality of the research. Sagan and his close friend the astronomer Frank Drake were becoming experts in identifying signals amid seas of noise picked up by radio telescopes. They agreed that the CRI's work set off alarm bells. Lilly appeared to be cherry-picking his data, carefully selecting the snippets of "dolphinese" that resembled human language and excluding the rest.

--------

The most unsettling feature of the CRI, however, was not Lilly's quixotic obsession with dolphin speech. It was the fact that his dolphins kept dying.

Among those who questioned Lilly's actions at the lab was Ted Nelson, a young Harvard graduate who had first met Lilly on a train in New Jersey. Nelson was intrigued by the idea of studying animal communication with computers. And he adored Gregory Bateson, with whom he shared an office ("*Oh, fuck! Fuck and damn!* he would say, with little provocation," Nelson recalled fondly). But Lilly struck Nelson as needlessly callous. True, the CRI's director was "brilliant and persuasive, and a fantastic self-publicist," Nelson admitted. Yet he was, at heart, "ruthless... a con man."

For a time, Nelson worked on a promotional video about the lab's work. It was meant to be a puff piece, but Nelson found the job unexpectedly difficult. The trouble was, he wrote, that the tropical splendor of the lab's setting kept swerving unexpectedly toward a more unsettling ambience:

> Through the windows, it was just before twilight and beautiful: the palm trees were blowing, pelicans scudded by. Atmosphere. Then I panned the camera across the room, showing concern on everybody's faces as they watched oscilloscope traces leaping. We were actually hearing what the scope showed: the shrieking and whistling of our dolphin pals in the tank on the floor below, coming out of different speakers as the sound was transposed from different audio ranges.

The truth was that Lilly's research with animals relied on what scientists at the time called "operant conditioning"—a polite term for using pain to control animal behavior.

## Summer 1964 to Spring 1965

By this time, an unlikely psychedelic subculture had emerged among a group of scuba divers, actors, and producers who worked on the television shows *Sea*

*Hunt* (1958–61) and *Flipper* (1964–67). Their mentor was Dr. Oscar Janiger, a UCLA psychiatry professor and friend of Sidney Cohen, Aldous Huxley, and Betty Eisner. Janiger believed that psychedelics must be "institutionalized" in a "supervised context"—but he did not mean psychiatric institutions. Instead, the freethinking Janiger cited the Eleusinian Mysteries of ancient Athens as his model. This volunteer initiation involved a transcendent experience that occurred deep underground as part of a secret rite. It appealed to Janiger because it was supposedly open to all—men or women, enslaved or free. But he also liked the theatricality of it. Janiger was particularly fond of working with the artists and writers of early 1960s Hollywood.

The divers and producers associated with *Sea Hunt* and *Flipper* began using LSD. And John Lilly, who consulted on both programs, found himself tripping for the first time in a Malibu beach house in early 1964. His guide was the actress Constance Tors, whose producer husband, Ivan, had hired Lilly to consult on the film version of *Flipper*. While Tors looked on, Lilly injected 100 micrograms of LSD into the muscle of his thigh (for reasons he never fully explained, he did so while completely naked). As the drug kicked in, Lilly wrote that he "suddenly realized that all of my previous training leading up to this point, all of my preparation, had been worth it." At the height of his trip, he stared at his face in a mirror for a long time. It seemed to him that his father's face was suddenly overlaid onto his own—and then his father's father's, and *his* father's, and so on. Each new face appeared at the pace of one per second. Then, after two thousand or so generations, "suddenly the face of a hairy anthropoid appeared on my face." A new kind of interspecies communion had been achieved.

Afterward, Lilly continued onward to a solo vacation in Hawaii, leaving his increasingly rocky relationships with Bateson and his wife, Elisabeth, back at the lab. He was hatching a plan. While returning from Hawaii on a United Airlines flight on April 28, 1964, Lilly earnestly quizzed his associate director. "Dear Gregory," he scrawled in schoolboy cursive on the airline's complimentary stationery. "Did you ever do any serious research with LSD? Its effect on communication? On yourself? With music, noises, etc? I'm getting some to try on a dolphin to see if it 'forgets to breathe' under its influence." He signed off with an invitation to join him: "I'm contemplating resuming the tank isolation work—are you interested? LSD in that milieu ought to push things very fast."

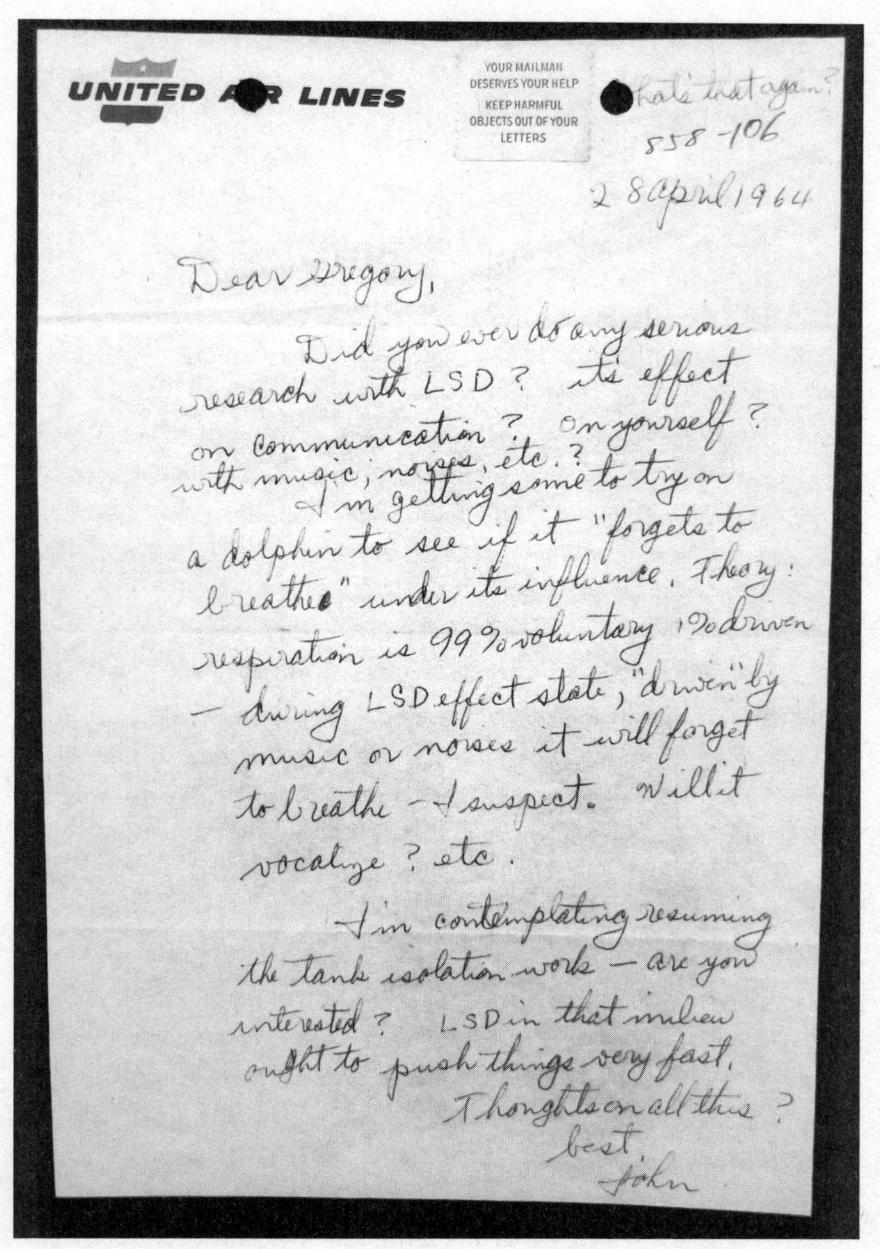
UNITED AIR LINES

YOUR MAILMAN DESERVES YOUR HELP
KEEP HARMFUL OBJECTS OUT OF YOUR LETTERS

What's that again?
858-106

28 April 1964

Dear Gregory,

Did you ever do any serious research with LSD? its effect on communication? On yourself? with music, noises, etc.?

I'm getting some to try on a dolphin to see if it "forgets to breathe" under its influence. Theory: respiration is 99% voluntary 1% driven — during LSD effect state, "driven" by music or noises it will forget to breathe — I suspect. Will it vocalize? etc.

I'm contemplating resuming the tank isolation work — are you interested? LSD in that milieu ought to push things very fast.

Thoughts on all this?

best,

John

Lilly's April 1964 letter to Gregory Bateson proposing LSD experimentation with dolphins.
*Estate of John C. Lilly and Stanford University Special Collections.*

Stopping back in Los Angeles en route to a conference in New York, Lilly then tripped for a second time with Constance and Ivan Tors. The next day, he boarded a flight to New York City to present a paper on dolphins at a conference of acoustic scientists. And then, after delivering his speech, exiting the conference room, and hitting the button of a hotel elevator, Lilly collapsed.

By his own account, he awoke three days later in a hospital bed, entirely blind. Slowly an explanation came together. When he injected himself with LSD, Lilly believed, tiny bubbles in the detergent used to clean the hypodermic syringe had entered his bloodstream, "cutting off the circulation to very critical parts of my brain, including the visual cortex." Lilly claimed he spent six weeks in recovery before returning to St. Thomas. By this time, news of the event was circulating throughout the scientific community—and, of course, among his employees. "The rumor went around that the episode had happened as a result of taking LSD," Lilly explained ruefully. After all, "LSD was found in my briefcase in the hospital."

Lilly resolved to extensively test the drug on the animals before he would try it again himself. But he was reassured about the risks, he wrote, when "the six dolphins tested apparently had very good trips."

From that point onward, Lilly began habitually tripping on LSD in his deprivation tank, floating motionless for hours at a time while he practiced his dolphin-inspired breathing technique. At a stroke, the CRI had become a drug research center.

--------

Lilly's near-death experience—what he called his "incident"—took place on May 7, 1964, a Thursday. That Saturday, Gregory Bateson could be found sitting in an auditorium at San Jose State University, participating in a symposium on the "problems and potentialities" of LSD. The event had been organized by one of Bateson's former colleagues from the Mental Research Institute. Frank Barron was also there, having fled the Leary circle by this time, but still a true believer in the transformative power of psychedelics. Richard Alpert was there, too. And watching from the audience was one of Bateson's former students at Stanford: Stewart Brand, a filmographer, writer, and technology enthusiast who would become an important figure in both the hippie movement and the digital revolution of the 1970s and 1980s.

Bateson's presence at the conference despite his decision to transition away from psychedelic psychiatry was emblematic of his larger role in the history of the field. He was a reluctant mentor, an accidental, even self-loathing, pioneer of a movement he was already beginning to disavow. But on that long weekend in May 1964, while Lilly lay in his hospital bed, Bateson remained at the heart of it.

Sagan had started the Order of the Dolphin as a fake secret society, a veneer of mystery that lent romance to the workaday world of midcentury science. But a real mystery sits at the center of this strange chapter in the history of psychedelics. Why did Lilly do it? And why didn't Bateson try to stop him?

--------

Lilly answered the question of why he gave LSD to dolphins in a half dozen ways, each contradictory. In May 1965, for instance, he made his way to a second Macy conference on LSD, organized once again by Harold Abramson. Here, Lilly described his work as a disciplined exploration of the dolphin psyche. The lab had instituted what he called "careful controls": one group of dolphins experienced the drug alone, another with companions; one received a placebo, and the other received a real dose; and so on. To Betty Eisner, who was in attendance, Lilly almost seemed to be portraying himself as a trans-species therapist.

At another conference a few months later, however, he sounded like an entirely different person. In October 1965, Lilly attended a secret symposium at the U.S. military's Edgewood Arsenal chemical weapons research center, where Frank Olson had once worked. "With man one is sometimes very much limited by some very stringent legal and social considerations," Lilly told the small audience of top-secret-cleared chemical weapon researchers and psychiatrists—a group that *also* included Harold Abramson:

> One can do certain things with animals that one couldn't possibly do with man at the present time. We are in this position with the

> dolphins. We do work of the type that one could not do with man without getting into severe moral, legal, or ethical problems. In other words, we can for example obtain brains under ideal conditions for neuroanatomical studies.

Given the context, Lilly's point was clear. He was arguing that his dolphins could stand in for human test subjects in experiments of LSD's role as a truth drug. And with dolphins, he boasted, you could dissect their brain afterward.

Whatever the true explanation for his actions, one thing is obvious: Lilly was putting an optimistic gloss on a bewildering mess. The surviving tape recordings of this period reveal Lilly and Margaret Howe struggling to make sense of the tripping dolphins' behavior—while also struggling to come to terms with what they themselves were doing to them. At one point, Howe describes how the dolphin Pam experienced distress after an LSD injection. This, in turn, provoked severe anxiety in Peter. "Somehow Peter knew that he must not leave Pam," Howe says. Howe and the CRI's two dolphin handlers—both young Black men who were St. Thomas locals—struggled to strap Peter down. But he kept thrashing.

"We shouldn't really put an animal through that kind of thing again," Lilly mutters a minute later. Margaret Howe agrees. "No," she says. "I don't like it either. This is Margaret Howe speaking, and *I don't like it.*"

The experiments with LSD kept coming over the months that followed. By the end of 1965, however, Margaret Howe had had enough. She left the program and disavowed the decision to use LSD on the dolphins.

Lilly never did.

--------

Lilly entered the world of psychedelics just as the door to their legal use was closing. In a July 1964 FDA memo, it was announced that Sandoz had cut off the LSD supply of several researchers in Silicon Valley as well as an indignant Harold Abramson. Frances Kelsey, the FDA official who had blown the whistle on thalidomide, was more favorable toward psychedelics, noting their "potential" value. But after a decade or more of testing, she believed that potential had not been shown to exceed their dangers. One by one in the fall of 1964 and early 1965, experimenters began to lose their licenses to buy and use psychedelics.

As Lilly spent more and more time tripping on a drug that was hurtling toward illegality, Bateson kept filling notebooks with notes on dolphin behavior. Given the circumstances, it would have been understandable if Bateson had simply quit. But Batesons were not quitters—not when science was concerned. "To close it now," Bateson wrote to one potential funder, "would be a terrible waste as well as a personal grief to me." In a draft of a letter to another, he admitted that it was "a depressing moment of my career." Then he scratched the phrase out.

But then came the opportunity to lead his own dolphin lab on the Hawaiian island of Oahu. "Was it Patrick Henry who said 'I regret that I have only one life to give for my country'? I want to remind you that I have only a few more years to give for science," he wrote curtly to Lilly, shortly before he quit the CRI.

"I must stop the dolphin research," Lilly remembered thinking around this time. "It was not being done in consonance with the new ethic." But this did not mean that he intended to stop working with dolphins. Far from it. Instead, Lilly began imagining an even more ambitious cetacean research project. To truly communicate, he decided, dolphins would need to live with a human family—*his* family—in a half-flooded house near the sea.

Lilly did not have a chance to implement this scheme. One by one, the CRI's dolphins—Peter and Pam, Elvar and Sissy—began to decline, mentally and physically. In the end, as Lilly put it, four out of their seven dolphins "committed suicide by refusing to eat or breathe."

--------

The work at the Communication Research Institute was about more than dolphins. The stated aim was profoundly utopian: a vision of a future that would be transformed by radical new advances in communication. For a little over a year, Lilly had deluded himself into thinking that psychedelics could somehow play a role in this project. Lilly's aspirations were lofty. But his methods were closer to those of the clandestine researchers at Edgewood than to those of idealists like Carl Sagan. He was, at his core, an elitist, and he saw his dead dolphins—and his unhappy employees—as acceptable sacrifices for his quest.

Bateson had quit his previous life and career in Silicon Valley out of disgust at the "manipulative" tactics of the psychiatrists he knew there. Now, with the dreamlike circularity of a horror film (or the absurd logic of a screwball comedy), he had moved his family across the country, attempted to master a new field, and endured all those damp, vexing, bewildering hours with dolphins... only to find that his new boss was doing *exactly* what he had tried to run away from.

Bateson found himself, once again, facing the sinister underbelly of psychedelic science. But he kept his misgivings to himself. "I had no idea how the dolphins would react to that," Lois Bateson remembered thinking, when asked about her own opinion of the LSD experiments. "I mean, humans didn't always react to it very well, you know."

She knew her husband loathed Lilly's decision to dose the dolphins with LSD. But she also knew that he found it almost impossible to talk about. It was too painful. "We did not get into that," she recalled. "We just got out of there."

CHAPTER 24

# Dialectics of Liberation (1965–68)

> Perhaps to some extent we have lost sight of the fact that [LSD] can be very, very helpful in our society if used properly.
>
> —*Senator Robert F. Kennedy, speaking at the congressional hearings on LSD, 1966*

## Winter 1965 to Spring 1966

IT IS THE EVENING of December 12, 1965, and Allen Ginsberg is backstage during the intermission of Bob Dylan's concert in San Jose, California. He is chatting with Dylan about Ginsberg's new portable tape recorder. "It looks groovy," says Dylan. "Is it worth it?"

"Oh yeah, it's an absolutely beautiful precision machine," Ginsberg boasts. "It can do anything."

Dylan is close to the recorder, and the mic picks him up clearly as he contemplates the rotating wheels of magnetic tape. "I don't know why the fuck I don't get one of those," he murmurs. A few minutes later, he and Ginsberg start gossiping about Marlon Brando.

Later that night, the same tape recorder captured another conversation—one between Ginsberg and Joe K. Adams, the Texas-born psychiatrist who had worked alongside Gregory Bateson and Don Jackson at Palo Alto's Mental Research Institute. By this time, Adams had quit science. Secure in the Big Sur campground he now owned, he dedicated himself to baking whole wheat bread, maintaining his vacation cabins, and writing playful essays about sexuality, consciousness, and the harms of psychiatric care. He no longer saw himself as a mental health professional with a Stanford affiliation. Instead, he was part of the psychedelic counterculture. As a contributing editor of a new journal, the *Psychedelic Review*, Adams shared duties with fellow editors Timothy Leary, Richard Alpert, and Alan Watts. He was also a core member of the Esalen Institute community, leading seminars and mingling with the institute's newly fashionable clientele. And, as happened to be the case on this night, hosting Allen Ginsberg and his lover, Peter Orlovsky, for a cup of tea.

On the tape, Adams has a friendly voice with a molasses-thick East Texas lilt. Ginsberg begins showing Adams his new gadget—how the recorder can be plugged into loudspeakers, radios, guitar amps, capturing everything like a roving electric consciousness. The thought brings Ginsberg back to six years earlier, to his first LSD trip at the MRI—the trip that inspired a poem called "Lysergic Acid" about feeling as if he were plugged into a vast electronic machine. The trip that Gregory Bateson had organized and Joe Adams had overseen.

"Hey, whatever happened to the tapes you made when I was high?" Ginsberg asks. "Was there anything interesting on them?"

"Well, I'm not sure if I have those tapes, or if they're back at the Mental Research Institute," Adams replies. "I would have to go through all my tapes and see."

They move on to other topics. But Ginsberg keeps circling back to the events of 1959 and 1960. At the poet's prompting, Adams recounts his catastrophic LSD trip, the one that contributed to the fracturing of the Mental Research Institute. "It was in January [1960] when I had my reaction," he remembers. "And I was high. For. *Months*... I was really crazy. I took my clothes off and started runnin' around the driveway."

"Here?" Ginsberg asks.

"No, no, nooo... it was in *Menlo Park*!" Adams says to sympathetic laughter from Ginsberg and Orlovsky. "A nice, suburban neighborhood. And I was intercepted by a psychiatrist in the driveway. My wife had called them." His voice is amused and detached, as if he's describing a scene from a play. "Three uniformed men came and tied me down on a stretcher and took me to the mental hospital." In his delusional state, Adams found himself dwelling on mass manipulation, psychological warfare, and World War II. "I realized how horrible this was," he explained, apparently referring to his experience of psychiatry during the 1940s and 1950s. He concluded that he was trapped in a sinister Cold War plot: "I tried to escape there and ran out and climbed over a fence and I yelled, '*This is not a hospital!*' . . . You see, there were two shifts of aides, and I thought there was a *U.S.* shift and a *Russian* shift. I thought it was a big experiment of some kind and I was a guinea pig. And the whole place—I thought the whole place had microphones in it. And, oh, well . . . it was very complex."

"What led you up to that?" Ginsberg asks, fascinated. "You know, very oddly, the first time I had LSD with you, I suspected that *you* were a Russian agent, a Russian spy. Remember that? Remember I mentioned that I thought it was a Russian plot? I knew it was involved in the Cold War somehow."

Adams laughs warmly. "No, no, I don't remember that," he says. "That's . . . uh . . . *huh*." He doesn't elaborate further.

--------

The two men had been brought together by Gregory Bateson at the end of the 1950s, during the height of psychedelic science's cultural prestige. Adams had once been a true believer. Ginsberg, in a sense, still was. It was this shared utopian sensibility, this deep idealism, that sparked their friendship.

When Ginsberg labeled the tape of his conversation with Adams, he gave it an ironic title: "Acid Test." One of the first acid tests—the name Ken Kesey and his followers gave to their parties—was in fact happening about three hours' drive north of Adams's cabin at the moment that they were discussing the strange events at the Mental Research Institute. Ginsberg's name was even on the poster. But he didn't show. Far more than his peers in the emerging psychedelic counterculture, Ginsberg was intrigued by the scientists, not the partiers. More than anyone else in the world in 1965, perhaps, Ginsberg was uniquely placed to understand the worlds both of psychedelic science and of the psychedelic counterculture. But as the winter of 1965 became the spring and summer of 1966, the poet grew increasingly disquieted by *both* worlds. He could not shake his sense that, as he put it, the psychedelic experiments at the MRI and elsewhere "had something to do with the Cold War." And he also believed that the new American messiahs of psychedelic spirituality were getting something fundamentally wrong.

Psychedelic science had always been a global project, winning adherents in places like Prague, Zurich, London, and even Baghdad years before it reached relative latecomers like Ken Kesey and Timothy Leary. The psychedelic counterculture was, at least at first, no less global. In Brazil, artist Hélio Oiticica created overtly "trippy" interactive artworks, inspiring a generation-defining cultural movement known as Tropicália. In Santiago, Chile, a young physician named Claudio Naranjo returned from a Fulbright scholarship at Harvard, where he took mushrooms with Frank Barron, believing psychedelics were tools for "a collective transformation of consciousness." And ancient psychedelic traditions, such as the use of ayahuasca, a potion made from two Amazonian jungle plants, found new followers amid rapid urbanization across the Amazon basin.

Stateside, however, the vivid diversity of global psychedelic culture was becoming monochrome in a distinctly midcentury American way. Mead had long called for a truly diverse and broad-based global cultural change. Yet while the psychedelic counterculture embraced "non-Western spirituality," it did so in a superficial and romanticized way that did not truly challenge Western cultural norms. Tom Wolfe memorably described Stewart Brand (who drives the Merry Pranksters' bus through the opening pages of *The Electric Kool-Aid Acid Test*) as wearing "just an Indian bead necktie on bare skin and a white butcher's coat with medals from the King of Sweden on it." Underneath the costumes, however, were bodies that were almost invariably American, white, middle class, and male. And in the new psychedelic spaces that emerged across the United States, from Esalen to Greenwich Village, it did not take long for all the usual ills of twentieth-century American life—alcoholism, political polarization, racism, greed—to reassert themselves.

## Summer 1966 to Spring 1968

"We are seeing accidents happen," Sidney Cohen lamented in March 1966. "We are frightening the public. We are getting laws passed [banning the drug]. We are not using the anthropological approach of insinuating a valuable drug of this sort into our culture" by "gradually demonstrating the goodness of the thing." For Cohen, "the anthropological approach" meant a science built on and with the cultural patterns of other societies, not an attempt to invent them from scratch—Margaret Mead's approach, in other words. But now, Cohen complained, the anthropological approach to drugs was being abandoned in favor of reckless experimentation—"these Acid Tests that go on with these bizarre individuals."

A few weeks later, on May 24, 1966, Senator Robert F. Kennedy sat down at a large table in the New Senate Office Building to convene a congressional investigation of LSD. "Here is a drug which has been available for well over 20 years," the senator said. "Yet suddenly, almost overnight, irresponsible and unsupervised use of LSD for nonscientific, nonmedical purposes has risen markedly." The dangers had, too. Kennedy feared that "without careful psychological screening, the drug will be used by some who suffer permanent damage as a result." In other words, "what was an experimental drug has become a social problem."

Given the decades of prohibition that began later—decades in which LSD and other psychedelic drugs would be lumped together with heroin as Schedule 1 drugs of abuse, and tens of thousands would spend years in jail for possession of psychedelic drugs—it is tempting to see Robert F. Kennedy's statement as an early salvo of what Richard Nixon, five years later, would label the "war on drugs." But 1966 was not 1971, and the future was not yet written. Senator Kennedy concluded his opening statement by describing the *benefits* of psychedelic therapy. "Experiments indicate that LSD may be useful in treating alcoholics—one of the largest groups of the handicapped," noted the man whose own family struggled with addiction. The "loss to the Nation if LSD were to be banned," Kennedy warned, "would be serious indeed."

This balanced tone was, however, already being drowned out by polarized commentary from both sides. In a parallel series of Senate hearings held the same month, a succession of narcotics officers and antidrug doctors shared horror stories about trips gone wrong. A jarring note of positivity came amid the scaremongering when one narcotics detective quoted an interview with Cary Grant in which the Hollywood actor credited LSD with improving his life and allowing him to "truly give a woman love for the first time." The thought seemed to disconcert one of the senators present, Thomas Dodd of Connecticut. "I think in fairness to Cary Grant," Senator Dodd ventured, "we do not know that he ever said any such thing." The detective agreed. The point he was making, he reassured the senator, was simply that the media was lying about psychedelics in such a way as to make the drug "attractive to our teenagers and our youth of today."

Allen Ginsberg spoke at the hearings, too. But the bearded poet did not exactly inspire confidence. "I don't think it is necessarily frightful or dangerous. It could be dangerous if we react dangerously to it," he told the senators. "But it is very hard to say yes, let the high school kids get it, because it scares everybody. The problem is not to scare everybody." Ginsberg tried to cite the members of the Macy circle as illustrious proponents of psychedelic therapy, but the details eluded him (Ginsberg cited the research of "Dr. Harold Abramson of New York, who is a very considerable figure, the head of some hospital or other").

And, worst of all, he lumped these distinguished scientists in with Timothy Leary. Because, when Leary himself rose to speak, the ex–Harvard lecturer informed the subcommittee that "the so-called peril of LSD resides precisely in its eerie power to release ancient, wise, and I would even say at times holy sources of energy which reside inside the human brain." Under its influence, he added, "you definitely go out of your mind, there is no question of that. To some people this is ominous." Leary's remarks were widely quoted and misquoted in newspapers. The senators were unimpressed. "I am trying to follow the best I possibly can," Robert F. Kennedy said to Leary, interrupting his rambling testimony. "And I find that I am completely unable to do so."

--------

By November, even Allen Ginsberg felt disenchanted. In a speech in Boston that month, he encouraged his audience to "try the chemical LSD at least once." But he also expressed fear of a "chemical dictatorship"—an allusion to the era's ever-present fear of brainwashing and mass mind control—and urged those in attendance not to assume that psychedelics were a solution for anything. They were simply one way of understanding the problem: "anger and control of anger."

Allen Ginsberg had a right to be angry. In an interview that fall, Leary had declared, "LSD is a specific cure for homosexuality." And then he had cited Ginsberg himself as an example of a gay man who had "turned on to women" thanks to LSD. This, despite the fact that Ginsberg considered his multidecade relationship with Peter Orlovsky to be a "marriage sealed by vows." Leary's claims were growing ever more outlandish. And the international media was reacting to them with equally outlandish horror stories. The *Saturday Evening Post* warned that "if you take LSD even once, your children may be born malformed or retarded," while the *Washington Post* ran an article with the headline "Worse Deformities Feared in LSD Than Thalidomide."

By the end of 1967, psychedelics had been banned by several U.S. states, and work was underway to draft a federal law that would entirely prohibit their sale or use. And while many scientists continued to be optimistic about the potential of psychedelic therapy, their hopes were fading fast. A turning point in the rejection of psychedelic therapy among scientists themselves had come at a packed auditorium at MIT in May 1967, when Timothy Leary publicly debated a young neuroscientist named Jerome Lettvin about the merits of psychedelics. Lettvin's mentors were long-standing members of the Macy circle—consciousness researchers like Walter Pitts and Warren McCulloch—and he saw himself as carrying forward their scientific legacy. The contrast was stark. First came a barefoot, bead-wearing Leary, standing before two different projections of trippy visuals as he told the crowd that "the real goal of the scientist is to flip out." Then Lettvin, speaking without props or visual aids, calmly urging the audience to seek a middle path. Lettvin denounced efforts to ban psychedelic drugs, but he also cautioned that utopia could not be found in a chemical.

Afterward, Lettvin corresponded with Gregory Bateson about the debate. They spoke of the "Faustian promise of a shortcut via drugs." Yet, despite his own ambivalence toward psychedelics, Bateson did not join the forces arrayed against the psychedelic counterculture. In fact, in the months that followed, he aligned himself with it.

--------

One of the only surviving clips from the two-week event that cemented Gregory Bateson's reputation among the 1960s counterculture captures Emmett Grogan, founder of the anarchist Diggers collective, delivering a passionate speech in praise of violence. It is mid-July 1967. Grogan is wearing enormous black sunglasses that reflect the circular banks of lights in a large London auditorium. "Violence is, and will always be, necessary to tear away the shreds of myth and absurdity that we've been encased in since we were born," he declares. The film then cuts to Black Power leader Stokely Carmichael, who throws his arm outward at the crowd, challenging their inaction in the face of anti-Black violence. Allen Ginsberg appears next, looking visibly tired (he had spent much of the previous week in the company of Paul McCartney, Mick Jagger, and John Lennon). "Don't escalate the hostility," he tells the raucous crowd. "Control your mind."

And then the camera cuts to a poster for the event itself, an event that has been called one of the defining cultural moments of the 1960s. "International Congress Dialectics of Liberation," the poster reads. Text below promises "a unique gathering to demystify human violence in all its forms." At the top of the list of speakers, beside "Stokely Carmichael," is the name of Gregory Bateson.

Dialectics of Liberation took place in London's Victorian-era Camden Roundhouse, an "enormous barn-like structure, its floor covered with sawdust," that "reeked heavily of marijuana," remembered one of the attendees, Angela Davis. "There were rumors that one speaker, a psychologist, was high on acid." Sheila Rowbotham, a young feminist activist who attended Dialectics, remembered the misogynistic tone of many of the speakers and the absence of women (with the notable exception of Angela Davis herself). Nevertheless, many of those onstage saw themselves as leading a vanguard of liberation—the transformation, as one of the organizers put it, of every social institution, from the family to the school to the mental hospital, "into a revolutionary center for transforming consciousness." A film crew roamed the crowd, shooting a BBC documentary that was later shelved because the broadcaster considered it too inflammatory. "Terror on a very high scale, far beyond the bounds set by Mao," a man with a mustache was heard to say to a small group. "Terrorism of consciousness—revolution of the psychedelic."

And there, in this charged environment, Gregory Bateson did a remarkably prescient thing. He stood up and delivered a speech about catastrophic global warming triggered by fossil fuels. Society-altering climate change would become irreversible, he warned, once the ice caps melted and triggered one of the runaway feedback loops he had long studied. As one historian writes, Bateson's speech on that day in the summer of 1967 "may well be the first instance of climate change being discussed before a lay audience." Amid this threat of looming catastrophe, Bateson continued, it was little wonder "that in the 1960s a very large number of people are looking to the psychedelic drugs" to heal what was clearly a broken society by achieving a kind of "systemic wisdom." But though he was not convinced that psychedelics were the best path—he now saw LSD as "a sort of short cut; and I'm not sure there are short cuts to wisdom"—he agreed that the goal was "understandable."

When Allen Ginsberg gave his own speech at the conference, he announced, "This week I've been impressed more than anything else by Bateson." The scientist, he said, had revealed the "apocalyptic aspect of the anxiety syndrome that we're suffering from. Did you all hear his thesis about the carbon dioxide layer over the planet?" Much of what followed was rambling, even incoherent. But it looped back repeatedly to Bateson's ideas. "Black Power, White Power, Porpoise Power," Ginsberg said at one point, having mentioned Bateson's work with dolphins. At times he sounded very much like a young Bateson and Mead at the peak of their scientific idealism in the 1930s—though when he spoke of "the Squares" he had a different object in mind, saying that his goal was to enlarge human consciousness so that everyone, even the people whom youth culture mocked as "squares," could participate in a new global culture in the making.

To change society's "whole fabric, top to bottom, where would you begin?" Ginsberg asked. To some extent, perhaps, he was asking Bateson himself.

--------

While the Dialectics conference was happening in London in July 1967, in New York City a group of unknown actors was rehearsing a musical theater production called *Hair*. At one point in the play, a character named "Margaret Mead" wanders onstage with a small notepad and identifies herself as "a visitor from another generation." As she observes the young hippies assembled onstage, Mead's character strikes a reassuring note: "Now remember this, kids, and don't forget it... I am your friend."

So far, so good. But there is a reason why the real Mead, who loved publicity, did not make much of her fictional role in the defining Broadway musical of the late 1960s: she was being mocked. The actor who played Mead was male. This fact reliably inspired gales of audience laughter. The character pulls aside her dress to reveal that she is wearing men's clothing underneath, then nods to the actor playing her husband and whispers, "He doesn't know." She then heads offstage, shouting, "Oh, fuck you, kids!"

Though Margaret Mead still saw herself as an ally to youth and an advocate for radical cultural change, there was no denying that she was now also a part of the vaguely defined elite that young people had begun referring to as "the establishment." Who could look at her daily life and claim otherwise? Mead spent much of the first half of 1967 advising a group of Christian elders on revisions to the Book of Common Prayer and teaching an introduction to anthropology class at Yale University, where her students were entirely male and overwhelmingly WASP. (The B-plus grade Mead awarded a young George W. Bush, who was among those students, would go down in history as the highest university grade he ever attained.)

In the same period, she also worked behind the scenes to defend the actions of the CIA. In February 1967, the *New York Times* published a front-page story that implicated over a dozen obscure charitable trusts as CIA "pass-through" organizations. One of them, the Knickerbocker Foundation, was a name Mead knew well. She still counted James Mysbergh as one of her closest friends, and Mysbergh—the former OSS commando turned career CIA officer—had used the Knickerbocker Foundation as a front to explain his presence in Jakarta, Indonesia, for two years in the mid-1950s. He had done so while *also* claiming that he was working as a researcher "on some project associated with Margaret Mead," according to the diary entry of a confused acquaintance who ran into him in Jakarta during this period. Aware that the journalists at the *Times* and other outlets had stumbled onto an explosive story, Mead scrambled to defuse the situation by contacting reporters behind the scenes. "I have put all my weight behind emphasizing the need to separate free money that wasn't whittled to pieces by Congress, from the specific spying activities," she wrote to Rhoda Métraux from a conference in Manila, soon after the story broke.

Gregory Bateson (identified as "B") and Margaret Mead ("C") agreed to have their body language photographed and analyzed—with an emphasis on interpersonal tension—as part of a study Mead coauthored in 1968. *From Margaret Mead and Paul Byers,* The Small Conference: An Innovation in Communication *(De Gruyter Mouton, 1968). Photographer: Paul Byers. Courtesy the estate of Paul Byers.*

Teaching privileged men at Yale, being mocked by young artists, and defending a shadowy intelligence service: none of this was how the radical Margaret Mead of the 1920s and 1930s had imagined her old age. So when a letter sent to *Redbook*—the women's magazine she now wrote a monthly column for—sought her opinion of LSD, Mead took the opportunity to be controversial. In the end, however, she managed to alienate both sides of an increasingly polarized debate. The reader's letter posed a question inspired by Timothy Leary's latest publicity campaign, which involved his rebranding as a religious leader who prescribed psychedelics to his followers as a religious sacrament.

Did Mead, too, believe LSD could facilitate mystical experiences?

By this time, even media outlets that had formerly shown interest in the potential of psychedelic science now filled their pages with accounts of psychedelic users who threatened civilization itself. Mead could easily have followed suit. But she didn't. Though LSD's users could be "solipsistic" and overly inward-focused, Mead wrote, she defended the use of psychedelics as tools of spiritual insight. The trouble was that the United States of the 1960s did not have a cultural background that allowed these insights to be meaningfully integrated into society at large. In the Balinese trance tradition, she explained, "people take a great many precautions in selecting and ritually training those who will engage regularly in trance and in controlling where and under what circumstances trance may be induced." To properly guide psychedelic users, she wrote, "very careful disciplines and rigorous forms of training would have to be developed."

While her provocative article sat on newsstands in January 1968, Mead penned a letter to James Mysbergh, who by this time was living in Saigon. To those who asked, the former OSS operative identified himself as a PR rep for an obscure regional airline. But he remained, as James Hamilton described him, a "deep cover CIA" officer. Throughout his posting in Saigon, Margaret Mead updated Mysbergh on the changing culture on the home front. And now, Mead told her old friend, things were moving fast. Mead believed that antiwar sentiment was reaching "a kind of climax" among both the "hippies" (whom she expressed no opinion of) and "Black Power advocates" (whom she supported).

Everyone, Mead said, was preparing for "the next long hot summer."

--------

The climax came sooner than expected. On April 4, 1968, a white supremacist assassinated Dr. Martin Luther King Jr. By this time, the Tet Offensive had drastically altered the state of the conflict in Vietnam, pushing antiwar protests to a boiling point and threatening the United States with the prospect of a shocking defeat at the hands of the Viet Cong. It had also nearly killed James Mysbergh, who wrote to Mead that he'd spent "two hours on my belly in a ditch" near the Saigon Airport as he sheltered from a Viet Cong rocket barrage. He wasn't angry at the North Vietnamese, he told his old friend. He was angry "at the social scientists back home... all those experts who remain in their damn ivory towers, walk in peace parades and haven't seen what goes on here. I am angry because I am helpless."

Similar sentiments energized student protesters around the world. In the summer of 1967, Bateson had sat alongside Stokely Carmichael and Herbert Marcuse at the Dialectics of Liberation conference. Now, less than a year later, Marcuse was the subject of a slogan—"Marx, Mao, Marcuse!"—that was appearing as revolutionary graffiti throughout France amid a vast general strike. Meanwhile, Carmichael had been targeted by J. Edgar Hoover's FBI in the aftermath of Martin Luther King's assassination. He was preparing for exile in Africa.

At the Dialectics of Liberation conference, Allen Ginsberg and Carmichael had gotten into a heated exchange. "The reason the hippies have taken on these beads, appurtenances, music, of shamanistic groups, of ecstatic trance-state types," Ginsberg said, "is because they are beginning to explore, for the first time, the universe of consciousness of other cultures beside their own."

"Mr. Ginsberg, I don't know much about the hippie movement, but I would like to beg to differ with you," Carmichael responded. The thousands of white Americans who journeyed to India and Nepal, who sought out ayahuasca shamans in Peru, who chanted from *The Tibetan Book of the Dead* or hung tribal artworks on their walls, were, he said, "confused little kids who have run away from their home and who will return to their culture within a year or two."

"There's no culture to return to," Ginsberg had replied. "*There's no viable group culture.*"

Margaret Mead's generation lived through the most rapid period of change in human history, before or since. People born in 1901, like her, had entered a world that was still lit by candles, gas lamps, and kerosene lanterns—a world in which heavier-than-air flight was reckoned "a vain fantasy," China was an ancient dynastic empire, the British Empire covered nearly one fourth of the planet's land area, no nation allowed women to vote, and hundreds of thousands still carried the living memory of being enslaved. And the youthful, utopian Margaret Mead of the 1920s and 1930s had been almost unique in predicting the scope of the changes to come. While her peers continued to imagine the future as one dominated by Western empires and by traditional social norms, she predicted an age of upheaval that would alter the *experience of being human itself*—an era of transformed consciousness.

Though she had predicted the problem with amazing accuracy, she had failed to find a solution.

*There was no culture to return to*, Ginsberg told Carmichael that day in London. It was an exaggeration, but it was not wrong. There would be no world federation, no global cultural evolution, no dolphins at the UN, no wonder drugs for curing schizophrenia or healing the Age of Anxiety. Instead, there was an age of fracturing, of what Gregory Bateson called "schismogenesis." Thousands of cultures existing in parallel, mutually hostile, staring over walls. And increasingly, Mead and Bateson found themselves on different sides.

CHAPTER 25

# "If Someone Throws Away the Box, Then What?" (1968–71)

When you put a probe into any human situation, the other end always sticks into you.

—*Gregory Bateson, as quoted in Richard H. Blum,* Society and Drugs *(1969)*

## 1968–70

WHATEVER HIS MOTIVES, ONE thing was clear about the Gregory Bateson who arrived at the castle in Germany in July 1968. Clear, at least, to his daughter Cathy. "He had started to care again," she realized.

Bateson had come to lead a conference organized by the Wenner-Gren Foundation, a private institution that funded interdisciplinary research on human evolution and culture. The conference was titled "The Effects of Conscious Purpose on Human Adaptation," and it gathered some of the most prominent scholars and thinkers of the time. The setting was a medieval castle surrounded by forests and lakes, where the participants lived and worked together for two weeks. Bateson was eager to share his ideas and challenge his colleagues on various topics, from cybernetics to ecology to schizophrenia. He often stayed up late into the night, arguing and debating with anyone who would listen. "What I'm mainly interested in is a formal description of the ways in which human planning and applied science tend to generate pathology," he said at the outset, "in the society or in the ecology or in the individual."

He was still in thrall to facts—but no longer to science. Or to spies.

"I think that Gregory tried to create spokesmen for the various aspects of his own psyche," one of the conference participants said—to gather "a group of people to speak for various parts of him. It was psychodrama." Bateson was not only seeking intellectual stimulation and validation from his peers; he was also trying to heal himself from his past traumas and regrets. He hoped, perhaps, that by creating a new circle of collaborators who shared his vision of a more harmonious and holistic world, he could redeem himself from a past that included not just psychological warfare and his close proximity to the inner circle of men responsible for MKULTRA, but also, more recently, the disastrous, acid-infused finale to his time at John C. Lilly's dolphin lab.

Amid the onrush of events that marked 1968 as a year of change, from the assassinations of Martin Luther King Jr. and Robert F. Kennedy to the chaotic antiwar protests at the 1968 Democratic National Convention in Chicago, one hit Bateson particularly hard: the death of his onetime protégé Don deAvila Jackson. The final year of the MRI's director had been a difficult one. An expert in family therapy, Jackson had himself undergone a painful divorce. And though he treated patients suffering from alcoholism and addiction, he found himself enduring these afflictions as well, drinking heavily while consuming what one associate at the MRI described as vast amounts of "dope"—prescription tranquilizers and sedatives. At the end of January 1968, Jackson's body was found in his Palo Alto apartment. An autopsy showed that he had consumed a high dose of Pentothal. Some called it suicide; others called it an accident. But one thing was obvious. For Jackson, the "new beginning" of psychedelics had been the beginning of the end. The decade he began with so much optimism was a decade that he would not survive.

The news of Jackson's death shook Bateson. He had introduced the younger man to the world of psychedelic research, hoping to unlock new insights into the human mind. But he had also unwittingly exposed him to the dark side of that world, where sinister agendas lurked behind the scenes. He had watched him fall prey to addiction, divorce, and despair, until he finally succumbed to a fatal overdose of the same drug the OSS had used to interrogate prisoners long before.

Bateson had maintained social ties to his friends in the CIA (notably his former OSS colleagues James Mysbergh and James Hamilton) up through the Dialectics of Liberation conference and in the months afterward. Now, however, he dropped out of touch with them for good.

They were ties that Margaret Mead refused to let go of.

--------

On a Monday morning in October 1968, Mead sat down for an interview with the first person who tried to write the history of the Macy circle, which had begun to wind down following the retirement of Macy Foundation executive Frank Fremont-Smith in 1960. In his other interviews, Steve Heims—a physicist who believed that the Macy conferences were far more important to the history of the twentieth century than anyone had yet realized—started off with probing questions that often caught his subjects off guard. Fremont-Smith, in particular, had struck Heims as oddly cagey. "He was so cautious that I got the impression he was anxious to keep something private," Heims later wrote of the interview. "Much later I learned of the CIA involvement, which might have been the source of his anxiety." Heims's interview with Margaret Mead went differently.

The transcript begins this way:

**Mead:** You better annotate it.
**Heims:** Oh, I will.
**Mead:** On the tape.
**Heims:** Oh, I will certainly do that. Yes.
**Mead:** Yes, but you better do it now. *This is October 14th*... Have you done that already?
**Heims:** Well, this is a new tape. I'll mark it when I...
**Mead:** No, I mean put it, *read it into the tape*. Read things into the tape.
**Heims:** Oh, I see...

> **Mead:** *Always read things into the tape.* Side 1, October 14, 1968. First tape, interview with Margaret Mead at 10:03. You know, have labels on things. If someone throws away the box, then what?

A minute later, Mead was still not letting it drop. Tapes might come in boxes, she said, "but you can't trust the box." The contents and the packaging didn't always match.

That same morning, three NASA astronauts were orbiting Earth in the first manned Apollo mission. In a dark room, an editor was piecing together *Midnight Cowboy*, a dystopian buddy film about two hustlers in a New York City that was rapidly falling apart. The costume shops of St. Mark's Place and Broadway were gearing up for Halloween, with strong sales projected for Casper the Friendly Ghost, Elvis, and—for the older crowd—Mia Farrow in *Rosemary's Baby*. And Margaret Mead was worrying about her legacy. Two years earlier, she and Rhoda Métraux had bought a luxury apartment in the building facing the American Museum of Natural History, so close that the cries of schoolchildren on field trips could be heard through the windows. She continued as always, her office virtually unchanged since the era of Prohibition, her typewriter still overflowing with her habitual five or ten thousand words a day, her mind still resolutely focused on the future of human consciousness. And yet, as the unseasonably warm days before the Halloween of 1968 gave way to a week of frigid rain and muddy puddles, as her electric typewriter clattered amid her assistants' murmured conversations about the worsening situation in Vietnam, Margaret Mead had entered her life's final decade.

She could not know this. But there was one fact she could not help noticing—a fact that ran deeper than the evidence presented each time she saw her face in press clippings or on television or in the mirror that hung in her office turret. She was becoming old. Not just *looking* old, but, far more dismayingly, being *perceived as* old by the culture around her. For a woman who once declared that eleven was her "true, spiritual age," this was a fact that was hard to face.

Her world was aging with her. Without anyone being aware of the change, the things that seemed so vital and important when she was younger—the shining "Democracity" of the 1939 World's Fair, the dream of a world federation, the urgency of salvage anthropology, the futuristic synthetic clothing and the modern slang and the transgressive books—had all taken on the faded quality of age. The flappers of the Silent Generation were now grandparents. (This was a group that Mead would shortly join, when the news came in the spring of 1969 that Mary Catherine was pregnant.) The sad, beautiful wartime songs that serenaded the red ballroom at the Hotel Statler—"We'll Meet Again," "Moonglow," "The Last Time I Saw Paris"—filled thrift stores. Many of the science-fiction authors of the genre's heyday, with their utopian dreams born out of the progressivism of the Great Depression era, had become bitter reactionaries, like Robert Heinlein, who had begun endorsing militarism and social Darwinism. L. Ron Hubbard, still leading the Church of Scientology, was in hiding from tax evasion charges.

And as the cultural divides of the 1960s deepened into unbridgeable chasms, Mead found herself in an unwinnable position, stuck between two extremes. Everywhere people spoke of a new opening up of society. But there was a closing, too. The pioneers of psychedelic science had lived in a different era, one marked by war, fear, and repression. They had been involved in covert experiments funded by the military and the CIA, testing the effects of LSD and other drugs on unsuspecting subjects. They had faced ethical dilemmas and moral compromises that haunted them for years. Many had also struggled with their own personal identities, especially their sexual orientations, in a culture that criminalized deviations from the norm. The world around them was opening up—but not for them. Their secrets stayed hidden, buried in boxes that no one would ever open.

--------

This was never more apparent than in October 1969, a few weeks after the Apollo moon landing, when Mead experienced what was probably the most violent public backlash of her career. The flashpoint was a drug that she had shown little previous interest in: cannabis. "It is my considered opinion that marijuana is not harmful," Mead told a Senate committee that month.

Speaking to reporters afterward, she sounded almost like a paid promoter when she declared, "If I were young today, I'm sure I would be smoking marijuana."

Although she had made similarly provocative statements for a long time, the setting brought national attention to her experimental approach to drugs. With it came heaps of scorn and public abuse. When pushed on the question, Mead said that she believed LSD was potentially dangerous, but she predicted that within fifteen years an as-yet-uninvented "mind-expanding" drug would be synthesized, one with the same effects as LSD but a better safety profile. And this hypothetical drug, she said, should be accepted.

The governor of Florida called her a "dirty old lady." Hate mail piled up. "The United States is becoming a nation of pill-takers," a grandmother from Fort Lauderdale opined, citing the damaging escapism of "the Hippies, the Leary group, and the college students"—a group that Mead, to her dismay, had apparently now joined. "All my life's hope for my children are now to be destroyed by you," a man from Willoughby, Ohio, wrote. "A dope crazed people cannot possibly survive."

--------

Alongside the publicity around Mead's public stance on drugs, a quieter storm was brewing around the ethics of drug research conducted by some of her closest friends and colleagues.

Early in the Cold War, scientists had seen the question of the ethics of human experimentation in simple terms: On the one side was the manipulative brainwashing and unspeakable cruelty of Soviet and Nazi drug tests. On the other, the unsavory but necessary "defensive" measures taken by researchers in the West seeking to match them. "It wasn't that we were Nazis," explained Louis Lasagna, a drug researcher who worked on psychedelic experiments in the 1950s.

But the ramping up of the civil rights movement in the mid-1960s had raised unsettling new perspectives. "Psychologists ought to stop investigating and examining people of color," Stokely Carmichael had told the crowd at the Dialectics of Liberation conference. "They ought to investigate and examine their own corrupt society." Mead, as ever, believed she was uniquely placed

# Margaret Mead's for Legal Pot

Washington, Oct. 27 (UPI) —Anthropologist Margaret Mead told senators today that marijuana should be legalized for anyone over 16 and that drinking and voting ages should match the draft age.

Prof. Mead, 67, said the harsh laws against marijuana use are damaging society much more than the prohibition on liquor in the 1920s and are forcing youngsters to turn to hard drugs such as heroin.

**"A New Form of Tyranny"**

She did not mention a minimum age for marijuana use in her testimony before Sen. Gaylord Nelson's Senate Small Business Monopoly subcommittee. But she told newsmen afterward that the minimum age should be "probably 16."

She said marijuana "doesn't have the toxic effects that cigarets have" and is milder than liquor. Therefore, she said, it should be permitted at a younger age than tobacco and alcohol.

Prof. Mead told the senators: "It is a new form of tyranny by the old over the young. You have the adult with a cocktail in one hand and a cigaret in the other saying: 'You cannot. . .' to the child. This is untenable."

**Book About the Gap**

She said youngsters switch from marijuana to hard drugs because of a philosophy that "you might as well be hung as a sheep as a lamb."

Prof. Mead, who recently wrote a book about the generation gap, told the senators that medical evidence leaves no doubt that marijuana is not addictive, does not by itself lead the user to hard drugs and is much milder in its effects than alcohol.

She said 99% of marijuana smokers never switch to harder drugs. Those who do, she said, are probably motivated by the harsh laws.

Prof. Mead is curator emeritus of ethnology at the American Museum of Natural History and professor of anthropology at Columbia University. She gained fame for her studies of natives in Samoa and New Guinea.

Anthropologist Margaret Mead testifying on use of marijuana.

MARGARET MEAD - YOU MUST BE MAD FOR ENCOURAGING MARIJUANA FOR YOUNGSTERS, WE MOTHERS HOPE YOU ARE PUNISHED - YOU MUST BE CRAZY A DOPE FIEND.

DIRTY LOUSE.

Mead received a thick stack of hate mail—including this angry note scrawled on a newspaper clipping—following her public call for cannabis legalization in 1969. *Courtesy Library of Congress.*

to intervene in this debate. The core of the issue was defined by the phrase "guinea pig," she wrote in a 1969 article on the ethics of human drug experimentation. The metaphor by which "human beings are reduced to the status of *experimental animals*" lurked behind public perception of such practices as "the creation of psychological states by hypnosis" or "injection of subjects with substances whose action is unknown," Mead wrote. "The mad scientist and evil scientist are typically portrayed with a set of caged animals in the background."

We cannot know if such thoughts were informed by her visits to Harold Abramson's lab at Cold Spring Harbor, with its rows of tanks containing LSD-spiked Siamese fighting fish and snails—or her visit to the St. Thomas dolphin lab overseen by Bateson and Lilly. But later in her article, when Mead wrote about the ethical vagaries of "field tests of new drugs," she noted what we might call the utopian's dilemma. The people responsible for ethical violations, she said, were not just "cynical" or malicious. They were often those "who take refuge behind the great good that [they are] doing." In such people, "the need to justify demeaning other human beings... is compensated for by another kind of delusion of grandeur: that of becoming someone who benefits mankind on a large scale." As an example of such a person, Mead made a thinly veiled reference to CIA-affiliated researchers like Harold Wolff, Harold Abramson, and James Hamilton: she wrote of "the scientist working for the intelligence agency of a particular government."

This was not the only public hint Mead dropped suggesting she was troubled by her own participation in covert Cold War drug research. In May 1970, Ralph Blum, Mead's closest confidant about psychedelic science during her LSD summer of 1954, published a thriller about a government mind control program closely modeled on MKULTRA. Blum's novel, *The Simultaneous Man*, interwove a standard techno-thriller plot with details torn from the real history of the CIA's early experiments with LSD. ("Boston Psychopathic Hospital needed volunteers for a drug experiment," a character explains at one point.) Blum confronted the legacy of racist psychedelic science from the 1950s, when people of color were disproportionately subjected to dehumanizing drug research—but he did so behind a science-fiction disguise. The centerpiece of his story is an African American soldier whose brain is manipulated using a fictional drug known as FLX and then "remade" with contents from the brain of a sinister "psycho-pharmacologist."

On the back of Blum's book, alongside an enthusiastic endorsement from Noam Chomsky, was a blurb from Margaret Mead. The book, she said, was not entirely fictional.

--------

Between the era's countercultural rebellion against "establishment" authority, the increasing recognition among Cold War researchers of their own past ethical improprieties, and the polarization of American society in the Nixon era, science itself was now falling from its pedestal. And as it did so, psychedelic science became something else—something distinctly more mystical. It was part of a larger cultural moment that, as the cultural historian Matthew Tribbe has argued, followed immediately in the wake of the Space Age. Many Americans were now disillusioned by the technocratic vision of progress embodied by NASA. Tarot cards, astrology, and crystals proliferated; the New Age movement found its footing; distrust toward institutionalized science grew.

Bateson was happy to change with the times. He now spoke of a transcendent pattern of mass consciousness comparable to God, a universal life force "immanent in the total interconnected social system and planetary ecology." There is no such thing as "Gregory Bateson," he said. There was, instead, a transient selfhood that existed as one small part of a *planetary* consciousness. "Under LSD, I have experienced, as have many others, the disappearance of the division between self and the music to which I was listening," Bateson explained. "The perceiver and the thing perceived become strangely united into a single entity. This state is surely more correct than the state in which it seems that 'I hear the music.' "

John C. Lilly, too, was pleased by the cultural shift underway. After the Communication Research Institute's funding ran out in 1968, he had found a position at a Maryland psychiatric research lab that allowed him to continue his LSD experiments. But early in 1969, he resigned from the position and moved to the Esalen Institute in Big Sur to take a series of self-exploration courses. After they ended, he stuck around. While the Apollo 11 mission was heading to the moon in July 1969, Lilly was inhabiting a succession of hot tubs, treehouses, and verandas, reassessing his life. At one point, he began dictating what he called "a series of metaprograms in blank verse" into his tape recorder while standing at dawn on a cliff overlooking the Pacific. "When I look back into my apparent past life, I find faulty recordings in my mind," he said to himself. "All my work was illusory."

Esalen's cofounder, Richard Price, had been a student of Gregory Bateson's at Stanford, and he envisioned the community he built in Big Sur almost as a new generation's take on the Macy conferences. The original Esalen seminars were academic in tone, featuring interdisciplinary discussions of anthropology, psychiatry, drugs, and religion that centered on consciousness expansion. By 1969, however, Esalen had become such a haven for young people suffering from bad trips that the Federal Bureau of Narcotics was sending over undercover agents. "We lost control of the place," said Price, remembering the crowds of hippies that descended in the late 1960s.

One of them was Charlie Manson. In the first week of August 1969, the would-be guru and three of his female followers began driving north on Highway 1 in a 1952 Hostess Bread truck. Manson hoped to impress Esalen's guests with his songwriting prowess. Instead, he ended up scaring one of the permanent residents, who sensed that, as he put it, the group was "bad news." Rebuffed, Manson and his acolytes headed back to Los Angeles. The Manson murders took place less than one week later.

Not long after the murders, Lawrence Kubie, the Manhattan psychoanalyst who had consulted for the Office of Strategic Services' truth drug experiments in World War II, wrote to Lilly urging him to rethink his stance on psychedelics. The son of a mutual friend, Kubie reported, had jumped from a balcony and died while on LSD. Lilly replied by speculating that the boy had died because the "parents did not understand . . . their children's problem." He decided to write a book on the value of LSD and other psychedelics from an "objective" point of view. He duly wrote a draft of the book. And then he decided to throw most of it away. What came in its place was something very far from an objective study of LSD. All throughout the spring and summer of 1971, Lilly labored on his manuscript. He lamented the "hysterical atmosphere" that now surrounded psychedelics. He wrote of his dolphins. He spoke of the universe as a great cosmic computer in which all conscious beings were mere simulations. He called it *The Center of the Cyclone*.

But when he sent the draft to Alan Watts, Lilly received a disquieting reply. Parts of the book were like "something out of a bad book on occultism dated 1910," Watts wrote. "I feel you have a responsibility for making it much more clear intellectually, why you passed from the scientific enterprise to the mystical."

Watts seemed especially worried about what Lilly's former colleagues would think. He singled out one possible response in particular: "Look what LSD did to a great mind!"

CHAPTER 26

# Esalen Sunset (1971–77)

> Is it in us really to alter after our 20s? I think not in fundamentals. I am sure that I shall go to my grave with my shoes untied, my trousers unpressed, my hair overdue for a cut, and probably imperfectly shaven. I shall still leave undone those things which I ought to have done—and shall still feel unhappy and guilty about it.
>
> —*Gregory Bateson in a letter to "Steve" (an ex-girlfriend), February 11, 1944*

THE PSYCHEDELIC SUBCULTURE OF the United States in the 1970s was, in large part, defined by an eclectic series of publications called the *Whole Earth Catalog*. In a commencement address at Stanford, Apple cofounder Steve Jobs remembered the series, created by Stewart Brand, as "sort of like Google in paperback form, 35 years before Google came along: it was idealistic, and overflowing with neat tools and great notions." Among those tools were DIY sensory deprivation tank kits, books explaining how to manufacture LSD, home birth manuals, and rudimentary personal computers. Taken together, the *Whole Earth Catalog* defined a cultural moment of "off the grid" communal living, concern for planetary ecology, and a melding of spirituality and science.

The catalogs were also shot through with adoration for Gregory Bateson, whose ideas Stewart Brand took every opportunity to promote and praise. The last and perhaps the most influential entry in the series, 1974's *Whole*

*Earth Epilog*, opened with a winsome photographic portrait of Bateson staring off into the sunset, his hair uncombed and his sweater wrinkled. "Gregory Bateson's insights lurk behind most of what's going on in this *Epilog*," Brand explained to his readers. The rest of that first page was dedicated to quotes from Bateson's writings—passages about addiction, consciousness, and human survival. Many of the early hackers who shaped the digital revolution were also avid readers of Stewart Brand's publications and, through them, imbibers of Gregory Bateson's ideas. They, too, were part of the psychedelic counterculture of the era, using LSD and psilocybin as they tinkered with circuit boards in garages.

Readers of the catalogs searched in vain for similar appreciation of Bateson's ex-wife. Despite Margaret Mead's repeated efforts to show solidarity with the psychedelic counterculture of the late 1960s and the 1970s, she found herself excluded from the movement she had helped inspire. And she was not the only one to feel left out. Many of the early pioneers of psychedelic science found themselves marginalized by an increasingly insular subculture. The fall of 1974, when the *Epilog* appeared, was a moment of peak polarization for both the United States and the world. The season had begun with the resignation of Richard Nixon amid a generation-defining global economic crisis. The national mood was one of outright paranoia—a paranoia that extended to the world of psychedelic science, as Richard Alpert, Allen Ginsberg, and Jack Leary called a press conference denouncing the now-imprisoned Timothy Leary as a "cop informant," "liar," and "schizophrenic." "I would not be surprised if he testified about my sister or myself if he could," Jack said.

In the 1950s and 1960s, Margaret Mead had numbered among the planet's most celebrated scientists. Now, however, it was impossible to say exactly what words like "celebrated" even meant. Celebrated *by whom*?

--------

Amid the charged social climate of the Nixon years, Margaret Mead's vision of science as a cross-cultural utopian project began to seem obsolete. Respect for scientific idealism was evaporating fast, and neither left nor right, squares nor "heads," knew where she stood.

Margaret Mead in her office at the American Museum of Natural History, 1971. *New York Public Library. Photographer: Martha Swope.*

Mead did her best to stay relevant—but the weight of her secrets made it difficult. An emblematic moment had come late in 1971. That year saw the publication of an unusual book, coauthored by Mead and James Baldwin and based on a long conversation between the two, called *A Rap on Race*. Since the 1920s, Mead had seen herself as passionately antiracist. And in African American newspapers throughout the 1930s and 1940s, she had been hailed as such. But Baldwin—Black, gay, brilliant, and proudly unconventional—had found his voice in the experimental 1950s. Though they discovered much common ground, Baldwin seemed frustrated both by Mead's unerring optimism and by her signature blend of surface-level openness and reflexive secrecy. Toward the end of their talk, their conversation took a turn away from race. Abruptly, as if aiming to establish mutual understanding, Baldwin suggested that he and Mead were both "exiles" in some other way—a *shared* way. "You see," he said, "*there is an area where we both were exiled.* You said you weren't, but you are because of what you know."

Baldwin, perhaps, had some inkling of Mead's own sexuality and was now inviting her to speak publicly about it for the first time. If so, his attempt failed.

"I am a what?" Mead asked.

"An exile," said Baldwin. Even as she tried to interrupt, he explained that he saw her as an exile "from the mainstream of life in this country."

"Oh, no, I am not," she replied. "I am not an exile. I am absolutely not an exile."

A note of desperation, of pleading, had snuck into a conversation that had earlier been free of it. It returned a few minutes later, when Baldwin rephrased his earlier point: "You and I are both in the same difficult hot seat. America doesn't want you any more than it wants me."

"Jimmy, it isn't true," Mead said.

"You think so?" was all James Baldwin said in reply.

--------

The lives of Margaret Mead and Ruth Benedict, notes the historian Naoko Wake, were both marked by "silence and ambiguity about their sexualities." Though Mead cohabited with her romantic and professional partner Rhoda Métraux for over two decades, her bisexuality did not reach public notice until several years after her death. There is strong evidence that Mead created intentional absences in the historical archive—lacunae not just about her own sexuality, but about her work with experimental drugs. These silences protected her. But at the critical moment of cultural change, a moment she had spent decades preparing for, it was the burden of her secrets that held her back.

Gregory Bateson shared many of Mead's secrets, but with two crucial differences. First, he was a man in a heterosexual relationship, and thus shielded from the misogyny and homophobia of the era. And second, he no longer desired power. He did not *want* to maintain his security clearance or his friendships with spies. And in showing this disdain, he gained the respect of thousands—as well as a new institutional home. In 1972, Bateson began working at a new research university nestled among redwood trees north of Esalen: the University of California, Santa Cruz. Officially, he was a lecturer in anthropology, but he also became something of a guru to the student body, who lined up for his office hours and recorded his wisdom for future reference. Bateson and Lois moved into a farmhouse on a semi-commune in the Santa Cruz Mountains, just down the road from Ken Kesey and his Merry Pranksters. It was rumored on campus that "there was some kind of connection" between them. Bateson's connection to another pioneer of psychedelic science turned UCSC professor—Timothy Leary's friend and research partner, Frank Barron—was clearer. The two men could often be found lunching at a strip-mall Polynesian restaurant called Zanzibar.

Only Margaret Mead knew the debilitating fear and guilt of the man who had returned from the jungles of Burma in 1945, unable to shake his memories of contributing to a manipulative science of consciousness. And only Lois Bateson could have understood the gut-wrenching conclusion of Bateson's time with John C. Lilly. But anyone who read the *Whole Earth Catalog* could see that Bateson's views about the dark side of science were finding an eager audience. In another of his publications, a magazine called *CoEvolution Quarterly*, Stewart Brand published letters to Bateson seeking advice, alongside Bateson's acerbic responses. It was as if he had become the reluctant therapist to an entire generation, a generation shaped by fears of carbon dioxide emissions, global overpopulation, ecological destruction... and the CIA.

In December 1974, news of the MKULTRA program broke. The first story in the *New York Times* was just a scrap, a hint of something larger. But soon enough, the world would learn about the CIA's illicit testing of psychedelic drugs. And amid these revelations, names long familiar to Bateson and Mead would surface—the names of friends.

--------

In the aftermath of the Watergate scandal, the media and the public began to uncover clues about other covert activities during the Cold War, including the existence of a group of previously unknown top-secret projects that the Central Intelligence Agency's leadership called "the family jewels." These included MKULTRA. In January 1975, just a few months into his presidency, Gerald Ford tried to get ahead of these ongoing investigations by appointing a commission to investigate CIA abuses, headed by Vice President Nelson Rockefeller. The Rockefeller Commission was supposed to restore trust and confidence in the government, but it failed. Senator Frank Church criticized the commission for classifying details about the CIA's most egregious acts—assassination plots and drug experimentation. In April 1975, Church chaired a Senate committee that conducted further investigations into these programs; before long, the Church Committee began collecting details of illegal CIA experiments on human subjects involving drugs, hypnosis, and torture. One of the best-publicized cases to emerge from the hearings was the sad end of Frank Olson, the biological warfare scientist who had been covertly dosed with LSD by his colleague Sidney Gottlieb in 1953 and died less than two weeks later after falling or jumping from a hotel window. Olson's family had been lied to about his death; now, thanks to this new public attention of 1975, they received a personal apology from President Ford in the Oval Office. A cash payment from the CIA followed.

For other victims of MKULTRA, especially the thousands of African American prisoners, inmates, soldiers, and civilians subjected to covert experimentation, compensation was harder to find. But at least there were now venues for learning the truth.

One of the most public of those victims was Edward M. Flowers, who appeared before the Church Committee's Senate hearings in November 1975. A small man in his midforties with a diffident smile and a quiet voice, Flowers had grown up in a large apartment block on 116th Street in Harlem. His father, an emigrant from Belize, did his best to maintain four children on a housepainter's salary, supplemented by Flowers's own work as a shoeshine boy. For a time, the family eked by. But after his father died, life got hard. By the time he was eighteen, Edward Flowers was addicted to opiates. Following a drug arrest, he lied about his age in order to be admitted to a special drug testing program in a government drug treatment facility in Kentucky known as the Narcotics Farm. The program, which by the time he entered it in 1953 was covertly funded by the U.S. Navy and MKULTRA, involved administering LSD and other experimental drugs to opiate-addicted patients in exchange for shots of morphine. As Flowers told the senators, he had heard rumors about its existence from other opiate addicts. He explained that "my whole reason for going into the program was just to get some drugs."

Now, twenty years later, Flowers found himself sitting across a table from Dr. Harris Isbell, the former medical director of the Narcotics Farm and a consultant for the MKULTRA program. When one senator asked whether Isbell had "authority" to test experimental drugs on young men like Flowers and whether the experimenters sought informed consent, Isbell did his best to explain—but came up short. "It was a different time, sir," was all he could muster.

--------

The revelations of the Church Committee congressional hearings triggered a public reckoning about the devastating consequences of the covert psychedelic experimentation program that had taken place earlier in the Cold War. But though the stories of men like Edward Flowers and Frank Olson inspired literally thousands of newspaper articles over the course of 1975 and 1976, the surviving members of the Macy circle had little to say about them.

Margaret Mead spent 1975 as president of the American Academy of Arts and Sciences, the nation's oldest and most prestigious scientific organization. Her presidential address included a reflection on how far the "science of the human" in the United States had come since the days of racial segregation. She spoke about science as a tool for "shared understanding." Citing everyone from Julian Huxley to Franz Boas to her own daughter (who was now an anthropologist herself), Mead made a pitch for a "cosmic" and "planetary" sense of our collective destiny, for a move away from nationalism and toward "a concern for the earth's atmosphere" and "new political forms" that could bring together the "irreversibly interdependent planetary community." She was looking to the stars and ignoring the newspaper headlines. In her lecture on the ethical progress and the cosmic future of science, there was no mention of the subject that was hitting international media and meriting mention in similar speeches from scientists at the time.

Above all, there was no mention of the troubling legacy of racism in psychedelic science—a theme that the Church Committee and subsequent hearings publicized for the first time. Mead did not publicly reflect on her own role in the imbalance of power between scientist and patient in an era before civil rights, an era when the brutality of the Nazi doctors was very much the stuff of living memory, and the infamous Tuskegee syphilis experiments (which studied the effects of syphilis on nearly four hundred African American men from Alabama, who were not treated for the disease or informed that they had it) were still underway. At the 1953 Macy conference on consciousness, for instance, without explaining why he was interested, Abramson had asked Mead about "candidates for experimentation from low-income groups." Rather than pausing to consider what Abramson was up to, Mead simply offered a suggestion. She told Abramson to look into Father Divine's Angels, a charitable organization in Harlem whose members were overwhelmingly poor and African American. It was a genuinely dangerous tip. Although there is no evidence that Abramson pursued the idea, there is abundant documentation of the MKULTRA program's predatory reliance on people of color like Edward Flowers as test subjects.

And as for Mead's ex-husband? In January 1976, Gregory Bateson was surprised to find himself the honored guest of California's newest governor, Jerry Brown; in fact, he was the keynote speaker at the governor's annual prayer breakfast. It was a long-standing tradition in the state. Throughout the early 1970s, then-governor Ronald Reagan had presided over sedate gatherings of government officials and churchmen. So newspapers were delighted by the odd scene that greeted them now. One reporter described "a bare-chested Indian dancer," a Sufi choir, and a spread of pineapples and yogurt in place of the traditional bacon and eggs. And then there was Bateson himself. "Peyote, snakes, pocketknives, and pregnant goats were the subjects chosen by the anthropologist-writer Gregory Bateson for the main address," as the *Sacramento Bee* newspaper put it.

In truth, Bateson was talking about guilt. He quoted the biblical book of Job. And then he read from Samuel Taylor Coleridge's famous poem *The Rime of the Ancient Mariner*: the key moment when the albatross tied to the mariner's neck as punishment for killing it finally falls and sinks "like lead into the sea." The mariner's symbolic release from guilt, Bateson explained, came from a moment of strange beauty, as, dying of thirst, he watches water snakes moving "in tracks of shining light" beneath the sea, then "blesse[s] them unaware."

It was the *unawareness*, the *unconsciousness*, of this act that captivated Bateson. If the mariner had been aware rather than unaware, conscious rather than unconscious, the symbol of guilt would never have fallen from his neck. The meaning of the poem, Bateson felt, derived from this inexplicable moment of redemption. But he never quite said this to the audience on that day in Sacramento in 1976. He just hinted that the strange stories he was telling had something to do with consciousness, transcendence...and guilt.

Bateson later wrote that his speech at the governor's prayer breakfast in 1976 had been partially inspired by his LSD experiences with his friend and MRI colleague, the psychiatrist Joe K. Adams. Specifically, he recalled Adams, at Esalen in 1962, "interrupting my psychedelic trip while he sets up a tape recorder." In trying to record a moment of sacred transcendence, Bateson said, Adams had destroyed it. This was what put him in mind of the albatross. The perplexing stories from that day in Sacramento, Bateson explained, all involved the danger of *too much* knowledge. They all pointed to the value, even the sacredness, of secrecy.

The central theme, as he put it, was not just concealing a secret, but "conceal[ing] *the fact that there is a secret.*"

--------

Throughout the rest of 1976, the investigation into MKULTRA appeared to have run out of steam. And then, early in 1977, hundreds of boxes of additional files were discovered in a CIA storage room. These files brought public attention uncomfortably close to the professional and personal lives of Mead and Bateson for the first time.

Several paths led there. The work of the neurologist Harold Wolff, long dead, was now being scrutinized. Attention fell on his CIA-funded Human Ecology program, a program that Mead had consulted for and that Rhoda Métraux had been an official employee of. At one point in the 1977 congressional hearings that the discovery of the new documents prompted, Senator Ted Kennedy began grilling an ex-CIA psychologist named John Gittinger about Wolff's program. When questioned earlier about the CIA's "west coast agent" for illicit drug testing, Gittinger had been careful to use George White's cover identity, Morgan Hall, rather than his real name. But now the questioning involved scientists who were not employed by the Agency. Scientists who did not have code names. And some, like Mead herself, who were *household* names.

Gittinger began haltingly. "The Society for the Investigation of Human Ecology, so-called, was actually a..." Then he stopped. "I am confused here now as to whether I should name you names," he explained to Senator Kennedy.

"We prefer that you do not," the senator replied, and the discussion moved on. One box, at least, had remained closed.

## October 1977, Sunset

One day in the fall of 1977, Allen Ginsberg sat in an airplane flying from Minnesota to San Francisco, thinking about the past.

His writings let us picture the moment in detail. It is 4:45 p.m. Ginsberg is staring out the window at the lengthening shadows of the Rocky Mountains. He is tripping on LSD. "To be high on acid in an airy transcontinental cabin and still have secrets!" he writes in his journal.

The poet thinks back once again to a spring day in April 1959. The day when, at the invitation of Gregory Bateson, he first stepped onto the creaky porch of the Mental Research Institute. In the weeks and months afterward, he begins reading the resulting journal entry at poetry readings. At one, Ginsberg asks, "So the question is, *how high up in CIA did acid get*? How much was acid spread through CIA? Early Contacts with Kesey or myself in 1959 Stanford Institute of Mental Health [*sic*], Dr. Joe Adams via Gregory Bateson. Was that not financed secretly by Army intelligence? So it is reported from Washington! Didn't I at that time register some perturbation that I was being eavesdropped by Big Brother ear? My talk and babblings were recorded including paranoiac suspicions that Army Intelligence and CIA were wired into *my* brain...so: who has those tape recordings *now*? I'd like to hear them!...eighteen years later, *I want to go back and hear the tapes.*"

When Ginsberg published this journal entry, the name had been changed to "Greyson Bateson." This could have been a transcriber's error. Or it could have been a nod to Bateson's desire to maintain some distance from his past. Bateson and Ginsberg, after all, were friends.

Ginsberg continued:

"What valuable history was destroyed? Was Leary their scapegoat? . . . Who remembers? . . . Am I 'Allen Ginsberg' the byproduct of one of the CIA's lamentable, ill advised, or triumphantly successful experiments in mind control?"

When he read it aloud at readings, he did so with humor in his voice, to an audience that occasionally broke into laughter—the poet was, to some extent, mocking himself and his own paranoia. But it was knowing laughter.

"Sun orange in tired heaven," he ended. "No blame in any direction."

No blame. But a persistent question, one that had been nagging at him for well over a decade. What *had* happened to those recordings that Bateson and Adams had made? What could they reveal about the origins of the psychedelic era? What other, similar boxes remained unopened?

Ginsberg never found the tapes. So far, no one else has either.

CHAPTER 27

# The Blue Glow of Life (1977 to July 4, 1980)

> And if you read between the lines
> You will find nothing there.
>
> —*from Gregory Bateson's unpublished poem "The Manuscript" (1978)*

ALLEN GINSBERG WAS NOT the only person in the fall of 1977 wondering about the Cold War secrets around psychedelics. International media from Sydney to São Paulo to London abounded with damning reports and lingering questions. And in Moscow, predictably, the state-run newspaper *Pravda* reported breathlessly on each new revelation.

Indirectly, one of the causes of it all was Albertine White, the elderly widow of the hard-drinking former narcotics officer and CIA consultant George Hunter White.

In retirement, White had become the fire chief of Stinson Beach, a small community in Northern California. To all who would listen, he argued strongly against drug legalization. He voted for Nixon. He missed the good old days before the "long hairs" had emerged. In other words, he seemed to resemble countless other men who came of age amid the sober, Christian, white masculinity of America during and after World War I. Sometimes, though, the mask fell off. "Where else could a red-blooded American boy lie, kill, cheat, steal, rape, and pillage with the sanction and blessing of the All-Highest?" he wrote in a nostalgic letter to his former CIA handler Sid Gottlieb. Even those without security clearances got the occasional hint that

White was something more than he appeared. In a letter to a Stanford professor about the menace of "dope," for instance, White mentioned that he himself had "served as a guinea pig from time to time." He added, "I did feel at times that I was having a 'mind-expanding' experience. But this vanished like a dream."

By the time he wrote these words, White was suffering from the cirrhosis of the liver that would kill him a few years later.

After her husband's death, Albertine White was left with approximately forty small leatherbound diaries filled with classified information. They were written in an almost illegible hand. At the suggestion of one of White's friends, she donated them to a most unlikely place: the Electronics Museum of Foothill Junior College in San Jose, California. It did not take long for journalists to discover that the diaries cracked open the story of MKULTRA. They led to an unspooling of an entirely new set of revelations involving the most salacious aspects of the program. These included records relating to drug experimentation that dated back to "the days of the Office of Strategic Services in 1943," a CIA spokesman announced. In other words, they had recovered the files relating to the truth drug experimentation of George Hunter White and James Hamilton.

One of the findings from this new information was that (as an anonymous source for the *New York Times* put it) James Hamilton was "a key guy" in MKULTRA. But there was far more on display in White's diaries—evidence not just of occasional lapses of judgment by scientists, but of a systematic program of unwitting testing of psychedelics spearheaded by a man with no medical degree, no training, and seemingly no moral qualms about the wreckage he left behind—a man who struck some as a dangerous sociopath.

White's diaries and the new boxes transformed the ongoing congressional investigation into MKULTRA. Throughout the fall of 1977, the Senate Select Committee on Intelligence summoned several former CIA officials and contractors to testify about the new details, including Sidney Gottlieb. Behind the scenes, the government investigations also compelled the CIA to put together its own internal team, the "MKULTRA Task Force," to attempt damage control. One member noted, accurately, that White's entries about drug testing were "cryptic, handwritten, and range from barely legible to meaningless scrawl." But there was enough to see that serious crimes had been committed.

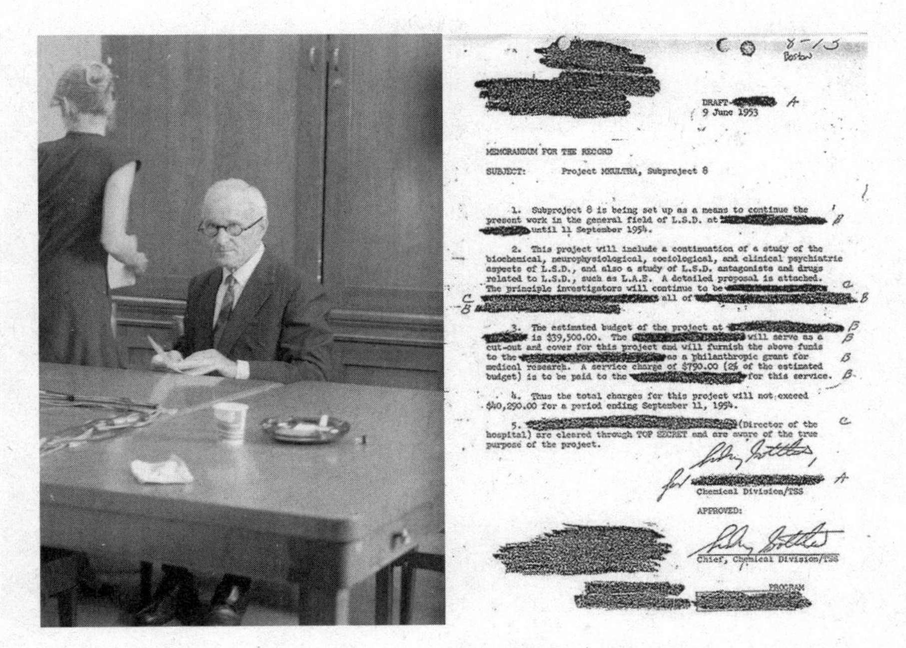

DRAFT-
9 June 1953

MEMORANDUM FOR THE RECORD

SUBJECT: Project MKULTRA, Subproject 8

1. Subproject 8 is being set up as a means to continue the present work in the general field of L.S.D. at until 11 September 1954.

2. This project will include a continuation of a study of the biochemical, neurophysiological, sociological, and clinical psychiatric aspects of L.S.D., and also a study of L.S.D. antagonists and drugs related to L.S.D., such as L.A.E. A detailed proposal is attached. The principle investigators will continue to be all of

3. The estimated budget of the project at is $39,500.00. The will serve as a cut-out and cover for this project and will furnish the above funds to the as a philanthropic grant for medical research. A service charge of $790.00 (2% of the estimated budget) is to be paid to the for this service.

4. Thus the total charges for this project will not exceed $40,290.00 for a period ending September 11, 1954.

5. (Director of the hospital) are cleared through TOP SECRET and are aware of the true purpose of the project.

Chemical Division/TSS

APPROVED:

Chief, Chemical Division/TSS

PROGRAM

*Left:* Sidney Gottlieb, September 21, 1977, at the MKULTRA hearings. *Right:* One of the internal CIA documents relating to MKULTRA made public as a result of the investigations. *Courtesy Wikimedia Commons.*

Members of the task force flew to the San Francisco Bay Area to interview White's former associates. They even visited the premises of the shuttered cabarets and strip clubs in the Barbary Coast district of San Francisco where White and Hamilton once trawled for unwitting test subjects. They attempted to find living victims but came up short. The world of the Barbary Coast had already faded away.

--------

As the new MKULTRA revelations were breaking, Margaret Mead was being feted at the first ever Margaret Mead Film Festival.

The lavish, multiday screening of hundreds of documentary films was held in thirteen venues throughout the American Museum of Natural History, including the enormous neoclassical auditorium at the museum's heart. The festival commemorated not just her seventy-fifth birthday and her fiftieth anniversary as a curator at the museum, but also what the *New York Times* article announcing the event called "her pioneering work in the study of human behavior." The festival organizers celebrated the power of film to document the diversity of human cultures.

The adulation she received from the thousands who attended the festival in the third week of September 1977 contrasted jarringly with the other stories that filled the newspapers on those days—revelations of sinister Cold War secrets involving scientists who had worked alongside Mead for decades. That contrast was likely on the mind of at least one of the Margaret Mead Film Festival's VIP guests: Gregory Bateson.

--------

By this time, Bateson had, almost by accident, attained real political power, having become a key mentor to California's young governor, Jerry Brown. Impressed by his unconventional speech at the 1976 prayer breakfast and encouraged by Stewart Brand, Governor Brown met with Bateson for a series of discussions that ranged from space colonies and climate change to Bateson's concept of the double bind. Brown, who recalled being impressed when Bateson sat on the floor at their first meeting, was part of a new generation of political leaders who embraced the counterculture. He had visited Esalen, earnestly studied cybernetics, even briefly dated John C. Lilly's daughter in Malibu. In 1977, Brown appointed Bateson a regent of the University of California. And that new position required yet another FBI security check—the last of many that had been conducted throughout Bateson's life.

This final investigation took place at the exact moment that the most damaging MKULTRA projects were first being reported. The day that Bateson signed the paperwork clearing the FBI to begin its investigation—October 13, 1977—was also the day a California newspaper reported that "CIA-financed mind control research" had been overseen at a state-run psychiatric facility for prisoners by "a prominent San Francisco psychiatrist, Dr. James Alexander Hamilton." Hamilton, who refused all comment, was revealed to have led the CIA's experimental testing of stimulants, tranquilizers, and hormones on well over a hundred inmates between 1962 and 1968.

Two days later came a reunion of psychedelic scientists on Bateson's home turf of UC Santa Cruz, where Albert Hofmann, LSD's creator, had been invited to lecture about his most famous discovery. Hofmann's "somewhat dry recitation," wrote a reporter in attendance, was "frequently interrupt[ed] with applause, cheers, and giggles." The audience, which included Richard Alpert and Timothy Leary, gave the elderly Swiss chemist a standing ovation. Afterward, brandishing newspaper clippings, Leary crowed that the news confirmed what he had long suspected. "The LSD movement was started by the CIA," he said. "I wouldn't be here now without the foresight of CIA scientists."

Amid the excitement, no one seemed to notice that one of the people who knew those scientists best was not there. Bateson had chosen to spend the weekend out of town.

Bateson and Mead could hardly have avoided the widespread press coverage of the fact that longtime friends such as Harold Abramson and James Hamilton had been key participants in the MKULTRA program. Nor would it have been news. Even if the details were not known to them, the archival record is clear that *something* had been, long before 1977. In two letters from January 1968, for instance, Bateson and Hamilton had discussed such things as the "unpleasant" nature of unwitting dosing with LSD and the "acute discomfort" caused by "hormonal stress." Their letters then had been couched in friendly terms, following a social visit between the two men and their wives. But when read with the knowledge that Hamilton was a veteran CIA consultant responsible for some of the most notorious MKULTRA drug experiments, they have a cryptic, even eerie subtext. "Best to Marjorie," Bateson signed off his letter, following a vague allusion to "those who get cubed (i.e., receive a dose without knowing it)."

By the fall of 1977, Bateson and Mead had turned away from the world of Cold War psychedelic science. But this world was not yet done with them, even as both their lives reached their end.

--------

In January 1978, after spitting up blood, Bateson went to a doctor for tests. He was told he had inoperable lung cancer.

Mead, meanwhile, was in high spirits, planning what she described as an "upbeat book" about the Soviet Union. But she had also begun losing a troubling amount of weight. Her twenty-two-year-old assistant, Amy Bard, was among those who noticed her decline. "MM sounded real bad today," she wrote that March. "Tired, unenthusiastic. I wonder, is she on her way out? Is she finally getting old?" Less than two weeks later, Bard discovered a note in Mead's handwriting that appeared to be transcribing a phone call. "Mass on the body of the pancreas most likely represents a malignancy," the note read.

Mead told no one. She explained her weight loss by saying that she had become anorexic. That summer, however, she began making plans to visit Esalen, where Bateson now lived as a semi-permanent resident, a kind of in-house philosopher, along with Lois and their infant daughter, Nora. And she also began planning a side trip to see a man who she knew could keep secrets well. She wanted James Alexander Hamilton to tell her if she really did have cancer.

--------

"An old rivalry with Gregory reasserted itself in her conviction that if he could beat cancer, so could she," remembered their daughter, Mary Catherine, about Mead's final illness in 1978. Yet she remembered, in their last meeting that summer, that Mead and Bateson "sat together like two lovers reconciled." They spoke of an ambitious new conference Mead was planning, a sequel to the Macy conferences of old: "She believed, I think," remembered Mary Catherine, "that the group she had chosen might in some way give birth to a crucial new idea."

Afterward, Mead arranged to meet Hamilton at his office. She asked him to examine her and to tell her the truth about her condition. Hamilton was a psychiatrist, not an oncologist, but her condition was obvious to anyone with medical training. She had pancreatic cancer, and it was terminal. Mead struggled on for ten more months, but in an increasingly diminished state, lapsing back into denial about her condition and placing her hopes on a faith healer whose ministrations so angered Rhoda Métraux that their relationship began to fray at the very end of a twenty-five-year romance.

Mead died, her biographer wrote, "furious" at having been taken away so soon: there was still so much to do. She died, in short, as she lived. Forever yearning for the breakthrough that would allow her to lead the creation of a new global culture. Forever living in the limited potential of the time allotted to her.

--------

In 1926, when she was twenty-four, Margaret Mead had a strange dream. She wrote about it to Ruth Benedict, from Samoa:

> I was in a laboratory with Dr. Boas and he was talking to me and a group of other people about religion, insisting that life must have a meaning, that man couldn't live without that. Then he made a mass of jelly-like stuff of the most beautiful blue I had ever seen—and he seemed to be asking us all what to do with it. I remember thinking it was very beautiful but wondering helplessly what it was for. People came and went making absurd suggestions.

Boas did his best to carry out each suggestion, "but always the people went away angry, or disappointed." One by one, the other people disappeared, leaving Mead and Boas alone together, illuminated by the blue glow. He turned to her with his wizened face. "I took some of the astonishingly blue beauty in my hand, and felt with a great thrill that it was living matter," Mead related. "I said 'Why it's life—and that's enough'—and he looked so pleased that I had found the answer—and said yes 'It's life and that is wonder enough.' "

Mead's dream of the glowing blue material was in some sense a dream of her own life. A life marked by strange dichotomies between openness and secrecy, selflessness and arrogance, universalism and elitism. Margaret Mead valued life above all, life in all its mystical, scientific, psychedelic glory. But though she treasured the idea that all humans were linked in a vast and interconnected pattern, she also saw herself as somehow *outside* of that pattern. In her dream, other people made "absurd suggestions" and then dwindled away, "angry" and "disappointed."

It was not until everyone else had left that she felt she could truly understand.

--------

One month after Mead died, on December 15, 1978, a letter reached Bateson's kitchen table at Esalen. He was invited by a British cultural institute to give a "last lecture": the talk he would give "if it were the ultimate statement" of his life. As the letter cautiously referenced, Bateson was himself dying. He had recovered from cancer, but now pain had returned in his lungs, and he was finding it difficult to walk. "LAST LECTURE," he scrawled on the request in large capitals. And then he wrote back to say yes.

As the lecture approached, Bateson returned to one of his favorite works of poetry, T. S. Eliot's *Four Quartets.* In it, he read about the "half-heard" voices of children in apple trees. He read about old patterns being "renewed" and "transfigured" in new ones. At the end of our lives, Eliot suggested, we do not reach somewhere new. Instead, we return to the place we came from, seeing it with fresh eyes. As he wrote notes for his speech, Bateson remembered his "boyish collecting of various invertebrates"; he remembered the early days of his love affair with Margaret Mead on the mirrored waters of the Sepik; he remembered the enormous optimism of his early years among the psychedelic psychiatrists of Silicon Valley; and, with regret but acceptance, he remembered John Lilly and their dolphins. Looking back, Bateson realized, "I never traveled far from where I started." His question, he said, had always been the same: what is the pattern of life, and what is the purpose of that pattern?

But lurking behind this was a more personal question, one he never fully articulated: *will becoming a great scientist allow me to redeem my family?*

As the final lecture approached, the organizer suggested he might want to visit Grantchester, the English manor house near Cambridge where he was born. The house where he and Martin and John once traveled the hedgerows in search of dragonflies, lizards, and fossils. But Bateson declined. He felt no need to revisit the place where his brothers' ghosts resided. After all, in all his wanderings, they had never left him.

--------

Bateson clung on for another two years, but by the first day of July 1980, he could feel it clearly. He was truly dying. And as he died, first at Esalen, then in a bed at the San Francisco Zen Center, his thoughts kept returning to an idea of his father's.

"We commonly think of animals and plants as matter," Bateson's father wrote in 1907, when Gregory was three years old. "But they are really systems through which matter is continually passing." The elder Bateson likened the stuff of life—the substance of a human mind, or of a bird and its egg, or of a parent and child—to "concentric waves spreading from a splash in a pool." Twenty-five years later in New Guinea, Gregory Bateson was startled to discover that the secret knowledge of the Iatmul elders, shared in a special, consecrated building set apart from their village along the banks of the Sepik, contained a remarkably similar insight. As Bateson put it, "It is said secretly that men, pigs, trees, grass—all the objects in the world—are patterns of waves."

The thought stuck with him. It was there during World War II, in his obsessive weeks in the back room at the Museum of Modern Art studying the "wavelike patterns" of manipulative media. And it was there one beautiful afternoon in 1962 when, with his friend Joe Adams, he had taken LSD in a garden in Big Sur. He had likened the experience to "the patterns of breaking waves." Now, at the end of his life, he described himself as a smoke ring, a momentary configuration of atoms in space, "endlessly turning upon itself." A circle spreading outward into nothingness.

Perhaps, in the moment of death, Bateson saw himself in that way. Not as a human being who had failed to achieve what he wanted—not as the old man at his birthday party at Esalen, surrounded by well-meaning sycophants, crying quietly and saying to himself, "It is so sad; it is *so, so sad*," amid the claims of achievement and meaning that he himself could never accept as true. Perhaps he finally saw himself as the thing he felt in his "flashes of awareness." Saw himself, in other words, as one of the evanescent waves that his father and the Iatmul elders had spoken of long before—one small part of the pattern that binds past, present, and future.

Bateson and Mead never achieved what they wanted. Like Rivera's painting, their vision of the coming decades and centuries was painted over and replaced by another. But the patterns they helped create are with us still, for good and for bad—renewed and transfigured, but still there.

--------

By the fall of 1980, it was not just Bateson and Mead who had died. So, too, had the ideas that inspired and drove the first generation of psychedelic science. Yet something of their sense of science as transformative and redemptive—and yes, even utopian—did survive. In September 1980, for instance, two months after Bateson's death, Carl Sagan and his wife, Ann Druyan, premiered the first episode of their television documentary *Cosmos*. In the years to come, it would go on to be viewed by over five hundred million people, a sizable percentage of the world's population.

"So much of what we did—*Cosmos*, *Contact*, and so many books—was stimulated, inspired, by many psychoactive substances," Druyan told me, relaying her experiences using substances from cannabis to MDMA to ayahuasca with Sagan. In 1964, Sagan had visited Bateson and Lilly at their dolphin lab, fascinated by their work, but in the end also troubled by it. Sagan's regard for Margaret Mead never faded, however. They were "kindred souls," Druyan said, because they were both "realistic utopians."

And their shared credo, she said, could be summed up quite simply: "Life is not perfectible. But it could be so much *better*."

# Acknowledgments

I am truly grateful to everyone who shaped the research and writing of this book. At my home institution of UC Santa Cruz, I thank my colleagues Jasmine Alinder, Bettina Aptheker, Dorian Bell, Muriam Davis, Jennifer Derr, Nathaniel Deutsch, Anna Friz, Alma Heckman, Kate Jones, Nidhi Mahajan, Marc Matera, Matt O'Hara, Greg O'Malley, Eric Porter, Juned Sheikh, Thomas Serres, Daniel Storey, Elaine Sullivan, and Alice Yang. For their support and friendship I would like to acknowledge Rose Andrade, Andrew Bailey, David Epstein, Madeleine Fairbairn, Meg Freitag, Sophia Gold, Robbie Guertin, Brian Jones, Charley Lanyon, Sara Madandar, Kate Miller, Tina Moeiny, Arash Mofakham, Dustin Neuman, and Neryvia Pilay. Since my first year of graduate school, Jorge Cañizares-Esguerra has been a steadfast mentor. I admire his fearlessness and originality as a historian, but more than this, I deeply appreciate his friendship. Henry Cowles was a witty sounding board and friend throughout the period of researching this book. At the very end of this project Hannah Zeavin provided encouragement and insights that I truly appreciated. Mike Jay has long been an ideal colleague as a fellow historian of drugs—I thank him warmly for his advice throughout every stage of research and writing. A special thank-you is in order for all my students at UCSC, especially the graduate students of the UCSC history department. Chris Heaney has been a true friend and a source of inspiration to me as a writer and person for over a decade now, and I hope for many more to come. And finally, *um grande abraço ao meu melhor amigo Felipe Cruz.*

The research and writing of this book was supported by the generosity of a National Endowment for the Humanities Award for Faculty (HB-273539-21), the Hellman Foundation Fellows Fund, and several grants from UC Santa Cruz. It is the hardworking archivists and librarians who truly make historical research possible. I salute Stephanie Schmidtz at Purdue University Archives and Special Collections; Shawn C. Wilson of the Kinsey Institute Library and Archives at Indiana University; Tim Noakes of Stanford University Special Collections (with special thanks for making the digitized copies of George Hunter White's appointment book and a key Allen Ginsberg recording available to me); Miriam Camacho Cabrera at UC San Diego Special Collections and Archives; and the staffs of the Library of Congress, the National Archives and Records Administration, University of Rochester Special Collections, the Drs. Nicholas and Dorothy Cummings Center for the History of Psychology, the Wellcome Library, the Columbia University Rare Book & Manuscript Library, Cambridge University Library, the Francis A. Countway Library of Medicine at Harvard University, the Harry Ransom Center at the University of Texas at Austin, the New York Public Library, and the Massachusetts Institute of Technology Libraries. Above all, I am profoundly grateful to the wonderful staff of UCSC Special Collections and Archives, especially Jessica Pigza, Teresa Mora, Kate Dundon, Rebecca Hernandez, and Luisa Haddad. My thanks also go to Philip Hansen Bailey and to the directors of the Bateson Idea Group. James Reidel played a key role in the research for this book by sharing Weldon Kees's diaries with me shortly after they resurfaced at an auction in 2020, an act of scholarly generosity that I greatly appreciate. Likewise, Christopher Riley generously offered me transcripts of several interviews conducted for the BBC documentary *The Girl Who Talked to Dolphins*. I was grateful to have the chance to present parts of this work at talks and seminars at Columbia University, UCLA, and UT Austin, as well as at a fascinating conference on John C. Lilly organized by Hannah Zeavin and Jeffrey Mathias. I thank the organizers and audiences at these events. Finally, I received invaluable help from my research assistants Josephine Kim, Alexander E. Taft, and Paloma Ruiz. A special thank-you to Piper Milton, who helped me complete the line edits for the manuscript.

I also want to acknowledge the generosity of those who made themselves available for interviews conducted by phone, email, or Zoom—often during the height of the COVID-19 pandemic, when many other things were on their mind. I am profoundly grateful to: Howland Abramson, Eliot Aronson, Jeff Bridges, Lois Bateson, Nora Bateson, Jonathan Beecher, Yvan Beaussant, Susan Block, Jerry Brown, Ann Druyan, Gary Dufresne, Robert Forte, John

Foster, Leo Goldberg, Jean Houston, Bob Houlihan, Joi Ito, Laurence Jarvik, Jeffrey Jarvik, Robert Janiger, Lisa Kornetsky, David Kornetsky, David Lipset, Steve Mahon, Michael Maccoby, Daniel Métraux, Eric Miles, Deborah Nightingale, Noah Oderberg, Phillip Oderberg, Forrest Robinson, Gary Stern, Rick Trout, Andrew Weil, Marianne Weltmann, and Irvin Yalom, and those who agreed to speak off the record. I particularly thank Susan Block, Jerry Brown, Ann Druyan, and Noah Oderberg for their generosity with their time, and Lisa and David Kornetsky for furnishing the striking "unconscious" version of the 1954 "Problems of Consciousness" conference photo that appears in this book and, in a modified form, on the cover.

The following people contributed advice or information that I drew upon while writing (though all errors remain my own): Amir Bar-Lev, Bob Bookman, Danielle Carr, James W. Chambers, James Clifford, Ken Dornstein, Don Lattin, Don Latin, Erik Davis, Carl Erik Fischer, Brian Goldstone, Bernard Geoghegan, Donna Haraway, Stephen Kinzer, John Lisle, Margaret McCaffrey, Max McGillivray, Rachel Moran, Margaret O'Mara, George Pendle, David Price, Joanna Radin, James P. Rathmell, Natasha Reichle, Rebecca Jo Plant, Michael Shermer, Simon Sadler, Courtney Stephens, Sharon Traweek, and Douglas Valentine. I thank them all.

My agent Yishai Seidman was by far my hardest-working reader. Colin Dickerman at Grand Central Publishing was an ideal editor, offering wise guidance and extremely sharp-eyed and judicious edits. A heartfelt thank-you to him and to Rachael Kelly, Ian Dorset, and the production, design, marketing, and sales team at Grand Central, who ably shepherded this manuscript through to publication.

My mother, Gloria Ataide Breen, died of cancer while I was in the midst of writing this book, in September 2021. Her courage during her illness and in her life as a whole has sustained me and inspired me. I honor her memory here. Thank you for everything, Mom. I also thank my other family—Roja, Ehsan, Eshrat, Heather, Nathan, and my father, John, for their support.

I dedicate this book, with inexpressible love and gratitude, to Roya Pakzad and to Yara Pakzad-Breen.

# Appendix

## Attendees of Two Macy Conferences

The participants in the 1954 Macy Conference on Problems of Consciousness, New York, NY.

1. Conan Kornetsky, 2. George Robinson, 3. Nathaniel Kleitman, 4. Hudson Hoagland, 5. Petty Ku
6. Frank Fremont-Smith, 7. Henry K. Beecher, 8. Ruth E. Rue, 9. Alexander Barry, 10. Ashley Montag
11. Seymour S. Kety, 12. Harold A. Abramson, 13. Mary Brazier, 14. Roy Grinker 15. Margaret Me
16. David G. Wright, 17. Eyleen A. Masser, 18. Harold Schlosberg

The participants in the 1953 Macy Conference on Cybernetics, New York, NY.

1. Leonard J. Savage, 2. Janet Freed Lynch, 3. Peggy Kubie, 4. Gerhardt von Bonin, 5. Henry W. Brosin, 6. Lawrence S. Kubie, 7. Gregory Bateson, 8. Lawrence K. Frank, 9. Frank Fremont-Smith, 10. Henry Quastler, 11. John R. Bowman, 12. George E. Hutchinson, 13. Donald G. Marquis, 14. Hans-Lukas Teuber, 15. Heinrich Klüver, 16. Julian H. Bigelow, 17. Claude Shannon, 18. Filmer S. C. Northrop, 19. Walter Pitts, 20. Heinz von Foerster, 21. Theodore C. Schneirla, 22. Yehoshua Bar-Hillel, 23. Margaret Mead, 24. Warren S. McCulloch, 25. Jan Droogleever Fortuyn, 26. Yuen Ren Chao, 27. W. Grey Walter, 28. Vahe E. Amassian

# A Note on Sources

Few people anywhere, on the mechanical level of words typed or scrawled or dictated, seem to have written as much as Margaret Mead. The finding aid for Mead's papers at the Library of Congress estimates that they include some 530,000 items. If every page in her archives were stacked, one on top of the other, the resulting tower of paper would mount 780 feet into the sky—taller than many skyscrapers in Manhattan. Bateson's own archive, now housed at the University of California, Santa Cruz, is smaller and stranger than Mead's. It contains such things as a torn-off cover of a James Bond novel on which is scribbled the home address of a famous scientist; paperwork lending William Blake's *Satan Exulting over Eve* (which had hung above Bateson's childhood breakfast table) to the Getty Museum in Los Angeles; the watercolor botanical diary of an eleven-year-old boy—Bateson's older brother, John—who would be shot dead on a World War I battlefield nine years later; an admiring letter from the French critical theorist Pierre Bourdieu; and correspondence with everyone from J. Robert Oppenheimer to Alger Hiss.

Readers looking to learn more about these two fascinating figures would do well to visit both archives, which house an incredible array of unstudied material. And they will also be rewarded by reading the available scholarly and biographical works. Below is a selective list of the books on Mead and Bateson that I have found most useful, as well as a few highlights of the rich and fast-growing historical literature relating to psychedelics and psychedelic science. I am profoundly thankful to all the historians and authors who have written them.

Margaret Mead's most comprehensive biography remains Jane Howard's *Margaret Mead: A Life* (New York: Simon & Schuster, 1984). Excellent studies that have appeared more recently include Maureen A. Molloy, *On Creating a Usable Culture: Margaret Mead and the Emergence of American Cosmopolitanism*

(Manoa: University of Hawaii Press, 2008); Nancy C. Lutkehaus, *Margaret Mead: The Making of an American Icon* (Princeton, NJ: Princeton University Press, 2008); Lois W. Banner, *Intertwined Lives: Margaret Mead, Ruth Benedict, and Their Circle* (New York: Vintage, 2010); Peter Mandler, *Return from the Natives: How Margaret Mead Won the Second World War and Lost the Cold War* (New Haven, CT: Yale University Press, 2013); and Elesha J. Coffman, *Margaret Mead: A Twentieth-Century Faith* (Oxford: Oxford University Press, 2021). For a convincing rejoinder to the many critiques of Mead's anthropological work in Samoa made by Derek Freeman, see Paul Shankman, *The Trashing of Margaret Mead: Anatomy of an Anthropological Controversy* (Madison: University of Wisconsin Press, 2009); and on Mead's relationship to race, see Mark Anderson, *From Boas to Black Power: Racism, Liberalism, and American Anthropology* (Stanford, CA: Stanford University Press, 2019). On Bateson, see David Lipset, *Gregory Bateson: The Legacy of a Scientist* (New York: Beacon Press, 1980); Noel G. Charlton, *Understanding Gregory Bateson: Mind, Beauty, and the Sacred Earth* (Albany, NY: SUNY Press, 2008); Anthony Chaney, *Runaway: Gregory Bateson, the Double Bind, and the Rise of Ecological Consciousness* (Chapel Hill: University of North Carolina Press, 2017); and Phillip Guddemi, *Gregory Bateson on Relational Communication: From Octopuses to Nations* (Cham, Switzerland: Springer Nature, 2020).

Charles King's *Gods of the Upper Air: How a Circle of Renegade Anthropologists Reinvented Race, Sex, and Gender in the Twentieth Century* (New York: Anchor, 2020) offers a deeply researched and sensitively drawn portrait of the intellectual community of Boasian anthropologists in which Mead first found her footing as a scientist. On anthropologists and the CIA, see David H. Price, *Cold War Anthropology: The CIA, the Pentagon, and the Growth of Dual Use Anthropology* (Durham, NC: Duke University Press, 2016); and on the Frank Olson case, see the excellent documentary *Wormwood* (directed by Errol Morris, 2017, Fourth Floor Productions and Moxie Pictures, a *Netflix* six-part series), which draws on original interviews. On Sidney Gottlieb and MKULTRA, see Stephen Kinzer, *Poisoner in Chief: Sidney Gottlieb and the CIA Search for Mind Control* (New York: Henry Holt, 2019). Ellen Herman's remarkable *The Romance of American Psychology: Political Culture in the Age of Experts* (Berkeley: University of California Press, 1996) is an indispensable resource for understanding the many military entanglements and strange bedfellows of American scientists during the Cold War. On how Mead, Bateson, and psychedelics fit into the larger history of twentieth-century science, see Rebecca Lemov, *World as Laboratory: Experiments with Mice, Mazes, and Men* (New York: Hill and Wang, 2005); and Rebecca Lemov, "Towards a Data Base of Dreams: Assembling an Archive of Elusive Materials, c. 1947–61," in *History Workshop Journal* 67, no. 1 (2009): 44–68; and Bernard Dionysius Geoghegan, *Code: From Information Theory to French Theory* (Durham, NC: Duke University Press, 2023).

Readers interested in the history of psychedelics should begin with the works by the distinguished scholars and writers cited in the third note of the prologue. For further reading on the history of LSD in particular, see Martin A. Lee and Bruce Shlain, *Acid Dreams: The Complete Social History of LSD* (New York: Grove Press, 1992); and Jay Stevens, *Storming Heaven: LSD and the American Dream* (New York: Atlantic Monthly Press, 1987). On the CIA's entanglements with psychedelic research, John Marks's *The Search for the "Manchurian Candidate": The CIA and Mind Control* (New York: Times Books, 1979) remains the most detailed source. Stephen Siff, *Acid Hype: American News Media and the Psychedelic Experience* (Champaign: University of Illinois Press, 2015) is a superb resource for understanding the impact of psychedelics in twentieth-century American media.

Recently, scholars of science and technology studies (STS), alongside historians of race and gender, have offered a particularly valuable perspective on the history of drugs and medicine during the Cold War. See, for instance, Danielle Judith Zola Carr, " 'Ghastly Marionettes' and the Political Metaphysics of Cognitive Liberalism: Anti-behaviourism, Language, and the Origins of Totalitarianism," *History of the Human Sciences* 33, no. 1 (2020): 147–74; Erika Lorraine Milam, *Creatures of Cain: The Hunt for Human Nature in Cold War America* (Princeton, NJ: Princeton University Press, 2020); and Keith Wailoo, *Pushing Cool: Big Tobacco, Racial Marketing, and the Untold Story of the Menthol Cigarette* (Chicago: University of Chicago Press, 2021). On changing conceptions of mental illness at midcentury, see Anne Harrington, *Mind Fixers: Psychiatry's Troubled Search for the Biology of Mental Illness* (New York: W. W. Norton, 2019); and Laura Stark and Nancy D. Campbell, "The

Ineffable: A Framework for the Study of Methods Through the Case of Mid-Century Mind-Brain Sciences," *Social Studies of Science* 48, no. 6 (2018): 789–820. The definitive scholarly study of John C. Lilly's ill-fated work at the Communication Research Institute can be found in D. Graham Burnett, *The Sounding of the Whale: Science and Cetaceans in the Twentieth Century* (Chicago: University of Chicago Press, 2012), chapter 6.

To learn more about the intellectual and social threads linking the Macy circle to contemporary "digital utopians," start with Fred Turner's excellent books *From Counterculture to Cyberculture: Stewart Brand, the Whole Earth Network, and the Rise of Digital Utopianism* (Chicago: University of Chicago Press, 2010); and *The Democratic Surround: Multimedia and American Liberalism from World War II to the Psychedelic Sixties* (Chicago: University of Chicago Press, 2013). On the history of technological utopianism in the twentieth century, see William E. Akin, *Technocracy and the American Dream: The Technocrat Movement, 1900–1941* (Berkeley: University of California Press, 1977); Howard P. Segal, *Technological Utopianism in American Culture* (Syracuse, NY: Syracuse University Press, 2005); and Richard Stites, *Revolutionary Dreams: Utopian Vision and Experimental Life in the Russian Revolution* (Oxford: Oxford University Press, 1988); on the history of Silicon Valley, see Margaret O'Mara, *The Code: Silicon Valley and the Remaking of America* (New York: Penguin, 2020); and Malcolm Harris, *Palo Alto: A History of California, Capitalism, and the World* (New York: Little, Brown, 2023), which touches on the LSD research around Stanford in the 1950s and early 1960s.

The literature on the history of gender, culture, and society in the mid-twentieth-century United States is vast, but in my reading, the following works stood out as particularly insightful: David K. Johnson, *The Lavender Scare: The Cold War Persecution of Gays and Lesbians in the Federal Government* (Chicago: Unversity of Chicago Press, 2004); Alan Petigny, *The Permissive Society: America, 1941–1965* (New York: Cambridge University Press, 2009); Rebecca Jo Plant, *Mom: The Transformation of Motherhood in Modern America* (Chicago: University of Chicago Press, 2010); Kate Brown, *Plutopia: Nuclear Families, Atomic Cities, and the Great Soviet and American Plutonium Disasters* (New York: Oxford University Press, 2013); and Elaine Tyler May, *Homeward*

*Bound: American Families in the Cold War Era*, rev. ed. (New York: Basic Books, 2017).

A note on citations: Mead's archives at the Library of Congress are organized by box and folder. Where possible, I have indicated both using the following citation style: "N120:5" means Box N120, Folder 5.

# Notes

## Abbreviations Used in the Notes

| | |
|---|---|
| *BW* | Margaret Mead, *Blackberry Winter: My Earlier Years* (New York: William Morrow, 1972) |
| GB | Gregory Bateson |
| GB papers | Gregory Bateson papers, University of California, Santa Cruz, Special Collections and Archives, Santa Cruz, CA |
| GHW papers | George Hunter White Papers, Stanford University Special Collections |
| JCL | John Cunningham Lilly |
| JCL papers | John C. Lilly Papers, Stanford University Special Collections, Stanford, CA |
| Kees diary | Unpublished diaries of Weldon Kees (1950–54), collection of James Reidel |
| LOC | Library of Congress, Washington, DC |
| MM | Margaret Mead |
| MM papers | Margaret Mead papers, Library of Congress, Washington, DC |
| MM FBI file | Margaret Mead FBI file, File #121-14450, accessible at https://archive.org/details/MargaretMead/mode/2up |
| NARA | National Archives and Records Administration, College Park, MD |
| NYPL | New York Public Library |

*NYT* — *New York Times*

RB — Ruth Benedict

*TCLW* — Margaret Mead, *To Cherish the Life of the World: Selected Letters of Margaret Mead*, ed. Margaret M. Caffrey and Patricia A. Francis (New York: Basic Books, 2006)

*WDE* — Mary Catherine Bateson, *With a Daughter's Eye: A Memoir of Margaret Mead and Gregory Bateson* (New York: William Morrow, 1984)

*WP* — *Washington Post*

White diary — Appointment books of George Hunter White, Stanford University Special Collections and University Archives, Stanford, CA

## PROLOGUE

1 **"Our Father, We thank Thee":** prayer offered by the Reverend Wilbur A. Korfhager, *Legislature of the State of California* (Sacramento: California State Printing Office, May 4, 1981), 1790.

2 **"a night-club scene":** Diego Rivera, *My Art, My Life: An Autobiography* (New York: Citadel, 1960), 126. On the utopian themes of Rivera's mural, see Robert Linsley, "Utopia Will Not Be Televised: Rivera at Rockefeller Center," *Oxford Art Journal* 17, no. 2 (1994): 48–62; for a detailed history of the work and its fate, see Catha Paquette, *At the Crossroads: Diego Rivera and His Patrons at MoMA, Rockefeller Center, and the Palace of Fine Arts* (Austin: University of Texas Press, 2017).

4 **The Macy circle was the seedbed:** For recent work on the history of psychedelic therapy, see Erika Dyck, *Psychedelic Psychiatry: LSD from Clinic to Campus* (Baltimore: Johns Hopkins University Press, 2008); Don Lattin, *The Harvard Psychedelic Club: How Timothy Leary, Ram Dass, Huston Smith, and Andrew Weil Killed the Fifties and Ushered in a New Age for America* (New York: HarperOne, 2010); Nicolas Langlitz, *Neuropsychedelia: The Revival of Hallucinogen Research Since the Decade of the Brain* (Berkeley: University of California Press, 2013); Beatriz Caiuby Labate and Clancy Cavnar, eds., *Peyote: History, Tradition, Politics, and Conservation* (Santa Barbara, CA: Praeger, 2016); Matthew Oram, *The Trials of Psychedelic Therapy: LSD Psychotherapy in America* (Baltimore: Johns Hopkins University Press, 2018); Lucas Richert, *Break on Through: Radical Psychiatry and the American Counterculture* (Cambridge, MA: MIT Press, 2019); Richert, *Strange Trips: Science, Culture, and the Regulation of Drugs* (Montreal: McGill-Queen's University Press, 2019); Michael Pollan, *How to Change Your Mind* (New York: Penguin, 2019); Ido Hartogsohn, *American Trip: Set, Setting, and the Psychedelic Experience in the Twentieth Century* (Cambridge, MA: MIT Press, 2020); and Danielle Giffort, *Acid Revival: The Psychedelic*

*Renaissance and the Quest for Medical Legitimacy* (Minneapolis: University of Minnesota Press, 2022).

4 **"She was truly interested":** Author interview with Jean Houston, August 11, 2021, 29:25 (audio).

5 **Mead's primary goal:** Margaret Mead, "This I Believe... the Human Potential," *Hartford Courant*, May 16, 1952, 16 (a transcript of her submission to Edward R. Murrow's radio program). On the self-perception among some Cold War scientists that they were engaged in a "battle over the future," see Jill Lepore, *If/Then: How the Simulmatics Corporation Invented the Future* (New York: Liveright Publishing, 2020).

5 **And it was Frank who:** Lawrence K. Frank, "Society as the Patient," *American Journal of Sociology* 42, no. 3 (1936): 335–44.

6 **Mead and Bateson saw themselves:** Margaret Mead, *Male and Female: A Study of the Sexes in a Changing World* (New York: William Morrow, 1949).

6 **Carl Sagan was once asked:** "Life Among the Stars," *Mosaic*, March/April 1977, 19.

6 **Mead saw herself as:** Mead as quoted in Harold A. Abramson, ed., *Problems of Consciousness: Transactions of the Fourth Conference* (New York: Josiah Macy, Jr. Foundation, 1954), 173 ("listening post") and 176 ("another level"). Science had expanded "our consciousness of being human beings," Mead explained at the conference: "A large amount of our behavior, which had been completely automatic and inaccessible to observation, became accessible, although we may have learned it unconsciously at the time that we learned it."

8 **"What we, as a group":** Harold A. Abramson, ed., *Problems of Consciousness: Transactions of the Fifth Conference* (New York: Josiah Macy, Jr. Foundation, 1955), 158.

9 **Timothy Leary's earliest:** Mervin B. Freedman, "Symposium: Interpersonal Circumplex Models (1948–1983)," *Journal of Personality Assessment* 49, no. 6 (1985): 623.

9 **And in one of his first:** Timothy Leary, "How to Change Behavior," speech at the International Congress of Applied Psychology in Copenhagen, Denmark, August 1961, reprinted in David Solomon, ed., *LSD: The Consciousness-Expanding Drug* (New York: G. P. Putnam's Sons, 1964), 104–6.

## 1. PROPHET OF THE LONG FUTURE

14 **Writing to Benedict:** *TCLW*, 159–60.

15 **"wide-spread Plains torture pattern":** Ruth Benedict, "The Vision in Plains Culture," *American Anthropologist* 24, no. 1 (1922): 4.

17 **"If you can get away":** MM, B14, as quoted in Hilary Lapsley, *Margaret Mead and Ruth Benedict: The Kinship of Women* (Amherst: University of Massachusetts Press, 1999), 68.

17 **"I need a companion in harness":** Lapsley, *Kinship of Women*, 69.

17 **After Mead married Cressman:** Luther Cressman, *A Golden Journey: Memoirs of an Archaeologist* (Salt Lake City: University of Utah Press, 1988), 92.

17 **"Anthropology had to be":** Virginia Yans-McLaughlin, "Science, Democracy, and Ethics: Mobilizing Culture and Personality for World War II," in *Malinowski, Rivers, Benedict and Others: Essays on Personality and Culture* (History of Anthropology, vol. 4), ed. George W. Stocking (Madison: University of Wisconsin Press, 1986), 193, citing Mead's unpublished rough draft of *Blackberry Winter*, 2.

**18 "The purpose of anthropology":** Ruth Benedict, *Patterns of Culture* (Boston: Houghton Mifflin, 1934), 17.

**18 "Ruth and I":** Lapsley, *Kinship of Women*, 124.

**19 "new kind of consciousness":** Margaret Mead, "The Future as the Basis for Establishing a Shared Culture," *Daedalus* 94, no. 1 (1965): 135–55, www.jstor.org/stable/20026899. This article, written in 1965, summarizes many of the ideas Mead had been developing for the previous thirty years.

**19 In the fall of 1926:** "Scientist Goes on Jungle Flapper Hunt," *NYT*, November 8, 1926, 190.

**19 "What constitutes courtesy":** Franz Boas, foreword to Margaret Mead, *Coming of Age in Samoa: A Psychological Study of Primitive Youth for Western Civilisation* (New York: William Morrow, 1928).

## 2. SOCIETY IS THE PATIENT

**23 "The peyote, Lord":** Antonio Saavedra Guzmán, *El peregrino indiano* (Madrid: Pedro Madrigal, 1599), 155. On the *bruja*'s magical potion in this poem, see Raúl Marrero-Fente, "De yerbas, animales y otras ponzoñas: Renovación mágica en *El peregrino indiano* de Antonio Saavedra Guzmán," *Hipogrifo* 5, no. 2 (2017): 199–214.

**23 Yet throughout the colonial period:** See, for instance, the case of William Lamport, an Irish exile in Mexico who was accused in 1642 of enlisting peyote-assisted divination to plan a rebellion against colonial authorities, cited in Alexander Dawson, "Peyote in the Colonial Imagination," in Labate and Cavnar, *Peyote*, 55–56. For other cases and the Inquisition's attempted bans, see Enrique Flores and Mariana Masera, eds., *Relatos populares de la Inquisición novohispana. Rito, magia y otras "supersticiones"* (Madrid: CSIS, 2010).

**23 "a marked effect":** Alfred Kroeber, *The Arapaho* (New York: American Museum of Natural History, 1902), 320.

**24 "false god":** "Peyote Used as Drug in Indians' 'Cult of Death'" *NYT*, January 14, 1923.

**24 "Peyote is just dope":** Peyote notes, Omaha fieldwork ("Fay, Miscelaneous [*sic*], Twins, peyote, marriage"), August 18, 1930, N120:5, MM papers.

**24 "It works in your brain":** MM interview with George Phillips, June 26, 1930, N120:5, MM papers.

**25 "Doesn't somebody warn the newcomer":** MM interview with Phillips, 2.

**25 But Mead quoted:** Margaret Mead, *The Changing Culture of an Indian Tribe* (New York: Columbia University Press, 1932), 27.

**27 There were stretches of happiness:** "Housekeeping in New Guinea," *The Sun* (Sydney), June 4, 1933, 24. The most detailed examination of this period in Mead and Fortune's life together can be found in Lise M. Dobrin and Ira Bashkow, "The Truth in Anthropology Does Not Travel First Class: Reo Fortune's Fateful Encounter with Margaret Mead," *Histories of Anthropology Annual* 6, no. 1 (2010): 66–128.

**27 In May, a woman in the city of Rabaul:** Edith Cobb to MM, May 5, 1932, MM papers.

**28 She had rushed into marriage:** *BW*, 177.

**28 Mead wrote to Ruth Benedict:** MM to RB, January 1, 1933, *TCLW*, 71–72. "The Baining," Bateson tried to explain, "are not an easy people to work among." See Gregory

Bateson, "Further Notes on a Snake Dance of the Baining," *Oceania* 2, no. 3 (1932): 337. Much later, the anthropologist Jane Fajans wrote an entire book, *They Make Themselves: Work and Play Among the Baining of Papua New Guinea* (Chicago: University of Chicago Press, 1997), that was partially inspired by Bateson's remark to Fajans that the Baining "broke my heart."

**28 Bateson once told his mother:** Jane Howard, *Margaret Mead: A Life* (New York: Simon & Schuster, 1984), 185.

**29 That fall, Mead wrote:** MM to Franz Boas, undated [fall 1932], N92:3, MM papers.

**29 "how surprised Bateson will be":** MM to RB, October 21, 1932, S4:1, MM papers.

**30 At the edge of the Iatmul village:** *BW*, 240.

**30 "Gregory Bateson is an awfully":** Sarah Chinnery, *Malaguna Road: The Papua and New Guinea Diaries of Sarah Chinnery* (Canberra: National Library of Australia, 1998).

**30 Porter once said:** Noel Porter to MM, November 1933, R10, MM papers.

**31 "By then," Mead recalled:** *BW*, 243.

**31 There was something "impish":** David Lipset interview with R. D. Laing, David Lipset papers, UCSC, 2.

**31 After he died:** Rodney E. Donaldson, "Gregory Bateson Archive: a Guide/Catalog," summarizing GB to Kirk Hogan, January 12, 1979, GB papers.

**31 Indigenous peyote users:** Ruth Benedict, "Anthropology and the Abnormal," *Journal of General Psychology* 10, no. 1 (1934): 59–82, 264 and 267. Drafts of this and related work reached Mead all through 1932 and 1933 via Benedict's letters.

**32 Benedict privately self-identified:** Lois W. Banner, *Intertwined Lives: Margaret Mead, Ruth Benedict, and Their Circle* (New York: Vintage, 2010), 10.

**32 Perhaps the society that excluded:** Benedict, "Anthropology and the Abnormal," 270.

**32 Mead's first attempt:** Margaret Mead, "Summary Statement of the Problem of Personality and Culture," 1933, N102:2, MM papers.

**33 historians of sexuality:** On the origins of the sex/gender split, see Shira Tarrant, *When Sex Became Gender* (London: Routledge, 2006); and Sandra Eder, *How the Clinic Made Gender: The Medical History of a Transformative Idea* (Chicago: University of Chicago Press, 2022). On the Squares, see Gerald Sullivan, "A Four-Fold Humanity: Margaret Mead and Psychological Types," *Journal of the History of the Behavioral Sciences* 40, no. 2 (2004): 183–206.

**33 "It was the closest I've ever come":** *BW*, 235.

**33 "All three of us together":** Howard, *Mead*, 161. Although Mead disavowed the Squares, she continued to tinker with the theory for several years afterward. See Margaret Mead, "Statement of the Squares Hypothesis, May 16, 1937," N12:2, MM papers.

**34 "dangerous madness":** Reo Fortune to Bronislaw Malinowski, January 12, 1937, LSE/Malinowski/7/8, as cited in Maureen A. Molloy, *On Creating a Usable Culture: Margaret Mead and the Emergence of American Cosmopolitanism* (Manoa: University of Hawaii Press, 2008), 120.

**34 His anger culminated:** For a full description of this episode see Banner, *Intertwined Lives*, 335. See also MM papers, "Notes to GCB (Gotthard Booth)," February 4, 1949, R10.

**34 "Threats of assault":** *BW*, draft, 6 and 15.

**34 "I feel singularly free":** MM to GB, June 12, 1934, *TCLW*, 89–90.

## 3. ARTIFICIAL PARADISES

**36 "Shamanism remained very much":** Mead, *Changing Culture*, 29.

**36 "remain below the surface":** See, for instance, Margaret Mead, *Culture and Commitment*, chapter 1, in which she summarizes twenty-five years or more of her fieldwork along these lines.

**37 "a new kind of consciousness":** Margaret Mead, "The Future as the Basis for Establishing a Shared Culture," *Daedalus* 94, no. 1 (1965): 135–55.

**37 In their place was a new freedom to experiment:** This era of what has been called "psychological modernism" was deeply rooted in cultural anthropology. Hormone science, writes the historian of science Michael Pettit, was part of a larger turn in the sciences immediately before World War II toward the study of the "hidden forces" underlying human behavior, with an emphasis on unexplainable edge cases, social deviance, and an "absorption with so-called 'primitive' culture.'" But it was also about intervening in culture through artificial means. Though originally used to reinforce social norms—male hormones for men suffering from impotence, or sedative drugs for mental patients—the new palette of experimental drugs would soon be applied to this goal: for instance, sex hormones used to subvert rather than reinforce societal expectation, or "psychosis mimicking drugs" to aid creativity. See Michael Pettit, "Becoming Glandular: Endocrinology, Mass Culture, and Experimental Lives in the Interwar Age," *American Historical Review* 118, no. 4 (2013): 1052–76.

**38 "Any successful cooperative":** Ralph J. Westlake, "Man Seen in Sight of the Goal of Brotherhood," *Dayton Daily News*, November 22, 1936, 15; and I. W. S., "Spontaneous Cooperation Among Primitive Peoples," ibid.

**38 In the Soviet Union:** See Richard Stites, *Revolutionary Dreams: Utopian Vision and Experimental Life in the Russian Revolution* (Oxford: Oxford University Press, 1988); and Asif A. Siddiqi, *The Red Rockets' Glare: Spaceflight and the Soviet Imagination, 1857–1957* (Cambridge: Cambridge University Press, 2010); and "Imagining the Cosmos: Utopians, Mystics, and the Popular Culture of Spaceflight in Revolutionary Russia," *Osiris* 23, no. 1 (2008): 260–88.

**38 Across Europe and Latin America:** On early Freudian psychoanalysis in a global context, see Hannah Zeavin, *The Distance Cure: A History of Teletherapy* (Cambridge, MA: MIT Press, 2021); and Mariano Ben Plotkin, *Freud in the Pampas: The Emergence and Development of a Psychoanalytic Culture in Argentina* (Stanford, CA: Stanford University Press, 2001).

**39 "a profane illumination":** Walter Benjamin, "Surrealism: The Last Snapshot of the European Intelligentsia" (1929); and Walter Benjamin, *On Hashish*, 52. See also Mike Jay, *Mescaline: A Global History of the First Psychedelic* (New Haven, CT: Yale University Press, 2019), 94.

**39 Among those volunteers**: Guttmann's "Artificial Psychoses Produced by Mescaline" appeared in the May 1936 issue of the *Journal of Mental Science* and was likely reporting on the series of studies in which Gorer had anonymously participated.

**39–40 "In the laboratory...I resent these visions":** Quotes in this paragraph are from the appendix to Geoffrey Gorer, *Bali and Angkor, or, Looking at Life and Death* (London: Michael Joseph, 1936), 216.

**40 Geoffrey Gorer arrived in Manhattan:** Howard, *Mead*, 180. Gorer had a knack for befriending interesting people: it was also in 1935 that he struck up a close friendship with George Orwell. Feral Benga, who remained Gorer's on-and-off partner, later followed him to Manhattan and became a participant in the Harlem Renaissance. A fascinating figure in his own right, Benga opened a bohemian nightclub in Paris after the end of World War II that featured a cabaret composed entirely of Africans resident in the city. Among the young people attracted to Benga's nightspot was a young Nico, of Andy Warhol and Velvet Underground fame.

**40 "the nearest approach to Utopia":** Gorer, *Bali and Angkor*, 52.

**41 "with doctors and motor buses":** Undated letter from MM to her parents, 1935, *TCLW*, 25–26.

**41 Mead found Beatrice Bateson:** MM to her mother, December 10, 1936, *TCLW*, 28.

**42 "from the ego-organized world":** Gregory Bateson, "Some Components of Socialization for Trance," *Ethos* 3, no. 2 (Summer 1975): 143–55.

**42 "would become for a time an animal":** Jane Belo, *Trance in Bali* (New York: Columbia University Press, 1960), 4–5.

**42 "great inter-disciplinary expeditions":** *BW*, 275.

**42 For all their theorizing:** Interview of I Madé Kalér by Geoffrey Robinson, as transcribed in Tessel Pollmann, "Margaret Mead's Balinese: The Fitting Symbols of the American Dream," *Indonesia*, no. 49 (April 1990): 1–35.

**42 "wanted to believe Bali":** Tessel Pollmann, interview with Ibu Gedong Bagus Oka, conducted in Denpasar, May 1986, in Pollmann, "Mead's Balinese," 21.

## 4. "A RESPECTABLE WORD FOR HYPNOSIS"

**46 Under the Third Reich:** Norman Ohler, *Blitzed: Drugs in the Third Reich*, trans. Shaun Whiteside (Boston: Houghton Mifflin Harcourt, 2017), 34; see also David Courtwright, *The Age of Addiction: How Bad Habits Became Big Business* (Cambridge, MA: Harvard University Press, 2019), 129–31. According to Ohler, Hildebrand chocolates sold in the Third Reich contained 14 milligrams of methamphetamine per portion; "*Hildebrand-Pralinen ergreuen immer*," went the slogan: "Hildegrand chocolates are always a delight."

**46 "tossed his pipe into the air":** *WDE*, 23.

**47 In an influential essay:** From her 1945 speech "Human Differences and World Order," in *World Order: Its Intellectual and Cultural Foundations* (New York: Harper & Brothers, 1945). See also Mead, "The Concept of Culture and the Psychosomatic Approach," *Psychiatry* 10, no. 1 (1947): 57–76. On Lawrence K. Frank's role in the history of science, see Rebecca Lemov, *World as Laboratory: Experiments with Mice, Mazes, and Men* (New York: Hill and Wang, 2005), 61.

**48 "mobilizing attitudes":** David Lipset, *Gregory Bateson: The Legacy of a Scientist* (New York: Beacon Press, 1980), 166. See also Mead's comments in *Proceedings of the Second Conference on Science, Philosophy and Religion*, Columbia University, September 8–11, 1941 (Ann Arbor: University of Michigan Press, 1942), 82, 84.

**48 It was the dawn of the Information Age:** On the role of big data in J. Edgar Hoover's rise to power, see Beverly Gage, *G-Man: J. Edgar Hoover and the Making of the American*

*Century* (New York: Penguin, 2022), chapters 7 and 21; and on the role of the Macy conferences in shaping the Information Age, see James Gleick, *The Information: A History, a Theory, a Flood* (New York: Vintage, 2011); and Ronald R. Kline, *The Cybernetics Moment: Or Why We Call Our Age the Information Age* (Baltimore: Johns Hopkins University Press, 2015).

48 **In Bali, Mead and Bateson:** As an example, one of Mead's field reports noted the exact moment when a dog licked the hand of a baby who had been born five minutes earlier: it was 10:45 a.m. on Tuesday, July 5, 1938. Two minutes later a child sitting nearby pulled apart a flower. "Film: Birth of Andoaiyanggen Baby," 1938/1946, N119, MM papers.

48 **On a spring evening in Lisbon:** See Nicholas Rankin, "Ian Fleming and American Intelligence, Part 1," Oxford University Press Blog, November 29, 2011, https://blog.oup.com/2011/11/fleming-1; and Nicholas Rankin, *Ian Fleming's Commandos: The Story of the Legendary 30 Assault Unit* (Oxford: Oxford University Press, 2011).

49 **"enigmatic man with slow eyes":** Ian Fleming, "The Secret of Edgar Hoover," *Sunday Times* (London), December 15, 1957.

49 **It was with a man:** John Pearson, *The Life of Ian Fleming* (London: Bloomsbury, 2013), 153–54.

49 **"a small informal conference":** Frank Fremont-Smith to Howard S. Liddell, April 8, 1942, Box 1, Folder 54, HSL papers, as cited in Robert G. Kirk and Edmund Ramsden, "Working Across Species Down on the Farm: Howard S. Liddell and the Development of Comparative Psychopathology, c. 1923–1962," *History and Philosophy of the Life Sciences* 40 (2018): 1–29.

50 **To Mead's delight:** Mead would later claim that a tape recording she made of Erickson hypnotizing her was so potent that it "was enough to send a whole bevy of secretaries, trying to transcribe, into trance" themselves. See MM to Erickson, May 23, 1939, in *The Letters of Milton H. Erickson*, ed. Jeffrey K. Zeig and Brent B. Geary (Phoenix: Zeig, Tucker & Theisen, 2000); and Margaret Mead, "The Originality of Milton Erickson," *American Journal of Clinical Hypnosis* 20, no. 1 (1977): 4–5.

50 **"I am not really anxious to destroy this world":** Peter Mandler, *Return from the Natives: How Margaret Mead Won the Second World War and Lost the Cold War* (New Haven, CT: Yale University Press, 2013), 70.

50–51 **They were driven:** Ellen Herman, *The Romance of American Psychology: Political Culture in the Age of Experts* (Berkeley: University of California Press, 1996), 305–6. See also Matthew Farish, *The Contours of America's Cold War* (Minneapolis: University of Minnesota Press, 2010), 78–88.

51 **"remained unaware of a lighted cigarette":** Erickson to MM, May 31, 1939, in *Letters of Milton H. Erickson*, 10.

51 **Afterward, an excited Lovell:** Stanley P. Lovell, *Of Spies and Stratagems* (Englewood Cliffs, NJ: Prentice Hall, 1963), 57–58. Lovell was evasive about the specifics of his meetings with hypnotists, but documents in Mead's archive clearly show that Milton Erickson began working as a hypnosis consultant for the OSS by the second half of 1942; see M35:6, MM papers ("Erickson hypnosis project, 1942–52"). This identification is strengthened by the fact that Lovell mentions discussing weaponized hypnosis with Dr. Lawrence Kubie, Erickson's research partner.

51 **One of them was sodium amytal:** On the role of sodium amytal and sodium pentothal as supposed "truth serums" in this period, see Alison Winters, *Memory: Fragments of a Modern History* (Chicago: University of Chicago Press, 2012), chapter 2.

51 **"Truth Drugs committee":** "Report on T.D.," 1, June 2, 1943, GHW papers.

51 **Of the four members:** Two were psychiatrists: John Whitehorn at Johns Hopkins and Edward Strecker at Penn. The other was Erickson's research partner and Mead's new friend, Lawrence Kubie. Trained as a neurologist, Kubie had converted to Freudian psychoanalysis in middle age. Though he was happy to consult with intelligence officers, he was also an eager seeker of publicity, a celebrity therapist whose clients later included Tennessee Williams and Ernest Hemingway. See Lawrence S. Kubie, "The Use of Induced Hypnagogic Reveries in the Recovery of Repressed Amnesic Data," *Bulletin of the Menninger Clinic* 7, no. 5 (1943): 172. On the role of neo-Freudian psychoanalysts like Kubie in 1940s and 1950s American culture, see Peter Mandler, "Totalitarianism and Cultural Relativism: The Dilemma of the Neo-Freudians," in *Psychoanalysis in the Age of Totaliarianism*, ed. Matt Ffytche and Daniel Pick (London: Routledge, 2016); and Dagmar Herzog, *Cold War Freud: Psychoanalysis in an Age of Catastrophes* (Cambridge: Cambridge University Press, 2017).

52 **"extremely menacing bowling ball":** John Marks, *The Search for the "Manchurian Candidate": The CIA and Mind Control* (New York: Times Books, 1979), 96.

52 **"most deadly and dedicated":** Lovell, *Of Spies and Stratagems*, 57.

52 **Another OSS colleague described him:** Mike Burke, interviewed in *ABC News Close-Up*, available on YouTube: "Mission Mind Control (1979)," at 4:55, https://youtu.be/DMH5WgGFxlc?t=303.

53 **Even the adventurous Major White:** One reason for the unconventional choice of the newly discovered compound THC (rather than mescaline, which had a longer history of testing as a truth drug) may have been the perception that Nazi scientists had already mastered mescaline research and hence developed antidotes for the drug. The year before, in the summer of 1942, the British cryptographic operation at Bletchley Park deciphered a top-secret message sent to SS Medical Headquarters. "Experiments to date of injecting parachutists with scopolamine were successful," the cryptic message read. "Therefore, experiments with mescaline are to be undertaken, since these injections produce and enhance effect through intoxication. The above is a request from SS Fuhrer in Ukraine for mind-altering drugs from SS medical headquarters (Berlin)." See Dominic Streatfeild, *Brainwash: The Secret History of Mind Control* (New York: St. Martin's Press, 2007), 35.

53 **Meanwhile, drug-induced hypnosis:** On the Macy conferences and the history of trauma in the twentieth century, see Jackie Orr, *Panic Diaries: A Genealogy of Panic Disorder* (Durham, NC: Duke University Press, 2006), chapter 3.

53 **"My fee is $25 an hour":** Roy Richard Grinker, *Fifty Years in Psychiatry: A Living History* (Springfield, IL: Charles C. Thomas Publisher, 1979), 10.

53 **"a therapeutically effective discharge of emotions":** Roy Grinker and J. P. Spiegel, "War Neuroses in North Africa: The Tunisian Campaign" (New York: Josiah Macy, Jr. Foundation, September 1943), 284.

**54 Traumatized soldiers were stuck:** Grinker and Spiegel, "War Neuroses," 286. For an earlier description of drug-assisted therapy for trauma, see W. J. Bleckwenn, "Narcosis as a Therapy in Neuropsychiatric Conditions," *Journal of the American Medical Association* 95 (October 18, 1930): 1168–71.

## 5. THE OFFICE OF STRATEGIC SERVICES

**55 "In battle, in forest, at the precipice":** Oppenheimer, who recited this poem to Vannevar Bush two days before the first atomic bomb test in New Mexico, was freely translating a passage from the *Śatakatraya* of Bhartṛhari, 1.99, circa fifth century CE.

**56 "the hypnotic fascination":** Gregory Bateson, "An Analysis of the Nazi Film *Hitlerjunge Quex*" (1943), in Margaret Mead and Rhoda Métraux, eds., *The Study of Culture at a Distance* (Chicago: University of Chicago Press, 1949), 312–14.

**56 "ripples and waves on the surface":** Gregory Bateson, *Naven*, 2nd ed. (Palo Alto, CA: Stanford University Press, 1958). The timing of the film's release held another meaning for Bateson, who obsessed over every detail of the propaganda film. *Hitlerjunge Quex* premiered on September 12, 1933, in Munich, with Hitler in the audience. That same day had been a pivotal one in Bateson's relationship with Mead, for it was on this day that she decided to choose Bateson over Reo Fortune (see the letters of both GB and Fortune to Mead on September 12, 1933, MM papers).

**57 They were years of:** See Jennet Conant, *A Covert Affair: Julia Child and Paul Child in the OSS* (New York: Simon & Schuster, 2011), 44–45; and Roger D. Stone, *The Lives of Dillon Ripley: Natural Scientist, Wartime Spy, and Pioneering Leader of the Smithsonian Institution* (Lebanon, NH: University Press of New England, 2017), 58, which claims that it was Bateson who first introduced Julia Child to her future husband, Paul Child.

**57 "the lenient, idiosyncratic atmosphere":** Conant, *A Covert Affair*, 45.

**57 A hastily scribbled note:** "Subject: Physical Examination," by Lieutenant Hampton H. Hill, March 16, 1943, O3:6; and small note (dated March 25?) attached to same, reading in part, "Tell M. M. Dr. Abrams will wire to Wash[ington]," in Papers of Colleagues: Gregory Bateson, MM papers.

**57 "the OSS has begun trying to pull me in":** GB to "Steve" (Elizabeth Stevenson Cubhold), May 11, 1943, 3:1, Papers of Colleagues: Gregory Bateson, MM papers.

**57 She met directly with an OSS officer:** MM to GB, untitled memo beginning with the words "Mr. Bateson's application," May 27, 1943, O4:10, Papers of Colleagues: Gregory Bateson, MM papers. Bateson's "Notebook 0" from 1943 (GB papers) contains a cleaner's bill for June 22, and notes regarding a hotel room in Washington, DC, which appear to be related to his OSS interview.

**58 When he was asked to list his aliases:** Undated typescript OSS application, 5:10, Papers of Colleagues: Gregory Bateson, MM papers.

**58 The complex adjoined a defunct hospital:** Elizabeth MacDonald, *Undercover Girl* (New York: Macmillan, 1947), 6.

**58 "a Chinese artist":** MacDonald, *Undercover Girl*, 9.

**58 The recruits learned how to fire:** Russell Miller, *Behind the Lines: The Oral History of Special Operations in World War II* (New York: St. Martin's Press, 2011), 59–60.

**58 "I teach what is called 'gutter fighting'":** John Whiteclay Chambers II, *OSS Training in the National Parks and Service Abroad in World War II* (Washington, DC: U.S. National Park Service, 2008), 192.

**58 "taught us how to roll":** Patrick O'Donnell, *Operatives, Spies, and Saboteurs: The Unknown Story of the Men and Women of World War II's OSS* (New York: Free Press, 2014), 11.

**58 As the narrator of:** "Gutter Fighting Training by OSS at Catoctin," U.S. National Park Service (1943), posted on YouTube: available at https://www.youtube.com/watch?v=rkvHoOH9I3w. This training film was directed by the legendary Hollywood director John Ford, who was an officer in the OSS Field Photographic Branch.

**59 "Everything about Area 'S'":** Roger Hall, *You're Stepping on My Cloak and Dagger* (New York: Norton, 1957), chapter 5.

**59 Only fragments of this work:** My account of the OSS's hypnosis interrogation project, which remains poorly understood and almost entirely unstudied, is based in part on a document found in both the Bateson papers at UCSC and Mead's papers at the Library of Congress that is described as "a stenotype report of an interview with Dr. Milton Erickson, participated in by Margaret Mead and Arnold Meadows, and based on a recording of the hypnotic seance of late December, 1943." Mead's archive also contains typewritten notes on the same event ("Followup of Jap. Hyp. Interview, December 18, 1943. Present: MM, GB, MEE"), along with numerous other references to this and related hypnosis sessions; see M35:6, MM papers ("Erickson hypnosis project, 1942–52"); and M37:2 ("Meadow, Arnold, study, 1944").

**60 A mimeographed transcript of one:** See Box 79, GB papers. In a letter to the MIT scientist Alexander Bavelas (July 31, 1951, M35:6, MM papers), Mead describes this as an "interview with a Japanese who had been trained in Psychology in the US and had been working with GB as an informant."

**60 One of his collaborators:** Kees diary, April 7, 1953.

## 6. OPERATION BITTERSWEET

**61 In wartime there is:** Gregory Bateson, "Cultural Problems Posed by a Study of Schizophrenic Process," in *Schizophrenia: An Integrated Approach*, ed. Alfred Auberback, American Psychiatric Association Symposium of the Havalian Divisional Meeting, 1958 (New York: Ronald Press, 1959).

**61 "an undetermined destination":** GB to "Steve" (Elizabeth Stevenson Cubhold), February 11, 1944, O3:1, Papers of Colleagues: Gregory Bateson, MM papers. On Mead's affair with Steve, see Banner, *Intertwined Lives*, chapter 11.

**61 Along with the orders:** The list of gear comes from Jane Foster, *An Unamerican Lady* (London: Sidgwick & Jackson, 1980), 111. As Foster and Bateson traveled to Ceylon together, departing Los Angeles in early March 1944 and arriving at the beginning of April, it's reasonable to assume he received the same standard-issue equipment. The booklet is described in MacDonald, *Undercover Girl*, 54.

**62 After a brief training:** September 8, 1944, OSS memo from George C. Demas to General Donovan ("Subject: Report of Trip to SEAC and CBI"), 5.

**62 The setting was picturesque:** Ann Todd Baum, "Betty and Jane: Two OSS Women who Waged Psychological Warfare in a Forgotten Theater" (PhD Dissertation, UT Austin, 2014), 140–41, citing Box 4, Folder 49, Child papers.

**62 One new arrival:** Wendy Law-Yone, *A Daughter's Memoir of Burma* (New York: Columbia University Press, 2014), 74–75.

**62 "Mr. Bateson is the only":** Among the other OSS anthropologists who worked with Bateson in Ceylon were Cora Du Bois (a close friend of Mead and Ruth Benedict) and Weston La Barre, who at thirty-three was already a leading expert on the anthropology of peyote. Among the plans La Barre and Bateson drew up was a scheme to forge luxurious menus for Japanese officers' messes, then hope that the officers' disaffected underlings somehow stumbled across them. September 8, 1944, OSS memo from Lieutenant George C. Demas to General Donovan ("Subject: Report of Trip to SEAC and CBI"), 10.

**63 "as a boy was brought up":** SSU SEA Plan (December 1945), 61–62, NARA.

**64 This was promptly rejected:** *WDE*, 40.

**64 "cool and scientific":** Mandler, *Return*, 159. On Mysbergh's background and for the source of the OSS quotes, see SSU SEA Plan (December 1945), 61–62, NARA.

**64 Then, toward the end:** GB and Carleton F. Scofield to David G. Mandelbaum, "Report on Expedition to Sat Tha Village, 26 January 1945," NARA; for further context, see Troy James Sacquety, "The Organizational Evolution of OSS Detachment 101 in Burma, 1942–1945" (PhD Dissertation, Texas A&M University, 2008), 253.

**64 "felt he could not":** Mandler, *Return*, 159.

**64 For weeks at a time:** *WDE*, 40. A fragment of a poem in Bateson's archive, entitled "Death of a Mongoose," may relate to this period.

**65 Documents declassified in 2015:** The project began at Stanford University before moving to Fort Detrick in January 1945. This correspondence is described in a declassified 1947 report on chemical and biological weapons in World War II: Rexmond C. Cochrane, "Biological Warfare in the United States, Vol II," Historical Section, Plans, Training and Intelligence Division; Office of Chief, Chemical Corps; November 1947, declassified 2015. Cochrane, in turn, cites a letter that is apparently still classified: Dr. Stanley Lovell to James A. Hamilton, June 23, 1944, "sub: Coccidioides Immitis, Introduction, Military Feasibility."

**65 "Hospital with Hamilton":** White diary, April 27, 1945.

**65 With Bateson's help:** "Ramree Town," OSS Arakan Field Unit, p. 3, Box 106, E 144, RG226, NARA.

**65 On May 8, when news:** White diary, May 8, 1945.

**66 "a vast eyeball-searing yellow":** Foster, *Unamerican Lady*, 136.

**66 "Within the week":** Arthur B. Darling, *The Central Intelligence Agency: An Instrument of Government, to 1950* (DCI Historical Series, HS 1, December 1953), chapter 2, p. 1, citing "G. Bateson's report, August 18, 1945 (OSS 12733 C "Suggestions"). Later published as Arthur B. Darling, *The Central Intelligence Agency: An Instrument of Government, to 1950* (University Park: Pennsylvania State University Press, 1990).

**67 "land and naval forces":** GB to General Donovan, August 18, 1945 (CIA.HSC/RG263/NARA), "Influence of Atomic Bomb on Indirect Methods of Warfare, Headquarters,

OSS, India-Burma Theater, approved for release by the CIA 2003/12/09, with typed note at top, "Transcribed for ABD [Arthur B. Darling] by CD, 6/1/53."

**67 A follow-up memo:** Redacted to General Donovan, September 4, 1945, released by the CIA on October 28, 2004.

**68 "interesting in patches":** GB to "Ar.B," August 21, 1946, 3:1, Papers of Colleagues: Gregory Bateson, MM papers.

**68 In his twenty-two months:** MM to Ralph Blum, January 30, 1954, C29, MM papers.

**68 "gradually sank, unconfronted":** Paul Boyer, *By the Bomb's Early Light: American Thought and Culture at the Dawn of the Atomic Age* (Chapel Hill: University of North Carolina Press, 1985), 194–95.

**68 The list ranged:** Donald B. Summers, "Final Summary Report of Toxic Materials," September 17, 1945, OSS Research and Development Branch, NARA (with thanks to John Lisle for sharing the document).

## 7. "A NOT UNPLEASANT, INTOXICATED-LIKE CONDITION"

**70 "Ergot is uncertain":** Joseph Taber Johnson, "The Proper Use of Ergot in Obstetrics," *Boston Medical and Surgical Journal* 107, no. 13 (September 28, 1882): 298–99.

**71 "I had the feeling":** From Stanislav Grof's interview with Albert Hofmann, Esalen Institute, Big Sur, California, 1984, available at https://maps.org/news-letters/v11n2/11222gro.html.

**72 In one trip report:** Albert Hofmann, "Boricht über 3 Selbstverguche mit d-Lysergaäure-ditthylamid," December 30, 1943, Hofmann papers, reporting on a trip in Claro, Switzerland, on October 2, 1943 (special thanks to Mike Jay for sharing this document).

**72 "A caged community of chimpanzees":** Hofmann, *Problem Child*, chapter 2.

**72 "Complaints that the soul has died":** Werner Stoll, "Lysergsaure-diathylamid, ein Phantastikum aus der Mutterkorngruppe" [Lysergic acid diethylamide, a phantasticum from the ergot group], *Schweizer Archiv für Neurologie und Psychiatrie* 60 (1947): 305–6.

**72 "Besides the hallucinations":** Stoll, "Lysergsaure-diathylamid," 302.

**73 It stretched back to some:** Jose Luis Martinez, ed., *El México antiguo* (Biblioteca Ayacucho, 1981), a scholarly addition of Sahagún's *Historia general de las cosas de Nueva España*. The cited passage is from the Codice Matritense del Real Palacio, cited in Martinez as "Adiciones: La Fiesta de los Mercaderes," 131, translated from Nahuatl into Spanish by Angel María Garibay.

**73 "wavy lines" of "brilliant colors":** Carlos P. Blacker, "Mescaline: 'Subjective experiences caused by mescaline (written on Saturday March 14th 1936),'" Wellcome Collection, PP/CPB/D.3, p. 3.

**73 Mescaline could draw:** True, the drug's naturally occurring form, ergot, had a long history in medieval European folklore, but this amounted to dim memories of affliction. A toxic fungus that grows on barley, ergot poisoning had plagued Mediterranean societies since ancient times. One chronicle from the year 857, the Annals of Xanten, described it as a "loathsome rot" that caused the limbs to become "loosened." An entire order of Catholic monks, the Order of Saint Anthony, was formed to treat the victims of what came to be known as St. Anthony's Fire.

**73 In a 1945 speech:** Mead, "Human Differences and World Order."

**74 "the fundamental fact of modernism":** Richard G. Wilson, D. H. Pilgrim, and D. Tashjian, *The Machine Age in America, 1918–1941* (New York: Brooklyn Museum/Harry N. Abrams, 1986), 23.

**74 "the first War of the Machines":** J. R. R. Tolkien to Christopher Tolkien, January 30, 1945, letter 96, in *The Letters of J. R. R. Tolkien*, ed. Humphrey Carpenter (London: George Allen & Unwin, 1981).

**74 "two or three monstrous super-states":** George Orwell, "You and the Atom Bomb," *Tribune*, October 19, 1945.

**74 "a great complicated machine":** MM quoted in "Remarkable Remarks," *Ventura County Star*, September 4, 1946, 12.

**75 "planning" a "world order"**: Mead, "Human Differences and World Order," 50.

**75 "constant flow" of "motifs":** Stoll, "Lysergsaure-diathylamid," 303.

**76 "default mode network":** Recent research has even found that psychedelics may increase "neurogenesis" (the formation of new connections between neurons) in brain regions associated with emotional processing. See Cato de Vos et al., "Psychedelics and Neuroplasticity: A Systematic Review Unraveling the Biological Underpinnings of Psychedelics," *Frontiers in Psychiatry* 12 (2021): 724606; and Ling-Xiao Shao et al., "Psilocybin Induces Rapid and Persistent Growth of Dendritic Spines in Frontal Cortex in Vivo," *Neuron* 109, no. 16 (2021): 2535–44.

**76 "a peculiar feeling of anticipation":** Hofmann, *Problem Child*, 35.

## 8. DEVIANTS LIKE MYSELF

**77 an ill-fated Hollywood film:** Greg Mitchell, *The Beginning or the End: How Hollywood—and America—Learned to Stop Worrying and Love the Bomb* (New York: New Press, 2020), 32.

**78 In February 1946:** Mitchell, *Beginning or the End*, 93.

**78 It was also in 1946:** In one bit of public advocacy for the Federation of American Scientists, Bateson returned to the "life insurance" theme of the letter he had begun writing on the day of Hiroshima itself, sitting across from Jane Foster at the OSS camp in Ceylon. He and Mead donated precisely $426.50 to the Federation, because this was the same amount they paid for life insurance. If the Federation failed in its efforts, Bateson wrote in an unpublished letter to the *New Yorker*, "then our daughter will almost certainly not benefit because either she herself, or the insurance company or (most probably) both of them will be destroyed at the same time that we are." (Bateson to *New Yorker*, November 18, 1946, Federation of American Scientists, GB papers.)

**78 "Anthropologists vs Atomic War":** See the original drafts for "The Pattern of an Armaments Race," in GB papers.

**78 "a sudden plunge":** Gregory Bateson, "The Pattern of an Armaments Race—Part II—An Analysis of Nationalism," *Bulletin of the Atomic Scientists* 2, no. 78 (October 1, 1946): 28.

**79 It was in the late 1940s:** Alan Petigny, *The Permissive Society: America, 1941–1965* (New York: Cambridge University Press, 2009); Joanne Meyerowitz, "Beyond the Feminine

Mystique: A Reassessment of Postwar Mass Culture, 1946–1958," *Journal of American History* 79, no. 4 (1993): 1455–82; see also Joanne Meyerowitz, ed., *Not June Cleaver: Women and Gender in Postwar America, 1945–1960* (Philadelphia: Temple University Press, 1994).

**80 "bisexuality, promiscuous love":** Massimiliano Mollona, "Seeing the Invisible: Maya Deren's Experiments in Cinematic Trance," *October*, no. 149 (Summer 2014): 176.

**80 It quoted her as saying:** "The Film That Dreams Are Made On," *Esquire*, December 1946, 187.

**80 "Until the very last minute":** Maya Deren's Haiti journals, archived in the Maya Deren Collection (MDC), Howard Gotlieb Research Center, Boston University, Box 1, Folder 15.

**82 "the part of him that wanted":** All quotes in this paragraph are from MM to Kelly, July 13, 1947, *TCLW*, 114.

**82 Bateson began filling:** Details in this paragraph come from notebooks dated July 14 [1947], pp. 1–2 and 11; and July 11–July 20 [1947], GB papers.

**83 "At first a sense of freedom":** Deren's 1947 journal as cited in Jamie James, *The Glamour of Strangeness: Artists and the Last Age of the Exotic* (New York: Farrar, Straus & Giroux, 2016), 268.

**83 "It has become increasingly":** James, *Glamour of Strangeness*, 268.

**83 "Gregory's dog tags":** VeVe A. Clark, Millicent Hodson, and Catrina Neiman, *The Legend of Maya Deren: A Documentary Biography and Collected Works*, vol. 1, part 2, *Chambers (1942–47)* (New York: Maple-Vail, 1984), 195.

**83 His eye-popping client list:** Peter Keating, "The Strange Saga of JFK and the Original 'Dr. Feelgood,'" *New York*, November 22, 2013, https://nymag.com/intelligencer/2013/11/strange-saga-of-jfk-and-dr-feelgood.html. A book on Jacobson, *Dr. Feelgood*, is poorly sourced, but the Kennedy link has been corroborated by many historians. For one example, see James M. Giglio, *The Presidency of John F. Kennedy* (Lawrence: University Press of Kansas, 2006), 80, which reports that Jacobson visited the Kennedy White House to deliver amphetamine injections at least twenty-seven times. The *New York Times* broke the story in 1972; see Boyce Rensberger, "Amphetamines Used by a Physician to Lift Moods of Famous Patients," *NYT*, December 4, 1972.

**84 "Ideas come at the speed of light":** C. David Heymann, *A Woman Named Jackie: An Intimate Biography of Jacqueline Bouvier Kennedy Onassis* (New York: Lyle Stuart, 1989), 307–8 (citing Heymann's interview with Truman Capote).

**85 "the extreme advocates":** Mead, *Male and Female*, 132.

## 9. "THE SCIENTIST IS *NOT* OUTSIDE"

**87 By one count, Kinsey:** Michael Yudell, "Kinsey's Other Report," *Natural History* 108, no. 6 (1999): 80–81.

**88 "a loincloth type of thing":** James H. Jones, *Alfred C. Kinsey: A Life* (New York: Norton, 2004), 287.

**88 "It is desperately strategic":** Wardell B. Pomeroy, *Dr. Kinsey and the Institute for Sex Research* (New Haven, CT: Yale University Press, 1982), 78.

**88 In March 1946:** Conference on Circular Causal and Feedback Mechanisms in Biological and Social Systems, held at the Beekman Hotel, March 8–9, 1946. In his speech, Bateson theorized that the Iatmul society, riven by internal conflict, avoids fracturing by allowing for inversions of the social order—such as socially accepted cross-dressing, which Bateson had focused on in the context of the gender-bending *naven* ceremony.

**88 It was in one such diner:** Allen Ginsberg, "The Hipster's Hipster," *NYT Magazine*, December 29, 1996.

**88 "We were part of Kinsey's":** From Allen Ginsberg tapes, Naropa University Digital Archives, "Basic Poetics I and II," recorded May 1, 1980, available at https://cdm16621.contentdm.oclc.org/digital/collection/p16621coll1/id/375.

**88 "I am very glad to have met":** Kinsey to MM, May 1, 1946, C15, MM papers.

**88 Had Mead herself agreed:** Nancy C. Lutkehaus, *Margaret Mead: The Making of an American Icon* (Princeton, NJ: Princeton University Press, 2008), 306, fn. 13.

**89 He hired a young anthropology:** Kinsey to MM, January 16, 1947, Kinsey Institute Archives, Bloomington, IN.

**89 "extraordinarily destructive of":** Margaret Mead, "An Anthropologist Looks at the Report," in *Proceedings of a Symposium on the First Published Report of a Series of Studies of Sex Phenomena by Professor Alfred C. Kinsey, Wardell B. Pomeroy and Clyde E. Martin* (New York: American Social Hygiene Association, 1948).

**89 Though it wasn't quite what:** "Kinsey Sex Report Is Called Destructive and Dangerous," *Richmond News Leader*, April 1, 1948.

**90 "something grim in the way":** Jones, *Kinsey*, 605.

**90 "I felt like my husband's":** Jones, *Kinsey*, 607.

**90 In August 1947:** Mead and Métraux were attending the Twenty-Eighth International Congress of Americanists, August 24–31, 1947. It is unclear when exactly Métraux and Mead initiated a romantic relationship; the editors of *To Cherish the Light of the World* believe it was in late 1948, immediately following Ruth Benedict's death, while Lois Banner (*Intertwined Lives*, 431) states that Métraux and Gregory Bateson had an affair earlier in 1947, after which "Rhoda became Margaret's partner for the rest of her life."

**90 "the extensive employment in":** David K. Johnson, *The Lavender Scare: The Cold War Persecution of Gays and Lesbians in the Federal Government* (Chicago: University of Chicago Press, 2004), 21.

**91 Benedict looked to her "almost supernatural":** Margaret Mead, ed., *An Anthropologist at Work: Writings of Ruth Benedict* (Boston: Houghton Mifflin, 1959), 438; and *WDE*, 117.

**91 The FBI began a loyalty:** See MM FBI file, p. 12, memo from S. J. Tracy to D. M. Ladd, December 10, 1948.

**91 The night after Benedict's funeral:** Lapsley, *Kinship of Women*, 301.

**91 For the rest of her life:** *WDE*, 126.

**92 On September 29, there was even:** White diary, September 29, 1948, and October 24, 1948. On December 4, 1948, White met with an Army captain and Hamilton in San Francisco to discuss "TD [truth drug] experimentation."

**92 two days earlier, the anthropologist:** "A Morning with Shock Therapy," October 22, 1948, "MSS, 1948–1949," GB papers. Later, a poet who worked with Bateson at Langley

Porter named Weldon Kees would dedicate a disturbing poem about electroshock, "The Clinic," to Bateson.

**93 The crucial contribution of the Langley Porter Clinic:** During World War II, for instance, a trans woman named Virginia Prince began working informally with Bowman and other doctors at Langley Porter and the UCSF medical school. A trained pharmacologist, Prince portrayed herself as a scientist taking a purely scholarly interest in "transvestism." But in fact, she identified strongly with the patients she worked with, especially a trans woman named Louise Lawrence; Prince later became an early transgender activist. See Virginia Prince, "The Life and Times of Virginia," *Transvestia* 17, no. 100 (1979): 11.

**93 This unique intersection of interests:** Around the time he was hired at Langley Porter, Bateson was interested in runaway feedback loops of violence, which he called "schismogenesis." He theorized that the Iatmul's gender-bending *naven* ritual played a role in preventing schismogenesis. The ceremony also involved the display of what Bateson called an "anal clitoris," which was in fact "an orange-coloured fruit." Bateson, *Naven*, fn. 21.

**94 "intellectual and emotional fragmentation":** Paul H. Hoch, Harry H. Pennes, and James P. Cattell, "Psychoses Produced by Administration of Drugs," Research Publications of the Association for Research in Nervous & Mental Disease (1953).

**94 Amid the experimental culture:** Later, in December 1949, Bateson and Kinsey met at Langley Porter; Bateson apparently screened his Balinese trance films, which piqued Kinsey's interest enough that he sought copies for his research (see Kinsey to GB, July 14, 1950, Kinsey Institute Archives).

**95 "our tools are people":** Gregory Bateson, "Social Planning and the Concept of Deutero-Learning" (originally published in 1942), reprinted in Gregory Bateson, *Steps to an Ecology of Mind* (New York: Ballantine Books, 1972), 163.

**95 a notably grumpy Frank Lloyd Wright:** The architect accepted the invitation with a telegram that read: "YOU ASKED FOR IT, I'LL BE THERE." Paul Venable Turner, *Frank Lloyd Wright and San Francisco* (New Haven, CT: Yale University Press, 2016), 81.

**95 "the enemy at the present time":** Western Round Table on Modern Art, 1949, Transcript A, p. 50.

**95 "trying to take the thing apart":** Western Round Table on Modern Art, 1949, Transcript A, p. 51.

**96 "When Martin switched from science":** Bateson to William Coleman (a historian of science at Johns Hopkins researching William Bateson), December 1, 1966, Box 7, UCSC Special Collections.

**97 "he was a copy boy":** "Wrong-Way Auto Tips Off Police to Narcotics," *World-Telegram*, April 23 [?], 1949.

**97 "Schizophrenia—Pseudoneurotic type":** Ann Charters, *Kerouac: A Biography* (New York: St. Martin's Press, 2011), 109–10.

**97 "apparition of an evil":** Allen Ginsberg, radio interview, "Discussion: Blake, LSD, meditation," August 20, 1981, Stanford Special Collections, Side 1.

**97 "longer than my own life":** Allen Ginsberg, *The Book of Martyrdom and Artifice: First Journals and Poems: 1937–1952*, ed. Juanita Lieberman-Plimpton and Bill Morgan (New York: Da Capo, 2008), Kindle Locations 3314–15. Ginsberg's vision was in the summer of 1948, but he wrote about it in an April 1949 diary entry.

**97 "a new state":** M. Mead, E. Chapple, and G. Brown, "Report of the Committee on Ethics," *Human Organization* 8, no. 2 (1949): 20–21, doi:10.17730/humo.8.2.30275r1p36h137nm.

**98 It was not until later:** Ginsberg directly connected his mystical experience to LSD, saying once that he had behaved "like an LSD person who tries to explain his trip to everybody and gets to be an old bore. Or to take LSD and take off your clothes and jump in front of the cars and say, 'Stop the machines, the world's coming to an end!'" Michael Schumacher, ed., *First Thought: Conversations with Allen Ginsberg* (Minneapolis: University of Minnesota Press, 2017), 199.

**98 "It was astounding":** Allen Ginsberg to Louis Ginsberg, May 20, 1959, in Allen Ginsberg and Louis Ginsberg, *Family Business: Selected Letters Between a Father and Son*, ed. Michael Schumacher (New York: Bloomsbury, 2001).

## Part II: THE PROJECT

**99 "The world is at":** "Expert Here to Lecture on the Sexes," *Los Angeles Daily News*, May 4, 1950, 12.

**99 "Note that I am":** Margaret Mead, "Preliminary LSD Memo," September 23–25, 1954, K65:5, MM papers (hereafter "LSD Memo"). Mead typed this memo in the style of running field notes, documenting events between April and September 1954, with the first three pages dated September 23, 1954, and the fourth dated September 25.

## 10. LIGHTS COMING ON IN WHITE ROOMS

**100 "The air is full of shimmer":** J. D. P. Graham and A. I. Khalidi, "The Actions of d-Lysergic Acid Diethylamide (L.S.D. 25). Part 2. Central Actions," *Journal of the Faculty of Medicine (Iraq)* 18, no. 1 (1954): 35.

**101 Rinkel experimented with administering methamphetamine:** J. Levine, M. Rinkel, and M. Greenblatt, "Psychological and Physiological Effects of Intravenous Pervitin," *American Journal of Psychiatry* 105, no. 6 (1948): 429–34, doi:10.1176/ajp.105.6.429.

**102 And so when a Viennese psychiatrist:** Kauders was an eminent psychiatrist who had published on the topic of sex hormones (1923's "On the Question of the Inner Secretion of the Testis and Sexuality") and hallucinations caused by head injuries. Lazaros C. Trairhou, "Otto Kauders (1893–1949)," *Journal of Neurology* 268 (2021): 3938–39, https://link.springer.com/content/pdf/10.1007/s00415-021-10420-0.pdf.

**102 "That was not Dr. Hyde's":** Max Rinkel, "Experimentally Induced Psychoses in Man," in *Neuropharmacology: Transactions of the Second Conference*, ed. H. A. Abramson (New York: Macy Foundation, 1956), 235–58.

**102 To gain some experience:** Meadow published on "the Discomfort-Relief Quotient" in 1952, coauthored with early LSD researcher Milton Greenblatt. H. Jackson DeShon and Max Rinkel credit Meadow with performing psychological tests on patients at Boston Psychopathic and Harvard who took LSD in group therapy; see "Mental Changes Experimentally Produced by LSD," 1952.

**102 "What's happening to you?":** MM to Arnold Meadow, November 1, 1949, MM papers.

**102 He had been trying to finish:** Meadow to MM, December 15, 1949, MM papers.

**102 He made plans to meet:** Margaret Mead, "Preliminary LSD Memo," September 23, 1954, K65:5, MM papers.

**103 By the end of 1949:** Based on a 1949 estimate of 471,200 patients in mental hospitals, which rose to 670,578 by the end of 1950. See "Statement of Dr. William C. Menninger," in "Mental Health Activities," February 20, 1952, Department of Labor-Federal Security Agency Appropriations for 1953, Hearings Before the Subcommittee of the Committee on Appropriations, House of Representatives (Washington, DC: U.S. Government Printing Office, 1952), 253, 264.

**103 At first, LSD followed:** "He said it was a drug which induced a psychosis," was all that Mead recorded in her notes of the meeting with Arnold Meadow when she first heard about LSD. It is not clear if Meadow himself tried the drug, but it seems likely. Rinkel and Hyde, his superiors, encouraged the use of LSD by clinicians at the hospital. After this work, his career shifted markedly away from Cold War topics and toward the study of creativity. See, for instance, Arnold Meadow and Sidney J. Parnes, "Evaluation of Training in Creative Problem Solving," *Journal of Applied Psychology* 43, no. 3 (1959), a paper that cites creativity and psychedelic researcher Frank Barron.

**104 As Ido Hartogsohn writes:** Hartogsohn, *American Trip*, 34.

**104 It was also the year:** Margaret Mead, "Character Formation and Diachronic Theory," in *Essays Presented to A. R. Radcliffe-Brown* (Oxford: Clarendon Press, 1949).

**106 The show trial of the Hungarian:** Streatfeild, *Brainwash*, 3–8 and 390. On the history of the concept of brainwashing, see Scott Selisker, *Human Programming: Brainwashing, Automatons, and American Unfreedom* (Minneapolis: University of Minnesota Press, 2016).

**106 "Somehow they took his soul":** David Seed, *Brainwashing: The Fictions of Mind Control: A Study of Novels and Films* (Kent, OH: Kent State University Press, 2004), 53, citing Paul Linebarger papers, Hoover Institution, Stanford.

**106 "special psychological technique":** Irving L. Janis, "Are the Cominform Countries Using Hypnotic Techniques to Elicit Confessions in Public Trials," U.S. Air Force Project RAND Research Memorandum, April 25, 1949.

**106 "the effectiveness of various drugs":** Janis, "Hypnotic Techniques," 17, 20.

**106 He warned that this research:** Janis, "Hypnotic Techniques," 20.

**106 In May 1949:** The group, known as the Haskins Committee, had been convened by the U.S. military and intelligence community to study biological, chemical, and psychological warfare. See D. M. Ladd to J. Edgar Hoover, FBI memorandum ("Subject: Biological Warfare"), May 31, 1949, available at https://drive.google.com/file/d/1MeOxLwRaYpzVbAdti8gkVNRdModI5k6b/view. The meeting was hosted by the physicist and lawyer Alfred L. Loomis at Mayfair House in New York City and attended by, among others, "Mr. G. C. Backster, Jr., CIA representative"—a reference to Grover Cleveland "Cleve" Backster, a longtime CIA interrogation specialist who took a special interest in the search for extrasensory perception. On the Haskins Committee, see Jacob Darwin Hamblin, *Arming Mother Nature: The Birth of Catastrophic Environmentalism* (New York: Oxford University Press, 2013), 39–41; and "A Global Contamination Zone: Early Cold War Planning for Environmental Warfare," in *Environmental Histories of the Cold War*, ed. J. R. McNeil (Cambridge: Cambridge University Press, 2010), 103–6.

**106 "People had quite a lot of fears":** John M. Crewdson et al., "Mind-Control Studies Had Origins in Trial of Mindszenty," *NYT,* August 2, 1977, 16.

**106 In his report, Janis directly cited:** Janis was likely drawing on classified research that he had conducted with Bateson, Mead, and Geoffrey Gorer in 1944. The four scientists had role-played in a series of "Experimental Socio-Dramas" that sought to analyze how unconscious or intentionally hidden information could be gleaned from conversations (see MM to GB, November 1, 1944, R3, MM papers; see also "Experimental Socio-Drama on Anglo-American Relationships," November 19, 1944, M24, MM papers, which featured Geoffrey Gorer and Janis role-playing as British and American army officers, with Mead and Rhoda Métraux observing).

**106 While Janis was researching:** A declassified CIA document about the Psychological Strategy board (CIA-RDP80R01731R003500150016-5) described Mead's Research in Contemporary Cultures project as among the "Psychological Warfare Research Studies within the Navy" that were "presently underway" (undated document, circa summer 1950).

**108 "Those who control the weapons":** Remarks of Harold Abramson in Heinz von Foerster, ed., *Cybernetics: Transactions of the Sixth Conference, March 24–25, 1949* (New York: Josiah Macy, Jr. Foundation, 1950), 103–5.

**108 That August, L. Wilson Greene:** L. Wilson Greene, CRL Report ETF 1500-1, "Psychochemical Warfare: A New Concept of War," August 17, 1949.

**108 Abramson appears to have started:** In May 1954, at a classified conference at Edgewood, Abramson said he had begun working with LSD "about three and one half or four years ago." Chemical Corps Medical Laboratories Special Report, MSLR No. 71, First Psychochemical Conference, 12 May, 1954, ed. Amedeo S. Marrazzi (September 1955), 33, available at https://apps.dtic.mil/sti/pdfs/AD0077032.pdf.

## 11. THE PSYCHEDELIC COLD WAR

**109 "Patients kept asleep longer":** George N. Raines and Lawrence C. Kolb, "Combat Fatigue and War Neurosis (Concluded)," *United States Naval Medical Bulletin* 41 (September 1943), p. 1307.

**110 "building up of a personality":** Memo from Morse Allen, "Subject: [redacted] Interview With," March 11, 1952, CIA document #C00140397.

**110 "No doubt the Rand people":** Draft of a letter in Notebook 14B, "A Science of Peaceful Adjustment," circa 1950, GB papers.

**110 "15 to 30 percent":** Notebook 13, "Abstract of speech to Fair Play Society—Palo Alto," January 12, 1950, GB papers. (The group was actually known as the Palo Alto Fair Play Council.)

**111 "a member of the family":** MM to Mysbergh, October 6, 1970, *TCLW,* 238. E. W. Kenworthy, "Triple Pass: How C.I.A. Shifts Funds," *NYT,* February 19, 1967. See also David H. Price, *Cold War Anthropology: The CIA, the Pentagon, and the Growth of Dual Use Anthropology* (Durham, NC: Duke University Press, 2016).

**111 "a deep cover CIA agent":** Jane Howard's interview with James Alexander Hamilton (undated, circa 1980–81), Box 36, Jane Howard papers, Columbia University Rare Book and Manuscript Library. In an interview, Rhoda Métraux's son Daniel Métraux confirmed that Mysbergh was a longtime CIA employee (interview with the author, July 28, 2021). Mysbergh's activities in the 1950s and 1960s are difficult to trace, but it appears he moved from Indonesia to Taiwan and finally to Saigon in this period. In 1955, for instance, Mysbergh appears in the diary of a Rockefeller Foundation official who was touring Java; the official, who was in the OSS with Mysbergh, is surprised to find him there. Mysbergh explained that he was doing research for Margaret Mead, but this was a shaky cover story: his formal employment in Mead's Columbia project ended four years earlier. See Charles B. Fahs diary entry, "Trip to the Far East," Jakarta, May 2, 1955, 41, Officer's Diaries, Rockefeller Foundation Archives. A 1951 letter Mead wrote to Mysbergh before his departure for Indonesia reads, in context, like an attempt to give her friend a vague cover as an academic researcher: Mead writes that she hopes he "may be able to follow up" on research "you began several years ago in Malaysia," but does not mention it again. MM to James Mysberg, February 13, 1951, B14:5, MM papers.

**112 But as the historian Elaine May:** Elaine Tyler May, *Homeward Bound: American Families in the Cold War Era*, rev. ed. (New York: Basic Books, 2017), 10.

**112 This double meaning of "containment":** On the RAND Corporation and the concept of containment, see Ron Robin, *The Cold World They Made: The Strategic Legacy of Roberta and Albert Wohlstetter* (Cambridge, MA: Harvard University Press, 2016). On Mead's contribution to RAND, see Pamela M. Lee, *Think Tank Aesthetics: Midcentury Modernism, the Cold War, and the Neoliberal Present* (Cambridge, MA: MIT Press, 2020), 89–115.

**112 Abramson's life was outwardly conventional:** Abramson's dedication of one of his books to his wife, Virginia "Beaver" Abramson, "who assisted me in all ways" during his early psychedelic research, implies that she, too, sampled the drug alongside their dinner party guests. See the dedication in Harold Abramson, ed., *The Use of LSD in Psychotherapy and Alcoholism* (Indianapolis, IN: Bobbs-Merrill, 1967).

**112 "When two males meet":** Mary Catherine Bateson, *Our Own Metaphor: A Personal Account of a Conference on the Effects of Conscious Purpose on Human Adaptation* (New York: Knopf, 1972), 5.

**113 "a tonic, a strange smelly medicine":** Mary Catherine Bateson, "Our Own Metaphor," pp. 5–6, 1968, Box 79, GB papers. A version of this essay was later published as the prologue to Bateson, *Our Own Metaphor*, but with different language. "The lives of those fish depended on the peace and continuity of our lives, just as our own peace depended on a wider political peace in those early years of the Cold War," Mary Catherine wrote of the aquarium Bateson bought her. "My mother, looking through the water past the water grasses at the early morning light, used to say it reminded her of the sun rising over a reef in the South Pacific" (*Our Own Metaphor*, 7).

**113 "sprinkling, sparkling, glittering spark":** *Problems of Communication*, transactions of the 1951 Macy conference, 142–43.

**113 "To isolate one drug":** *Problems of Communication*, transactions of the 1951 Macy conference, 144.

**114 Mead was now fast friends:** Howard, *Mead*, 332.

**114 by 1951, James Hamilton, too:** It is unclear when Hamilton joined the CIA as a consultant, but based on existing documentation, it likely took place between 1950 and 1952; it may have been as early as the spring of 1949, when he exchanged two cryptic letters relating to biological weapons with the CIA's then director; the previous fall, Hamilton had complained to Bateson that he found life as a civilian psychiatrist boring and hoped to find a more interesting line of work. (Transcript of a conversation between GB and Hamilton, November 4, 1948, Box 46, GB papers.)

**116 "that loveable exponent of Viennese":** Weldon Kees to the Myrers, August 18, 1951, in Robert E. Knoll, ed., *Weldon Kees and the Midcentury Generation: Letters, 1935–1955* (Lincoln: Bison Books/University of Nebraska Press, 2003), 156.

**116 He began inviting:** *WDE*, 56.

**116 L. Ron Hubbard also claimed:** This description is based on firsthand accounts from L. Sprague de Camp, "El-Ron of the City of Brass," published in *Fantastic* (August 1975); and Michael Sherman's 2011 interview with Harlan Ellison (with thanks to Michael Shermer for making his interview notes available to me).

## 12. "NARCOSYNTHESIS IS NOT VERY DIFFICULT"

**119 "the era of the expert":** May, *Homeward Bound*, 29.

**119 "took over the role of psychic healer":** Joseph Veroff et al., *The Inner American: A Self Portrait from 1957 to 1976* (New York: Basic Books, 1981), 194.

**120 "The goal of Dianetics":** L. Ron Hubbard, *Science of Survival: Prediction of Human Behavior* (Hubbard Dianetic Research Foundation, 1951), 1.

**120 "Psychology—no, I'm afraid":** Tony Hitchman and L. Ron Hubbard, "An Introduction to Scientology: An Exclusive Filmed Interview with L. Ron Hubbard," filmed for Rhodesian television, May 1966, www.worldreligionnews.com/wp-content/uploads/2015/05/LRHHitchmanInterview.pdf.

**120 Campbell ran a long article:** "Varieties of Culture," *Astounding Science Fiction*, December 1950, 92–96. The letter about Dianetics appears on p. 152.

**120–21 "absence of jealousy":** L. Ron Hubbard, "PTP's, Unknownnesses," p. 18, lecture delivered on August 22, 1961, transcribed in SHSBC Tapes 001-100; on ethnology being "a hobby of mine," see L. Ron Hubbard, "Question and Answer Period: Goals Search," August 9, 1961, p. 19, transcribed in SHSBC Tapes 001-100.

**121 "the sodium pentothal system":** Campbell to Heinlein, July 26, 1949, 5, Heinlein papers, UCSC. Heinlein responded by saying that Hubbard should get in touch with his friend Cal Laning of the U.S. Navy. At that time, however, Laning was busy working on a classified study of Soviet psychology; soon after, he began collaborating with Margaret Mead on the same topic. See Caleb B. Laning to MM, September 15, 1950, LOC, and the discussion in Mandler, *Return*, 249–50.

**121 "working a patient with narcosynthesis":** L. Ron Hubbard, "Psychotics, Parts 1 and 2," June 23, 1950, originally published by Hubbard Dianetic Research Foundation, New Jersey, 1950, accessible at https://archive.org/details/institutionaldianetics23jne1950researchanddiscoveryseriesvolume1.

**121–22 "the one-shot clear":** Hubbard was studiedly vague about what exactly "GUK" was. He often described it simply as a compound of glutamic acid (a neurotransmitter and food additive) and vitamin B1. At other times, he references amphetamines. This would explain why Hubbard sometimes made outlandish claims about Guk's psychoactive effects. Glutamic acid and B vitamins have no real psychoactive properties, but Hubbard warned in one lecture that the ingredients in Guk could potentially "produce hallucination." See L. Ron Hubbard, "Uses of Validation Processing," lecture on June 27, 1951.

**122 One of them:** Hank P. Albarelli, *A Terrible Mistake: The Murder of Frank Olson and the CIA's Secret Cold War Experiments* (Springfield, OR: Trine Day, 2009), citing Gottlieb's deposition in which he recalled "one meeting with a hypnotist named Erickson in New York around 1952 or 1953." Albarelli's historical analysis is often questionable, but I have been able to verify a number of his factual claims about both Sidney Gottlieb and Harold Abramson through archival research and interviews.

**123 "opposition intelligence services":** Memo for Director of Central Intelligence, Subject: Report of Inspection of MKULTRA, July 26, 1963, https://edgewoodtestvets.org/court-filed-documents/pdfs/20100915-Bowen-Dec-ISO-Opp-To-30b6-Exh-J.pdf ("biochemical controls of human behavior" appears on p. 3). The report says that Allen Dulles officially approved MKULTRA on April 3, 1953. However, based on Gottlieb's testimony and White's diary, it appears the groundwork for the project was already being laid over a year before this. The cryptonym MKDELTA (the sister project of MKULTRA, involving R&D and testing of psychochemical compounds) was approved on October 20, 1952.

**124 "I drove up to Boston":** Division 19 was a small team of researchers during World War II, led by Gottlieb's boss, CIA science director H. Marshall Chadwell, who worked with Lovell and the OSS to develop covert spy tech, including truth drug preparations. Gottlieb had clearly been speaking to Chadwell about the truth drug experiments from 1943.

**125 "Both interrogation and provoking":** See CIA memo, "Subject: Telephonic Response of Dr. Gottlieb to Our Letter of 30 April 1979," approved for FOIA release January 2000, https://ratical.org/ratville/JFK/MKULTRA/19790430-GottliebResponse.pdf.

**125 "It turned out to be a bad thing":** Gil Fox's recollection is quoted in Douglas Valentine, "Sex, Drugs, and the CIA, Part 2," *Counterpunch*, June 19, 2002, available at www.counterpunch.org/2002/06/19/sex-drugs-and-the-cia-3/. For other examples of White's covert dosing of both friends and strangers in these months, see "Memorandum for: Director of Central Intelligence, Subject: MKULTRA—notification of Unwitting Subjects, April 25, 1979, p. 96. Memo, Subject: MKULTRA Subprojects 3, 14, 16, 42, 132, 149, and MKSEARCH 4," p. 9.

## 13. THE TELEPHONE AT THE END OF THE WORLD

**127 "We were literally terrified":** As quoted in "Project MKULTRA, the CIA's Program of Research in Behavioral Modification," Joint Hearing before the Select Committee on Intelligence, August 3, 1977 (Washington, DC: U.S. Government Printing Office), 393.

**127 Then he spotted him:** Oral history interview with A. Robert Abboud conducted by Mark DePue on September 26, 2007, Abraham Lincoln Presidential Library, Springfield, IL, p. 27 (available at: https://presidentlincoln.illinois.gov/oral-history/collections/abboud-a-robert/interview-detail/).

**128 "Nobody will probably ever know":** As quoted in Gary Lachman, *Jung the Mystic: The Esoteric Dimension of Carl Jung's Life and Teachings* (New York: Penguin, 2010), 179.

**128 Clover came to look on:** "You rebuked me once for calling you God," Clover Dulles once wrote to the neurologist, "but if you will not allow your expenditure of time to be reimbursed in some fashion then I consider that it will be you yourself who calls yourself God." Clover Dulles to Harold Wolff, November 5, 1954, Box 1, Folder 15, Isabel Bishop papers, Archives of American Art, Smithsonian.

**129 "hidden empire":** Daniel Immerwahr, *How to Hide an Empire: A History of the Greater United States* (New York: Farrar, Straus & Giroux, 2019).

**129 They often spoke about:** Mark Zborowski had once been a covert agent for the Soviet Union's secret intelligence service, the NKVD. Allegedly a member of a Soviet assassination team active in 1930s Europe, Zborowski had by the early 1950s switched both his affiliation and his career; he became an anthropologist of Jewish shtetl life and, along the way, a close associate and employee of Mead in her "culture-cracking" project at Columbia. Reading between the lines, there was almost certainly a covert angle to this work—perhaps American intelligence services tolerated Zborowski's presence due to his willingness to provide information about Soviet spies of his acquaintance who were still active on the other side of the Iron Curtain. Little is known with certainty about this mysterious character, but we do know this: Margaret Mead counted him as a close personal friend in the early 1950s, and introduced him to Harold Wolff as an expert in the study of pain.

**130 In the same period:** White's diary records a package pickup from Abramson's office on August 10, 1953.

**130 The CIA funded an eclectic range:** See Frances Stonor Saunders, *The Cultural Cold War: The CIA and the World of Arts and Letters* (New York: New Press, 1999); and Joel Whitney, *Finks: How the CIA Tricked the World's Best Writers* (New York: OR Books, 2017). On the CIA's remote viewing experiments, see Annie Jacobsen, *Phenomena: The Secret History of the U.S. Government's Investigations into Extrasensory Perception and Psychokinesis* (New York: Little, Brown, 2017).

**130 In that year alone:** Marks, *Manchurian Candidate*, 61–62.

**131 "For one reason or another":** Albarelli, *Terrible Mistake*.

**133 "every sort and kind of American":** Margaret Mead, *New Lives for Old: Cultural Tranformation—Manus, 1928–1953* (New York: William Morrow, 1956), 168.

**134 "Science is a cooperative task":** MM to Lenora and Theodore Schwartz, May 18, 1954, MM papers. On her plans for the trip, see MM to Herbert Dinerstein, April 12, 1953, C28, MM papers.

**134 "I realized for the first time":** All quotes in this paragraph are from Mead, *New Lives*, 24–26.

**135 "everything belonging to the past":** Mead, *New Lives,* xii.

**135 "through diviners and mediums":** Mead, "Different Concepts of Immortality," a lecture delivered in 1956 republished in *In Search of God and Immortality* (1961), 95.

**135 By the time of Mead's arrival:** Mead, *New Lives*, xx.

**136 "What brand of cigarette":** Mead, *New Lives*, 498.

**136 These entries suggest:** See Kees diary, January 6, 1953; January 27, 1953; July 29, 1953.

**136 "Betty looking like hell":** Kees diary, May 11, 1953.

**138 By the late 1950s:** For instance, see *Theodore Brameld, Patterns of Educational Philosophy* (Yonkers-on-Hudson, NY: World Book Company, 1950), 59; and "Control and Reduction of Armaments," Hearing Before a Subcommittee of the Committee on Foreign Relations, United States Senate, Eighty-Fourth Congress, Second Session, Pursuant to S. Res. 93, January 25, 1956 (Washington, DC: U.S. Government Printing Office, 1956), 696, www.google.com/books/edition/Control_and_Reduction_of_Armaments/1DEvAAAAMAAJ?hl=en&gbpv=1&dq=%22schizophrenic+society%22&pg=PA696&printsec=frontcover.

## 14. NEMBUTAL AND THE SIAMESE FIGHTING FISH

**139 "Do you feel drowsy?":** Harold A. Abramson, Murray E. Jarvik, M. R. Kaufman, Conan Kornetsky, A. Levine, and M. Wagner, "Lysergic acid diethylamide (LSD-25): I. Physiological and perceptual responses," *Journal of Psychology* 39, no. 1 (1955): 3–60.

**139 In the early morning of November 28:** This account is drawn from Lashbrook's own testimony on the day after the death of Frank Olson, when he was questioned by internal CIA security officers; see the security office memo from December 3, 1953 (Case No. 73317), which appears in a tranche of declassified documents released to Olson's family as part of a 1977 settlement with the U.S. government, available at https://frankolsonproject.com/documents/. See also Brian Mooar, "Tests Contradict U.S. Story of Man's Suicide; Family Suspects CIA Killed Researcher," *WP*, July 12, 1994, B1.

**140 In Lashbrook's suit pockets:** The initials may have been a reference to a visit to George White's 81 Bedford Street safe house, which he maintained under the pseudonym Morgan Hall. At the time, White was out of town nursing his ailing mother in California; his diary records that a CIA "tech squad" was updating the safe house's hidden recording equipment while he was gone.

**140 Detective Mullee's best guess:** CIA Field Office Report, December 3, 1953, p. 3, reproduced in "Project MKULTRA," Joint Hearing Before the Select Committee on Intelligence, August 3, 1977 (Washington, DC: U.S. Government Printing Office, 1977), 398.

**140 "he did not know of any":** CIA Field Office Report, p. 3.

**141 "engaged in psychiatric research":** Agent Jeremiah to Sheffield Edwards, December 4, 1953, Case No. 74150, as cited in Albarelli, *Terrible Mistake.*

**141 He was a colleague:** Likewise, George Hunter White's diary records several meetings with "Olson" and "Frank O" in February of 1944, indicating that Olson may also have worked with White and Lovell, the research head of the OSS, during their truth drug research.

**141 In the golden light:** All information in this paragraph is from CIA memo, "Memorandum for: General Counsel. Subject: Observations on Mr. Frank Olson from 24 November through 27 November 1953," dated December 4, 1953.

**142 On that afternoon in November:** CIA memo, December 4, 1953, Investigative Division, referencing an internal report of an interview by "Agent ILIFF with Abramson" on May 22, 1953.

**143 He recommended that Olson:** So secure was Chestnut Lodge that both CIA director Allen Dulles and his deputy, Richard Helms, apparently sought psychiatric treatment there themselves; see Hank Plante, "Inside Chestnut Lodge," *Washington Sentinel*, October 24, 1974. (A clipping is preserved in the CIA's FOIA archive: https://archive.org/details/CIA-RDP91-00901R000500110034-7.)

**143 "I was awakened":** "Observations on Dr. Frank Olson," by Robert V. Lashbrook, December 7, 1953. Olson and Lashbrook checked in to Room 1018A, but as the first three stories of the hotel were not counted in the room number, this was evidently the thirteenth, not the tenth floor.

**143 "wanted to be kept out":** The Special Agent report explained that at 5 p.m. on November 28, Agent "Walter, P.T., Jr met reporting agent and Robert Lashbrook" at Penn Station; after dinner and a movie (*Cease Fire* at the Criterion theater) they went to Abramson's office. "Project MKULTRA," Joint Hearing Before the Select Committee on Intelligence, August 3, 1977 (Washington, DC: U.S. Government Printing Office, 1977), 398, citing CIA Field Office Report, December 3, 1953, p. 3.

**145 "Learned its non-addictive character":** Mead, "LSD Memo," 1.

**145 "a chemist working with":** Mead, "LSD Memo," 1.

**145 "He told me":** Judge Howland Abramson, personal communication, February 15, 2022.

**145 "very smart but somewhat peculiar":** Interview with Murray Jarvik (2001), in *Addiction: Evolution of a Specialist Field*, ed. Griffith Edwards (New York: Wiley, 2008), 76.

**146 "My father thought":** Author interview with Laurence Jarvik, June 29, 2021.

**146 "a physician who combines":** Harold A. Abramson, *The Patient Speaks* (New York: Vantage, 1956), reviewed by Margaret Mead in *The Psychoanalytic Review* 46B, issue 2 (1959): 126–27.

## 15. THE LSD SESSIONS

**147 "prophetic dreams" . . . "a utopia":** Mead, *New Lives*, 38, 453.

**147 "the apocalyptic cult aspect":** Margaret Mead, *Continuities in Cultural Evolution* (New Haven, CT: Yale University Press, 1964), xxxiii.

**148 Perhaps, as she wrote:** Mead, "LSD Memo," 1. On the idea of curing a sick society as a prevalent theme in postwar psychiatry, see Ran Zwigenberg, "Healing a Sick World: Psychiatric Medicine and the Atomic Age," *Medical History* 62, no. 1 (2018): 27–49, www.ncbi.nlm.nih.gov/pmc/articles/PMC5729876.

**148 best-known living scientist:** Google Books Ngram Viewer, accessed May 26, 2023, https://books.google.com/ngrams/graph?content=Albert+Einstein%2CMargaret+Mead%2CJ.+Robert+Oppenheimer%2CJonas+Salk%2CNiels+Bohr%2CEnrico+Fermi%2CLinus+Pauling&year_start=1930&year_end=1960&corpus=en-2019&smoothing=0.

**148 "tremendous optimism":** Carl Sagan interviewed by Shirley Thomas, October 14, 1962, "Men of Space," Reel 2, Tape 1, Lilly Library Special Collections, Shirley Thomas papers, Indiana University Bloomington.

**148 scoring an interview:** Keay Davidson, *Carl Sagan: A Life* (New York: Wiley, 1999), 47.

**150 "new set of optics":** Marks, *Manchurian Candidate*, chapter 7.

**150 "interested observer":** Coincidentally, a few months earlier, Gregory Bateson and Weldon Kees had applied directly to Eberhardt for funding, and been turned down. See Kees's diary, June 18, 1954.

**150 "too serious":** Author interview with Marianne Weltmann, July 26, 2022.

**150 "Graceful dark little German":** Mead, "LSD Memo," 2.

**151 "in a state of giggles":** Mead, "LSD Memo," 2.

**151 "Things are now under way":** MM to Ralph Blum, September 24, 1954, MM papers.

**151 "Drugs, such as mescal or LSD":** Mead, *New Lives*, 522–23.

**152 an MKULTRA conference of sorts:** White diary, April 15, 1954.

**152 All of Hoch's test subjects:** Another attendee, an in-house CIA pharmacologist named Dr. Edward Pelikan, has left little trace in the historical record, except for this one: In the late 1950s, the CIA experimented with deploying three-man units for drug interrogations of prisoners and defectors. They were called "Pelican teams." See Peer de Silva, *Sub Rosa: The CIA and the Uses of Intelligence* (New York: Times Books, 1978), 48–50.

**153 "will send list":** Handwritten note of telephone call with Paul H. Hoch, K65:5, MM papers.

**153 "Margaret Mead was a realist":** Rhoda Métraux, "The Study of Culture at a Distance: A Prototype," *American Anthropologist* 82, no. 2 (1980): 362–73.

**153 "I think I'll have to put off":** MM to Gotthard Booth (draft letter), November 8, 1954, K65:5, MM papers.

**153 "I've not forgotten":** MM to Harold Abramson (draft letter), November 8, 1954, K65:5, MM papers.

**154 In June 1953:** John D'Emilio, *Sexual Politics, Sexual Communities*, 2nd ed. (Chicago: University of Chicago Press, 1998), citing Mattachine Society Coordinating Council minutes, June 26, 1953, and July 10, 1953, in Don Lucas papers, Stanford University, p. 83.

**155 Then one day the scorned husband:** This event is described in Banner, *Intertwined Lives*, citing MM to RB, February 29, 1940; March 2, 1940, S5, MM papers; MM to GB, February 27, 1940, S1, MM papers.

**155 "I have become increasingly conscious":** The letter is reprinted in *WDE*, 15–16.

**155 Her daughter, Mary Catherine, wrote:** *WDE*, 119.

## 16. GADABOUT

**157 At the end of March:** White diary, March 31, 1955. For unpacking, see April 3, 1955, entry.

**158 He hired Maurice Sands:** White diary, May 6, 1955.

**158 They aimed for a bohemian look:** Item list is based on two invoices: MKULTRA 42-144, "Inventory of [censored]," and MKULTRA 42-179, June 30, 1955. White's expenditures would later prompt an internal CIA audit, which complained that he had spent over a thousand dollars on liquor alone; see MKULTRA 42-150, December 21, 1955.

**158 The first experiments with LSD began:** See White's diaries for May 2, July 9, and July 20, 1955. White's colleague on the case, an undercover agent named Ira Feldman, later testified in a sworn deposition that he personally drugged between ten and twelve people with LSD at 225 Chestnut that summer, some of whom were recorded engaging in sex acts while under the influence; Feldman's testimony is available here: https://caselaw.findlaw.com/us-9th-circuit/1350982.html.

**159 "fantasy and imaginative expression":** Timothy Leary, "A Theory and Methodology for Measuring Fantasy and Imaginative Expression," *Journal of Personality* 25, no. 2 (1956): 159–75.

**159 He prescribed Ann:** James Reidel, *Vanished Act: The Life and Art of Weldon Kees* (Lincoln: Bison Books/University of Nebraska Press, 2003), 304–5.

**159 "He's the answer":** Reidel, *Vanished Act*, 294.

**160 "from the world of big":** Weldon Kees and Vincent McHugh, "Gadabout: An Original Screen Story," incomplete screenplay, as cited in Reidel, *Vanished Act*, 324. The earliest printed work to mention LSD that I have been able to locate was *The Splintered Man* (1955), a pulpy thriller that claimed to be drawing from secret government sources. Vincent McHugh, Kees's writing partner on the *Gadabout* script, had a long-standing interest in the topic as well. His novel *I Am Thinking of My Darling* (1943), about a pandemic in New York City that causes sexual disinhibition, includes a discussion between two scientist characters trying to diagnose the cause: could it be drugs? One character says he has checked everything: amphetamines, sedatives, even a South American drug called "Nyopo." "We tried peyotl too," the scientist character adds. "I was startled by one account. A man who'd taken it experimentally said he had a feeling of universal brotherhood. Said he felt as if the millennium had come." See Vincent McHugh, *I Am Thinking of My Darling* (New York: Simon & Schuster, 1943), 71–72. A transcript of recordings of Kees discussing *Gadabout* can be found in Kees, *3 Entertainments* (Newton-le-Willows, UK: Knives Forks and Spoons Press, 2012).

**161 "violence and irrationality":** Weldon Kees, "How to Be Happy #1053," *New Republic*, July 1955. The essay was a review of Arnold A. Hutschneker's *Love and Hate in Human Nature.*

**162 The same afternoon:** Knoll, *Midcentury Generation*, 192.

**162 Kael remembered the Kees:** Knoll, *Midcentury Generation*, 186.

**162 Weldon Kees had just missed:** "His death seemed to precede any poetic outbreak of ours," Allen Ginsberg once wrote to Kees's biographer, James Reidel (July 2, 1993). Accessible at www.bonhams.com/auctions/20906/lot/2115/.

**162 "to overcome a block in his writing":** Jonah Raskin, *American Scream: Allen Ginsberg's Howl and the Making of the Beat Generation* (Berkeley: University of California Press, 2004), 154–56.

**162 "a Golgotha-robot—eternal":** Ginsberg to Jack Kerouac, November 9, 1954, in *The Letters of Allen Ginsberg*, ed. Bill Morgan (New York: Da Capo, 2008).

**163 "I cannot survive if I do":** Bateson's comments in Carl A. Whitaker, ed., *Psychotherapy of Chronic Schizophrenic Patients* (New York: Little, Brown, 1958), 42.

**163 "invented this theory without":** Robert Kolker, *Hidden Valley Road: Inside the Mind of an American Family* (New York: Doubleday, 2020), 36.

**163 "constellation . . . of trauma":** Bateson, "Epidemiology of Schizophrenia," in *Steps to an Ecology of Mind*, 204, an edited version of a talk, "How the Deviant Sees His Society," given in May 1955 at the University of Utah.

**163 Such a person, Bateson wrote:** Bateson et al., "Toward a Theory of Schizophrenia," in *Steps to an Ecology of Mind*, 216.

**164 Bateson wrote a letter:** Red notebook, draft of a letter addressed to "Darling," dated July 11 [1947], GB papers, donation of Rodney Donaldson.

**164 He *lived it*:** And so, too, did Weldon Kees, whose disappearance was still fresh in Bateson's mind that day on Sea Island. In the moment when he first introduced the double bind to his colleagues, Bateson used a telling phrase: "Let me make another jump." Bateson's comments in Carl A. Whitaker, ed., *Psychotherapy of Chronic Schizophrenic Patients* (Boston: Little, Brown, 1958), 39.

**164 Writing for the *New York Times*:** Margaret Mead, "One Vote for This Age of Anxiety," *NYT*, May 20, 1956, 227 and 249. On the concept of the Age of Anxiety and its relation to midcentury science, see Jessica Wang, *American Science in an Age of Anxiety: Scientists, Anticommunism, and the Cold War* (Chapel Hill: University of North Carolina Press, 1999).

**165 The first approved drug:** On Miltown's commercial triumph in the late 1950s, see Andrea Tone, *The Age of Anxiety: A History of America's Turbulent Affair with Tranquilizers* (New York: Basic Books, 2008), 59–62.

**165 By 1965, 25 percent:** Lexis-Nexis, Reference (Polls and Surveys), April 28, 1965, as cited in Petigny, *Permissive Society*, 25.

**165 This meteoric rise:** Patrick Radden Keefe, *Empire of Pain: The Secret History of the Sackler Dynasty* (New York: Anchor, 2021), chapters 3 and 4. Intriguingly, the Sacklers took an early interest in LSD before their involvement with Valium. Raymond Sackler was in the audience when Max Rinkel first reported on the results of the LSD tests at Boston Psychopathic, the first LSD experiment conducted in North America. "What were the physical manifestations?" asked Sackler. "Were there any?" See Discussion, American Psychiatric Association Meeting (Detroit), May 1950, Box 2, Folder 66, Max Rinkel papers, Countway Library, Harvard. Later, Arthur Sackler would coauthor a paper on LSD's possible role as a stress reducer; see Stanley A. Weltman and Arthur M. Sackler, "Effect of Lysergic Acid Diethylamide (LSD-25) on Growth Metabolism and the Resistance of Male Rats to Histamine Stress," *Journal of Pharmaceutical Sciences* 54, no. 9 (1965): 1382–84.

**166 In one notepad from 1956:** Bateson notebook, VA notes (circa 1956), Box 71, GB papers.

## 17. THE LEARYS AND THE BATESONS

**167 "The deafness isolated him":** Robert Greenfield, *Timothy Leary: An Autobiography* (Boston: Mariner Books, 2007), 73, citing an interview with Gerald Kasin.

**167 "I fell in love on the spot":** Greenfield, *Leary*, 64.

**169 "everyone was aware of it":** Greenfield, *Leary*, 77–78.

**170 He and Marianne went:** Box 14, Leary papers, NYPL.

**170 "Psychologist's Wife Ends Life":** *Long Beach Independent*, October 23, 1955, 30.

**170 "Housekeeper wanted to live":** Box 10, undated scrap of newspaper, Leary papers.

**171 The dean thanked Leary:** Box 10, Leary papers.

**171 "Written when blind drunk":** Box 19, Leary papers.

**171 "Leave then come back":** Box 19, Leary papers.

**171 "Tim drank all the time":** Charles Slack, the colleague in question, himself descended into opiate and alcohol addictions that ultimately led him to "nights in the drunk-tank of the Birmingham City Jail." See "My History of Drug Abuse and Recovery," www.leaderu.com/socialsciences/slack2.html.

**171 According to Freedman:** Freedman, "Symposium," citing Jurgen Reusch and Gregory Bateson, *Communication: The Social Matrix of Psychiatry* (New York: Norton, 1951), 105.

**171 "He wanted a model where":** Ram Dass, *Being Ram Dass* (Louisville, CO: Sounds True, 2022).

**172 "observed by Gregory Bateson":** Jack Block, "My Unexpected Life," in *Pioneers of Personality Science: Autobiographical Perspectives*, ed. Stephen Strack and Bill N. Kinder (New York: Springer, 2005), 11–12. For "enrich and remarkably broaden" quote, see SRCD Oral History Interview with Jack Block, May 1, 1994 (interviewed by David Harrington in El Cerrito, CA, available at www.srcd.org/sites/default/files/file-attachments/block_jack_interview_0.pdf).

**172 "I wish we could have":** GB to Aldous Huxley, April 17, 1957, GB papers. Bateson also referenced Block's mescaline experiments in a December 4, 1957, letter to Max Levin.

**173 He remembered thinking:** Block, "My Unexpected Life," 11.

**173 The scientist should be a participant:** Greenfield, *Leary*, 106.

**173 "I thought I knew":** Timothy Leary, *Flashbacks: An Autobiography* (Los Angeles: Jeremy P. Tarcher, 1990), 16.

**173 "A phony persona":** Frank Barron, page of notes titled "Tim Leary's hell," undated, Barron papers, Folder 2, Nicholas and Dorothy Cummings Center for the History of Psychology, Akron, OH.

**173 "One of Tim's attributes":** Barron's notes for a memoir, undated, Folder 2, 15, Barron papers.

**175 "a derivative of the":** Block, "My Unexpected Life."

**175 "I even wondered":** Frank X. Barron, *No Rootless Flower: An Ecology of Creativity* (Cresskill, NJ: Hampton Press, 1995), 182.

**175 "There was a feeling":** Author interview with Dr. Susan D. Block, June 8, 2022, 27:00 to 29:30 (audio).

**175 "dissolving of self":** Susan Block interview, 7:07 to 8:45.

**176 "He was intrigued":** Jack Block interview, www.srcd.org/sites/default/files/file-attachments/block_jack_interview_0.pdf.

**176 "sweating in a small room":** Leary, *High Priest* (New York: New American Library, 1968), 4–5 (and see Greenfield, *Leary*, 101–3).

**176 "I died. I let go":** Leary, *High Priest*, 9.

**177 "revise and integrate":** All quotes in this and the following two paragraphs are from Mead, *New Lives*, 4, 524–25.

## 18. OUT OF THE LAB

**178 "Psychoanalysis resembles Galileo's":** Humphry Osmond, "A Review of the Clinical Effects of Psychotomimetic Agents," *Annals of the New York Academy of Science*, 66, no. 3 (1957): 418–34.

**179 By the late 1950s:** Dyck, *Psychedelic Psychiatry*, 39.

**181 "It was as though I was":** Betty Eisner, *Remembrances of LSD Theory Past* (unpublished memoir, August 7, 2002), 7–8, available at www.erowid.org/culture/characters/eisner_betty/remembrances_lsd_therapy.pdf. Eisner was a complex and influential figure who deserves a more prominent role in the history of psychiatry and psychedelic science. To date, the most comprehensive study of her life has been made by Tal Davidson, in "The Past Lives of Betty Eisner: Examining the Spiritual Psyche of Early Psychedelic Therapy Through the Story of an Outsider, a Pioneer, and a Villain" (MA thesis, York University, Canada, 2017).

**181 "I am certain that":** Don Lattin, *Distilled Spirits: Getting High, Then Sober, with a Famous Writer, a Forgotten Philosopher, and a Hopeless Drunk* (Berkeley: University of California Press, 2012), chapter 7.

**182 "an extremely serious person":** Author interview with Phillip Oderberg, July 15, 2022.

**183 a *Life* article:** On Wasson's article as a "publicity campaign," see Stephen Siff, "R. Gordon Wasson and the Publicity Campaign to Introduce Magic Mushrooms to Mid-Century America," *Revue française d'études américaines* 156 (2018): 91–104. See also Valentina Pavlovna Wasson, "I Ate the Sacred Mushroom," *This Week*, May 19, 1957, 8–10, 36.

**183 "oceanic experiences of identity":** L. G. de Alvarez de Toledo, A. Fontana, and F. P. Morales, "Psicoanálisis y dietilamida del ácido lisérgico (LSD-25): Fundamentos para una técnica terapéutica combinada," *Acta Neuropsiquiátria Argentina* 4 (1958): 28 and 34 ("vivencias oceánicas de identidad... con el cosmos").

**184 By the summer of 1957:** Eisner, *Remembrances*, 40.

**184 Even then, he recalled:** Bateson described his experience in a September 9, 1958, letter to Humphry Osmond, Box 26, GB papers.

**185 not just witnessing but shaping:** On the role of military funding in the origins of Silicon Valley, see O'Mara, *The Code*, chapter 1; and Harris, *Palo Alto*, passim. Bateson influenced the internet's development in characteristically eccentric ways. For instance, in 1964, he shared an office with Ted Nelson, who was in the process of developing the concepts of "hypertext" and "links" between digital pages. That same year, Bateson was discussing communication theory with J. C. R. Licklider of MIT during the period when Licklider was laying the intellectual foundation for the World Wide Web. Much of his conversations with both men, however, seem to have centered on communication between octopuses rather than humans.

**186 The MRI was:** Although the MRI's own institutional history describes Don Jackson as the sole founder of the organization, Bateson was initially described as a co-leader of the project ("P.A. Medical Research Group Gets $20,000 Study Grant," *Peninsula Times Tribune* March 26, 1958, 48). By June of that year, the MRI had formally come into being, but it was now led solely by Jackson, with Bateson relegated to the role of an advisory board member and consultant. On the early MRI, see "Institute Probes Treatment, Causes of Mental Illness," *Peninsula Times Tribune*, September 15, 1959, 44, which describes its locations as a "frame house" for research space and a suite of offices at 860 Bryant Street in Palo Alto.

**186 "one of the cleverest men":** Joe Kennedy Adams, *Secrets of the Trade: Madness, Creativity, and Ideology* (New York: Viking, 1971), 37.

**186 "creative individuals while":** Frank Barron to Efrén del Pozo, July 15, 1958, Frank X. Barron papers, Drs. Nicholas and Dorothy Cummings Center for the History of Psychology. Del Pozo had influential mentors, including the Mexican physician and cybernetics pioneer Arturo Rosenblueth, who had been a member of the Macy circle from its earliest days. An expert in tropical diseases and drugs, del Pozo had been studying Aztec and Mayan medicines since 1945 and was already conducting human trials with at least one ethnobotanical specimen by 1949, an obscure Mayan herb that was thought to increase the supply of mother's milk. See Efrén C. del Pozo, "Estudios farmacológicos de algunas plantas usadas en la medicina azteca," *Boletín Indigenista* (1946): 350–64. For a short biography, see Gabriela Castañeda, "La Sociedad Mexicana de Ciencias Fisiológicas a cincuenta años de su creación," *Revista Mexicana de Ciencias Farmacéuticas* 39, no. 3 (2008): 32. On the Merck partnership, see "A Neglected Mayan Galactagogue: Ixbut," *Botanical Museum Leaflets*, Harvard University, November–December 1978, https://archive.org/stream/mobot31753003541049.

**188 In 1958, a vice president:** Harold Abramson to Carlo Henze, February 4, 1958, Eisner papers, Stanford.

**188 "rumors . . . of suicides caused":** Harold Abramson, ed., *The Use of LSD in Psychotherapy* (New York: Josiah Macy, Jr. Foundation, 1960), 27.

**189 Ginsberg was invited:** Bill Morgan, *I Celebrate Myself: The Somewhat Private Life of Allen Ginsberg* (New York: Penguin, 2006), 296. The MRI had two facilities at this time—one was a midcentury office park that it shared with its parent organization, the Palo Alto Medical Foundation, and the other was a wood-framed house a few blocks north of Stanford in which psychedelic therapy sessions were held.

**189 "I went down to Stanford":** Allen Ginsberg to Louis Ginsberg, May 20, 1959, in Ginsberg and Ginsberg, *Family Business*, 121–22.

**189 The experience inspired:** A few months later, Ginsberg appeared on a radio panel to discuss a newly coined term, "the Beat Generation." Ginsberg's fellow panelist was Margaret Mead. One wonders whether they discussed psychedelics—or, for that matter, Gregory Bateson. All that is recorded of their meeting is that they spoke about the cultural experimentation of the 1940s. "The word beat is a word from the 'hip' vocabulary, which Dr. Mead and I were talking about before," Ginsberg said at one point. It was, he said, a concept "really from the late forties rather than now, actually." See https://writing.upenn.edu/pennsound/x/Ginsberg.php. The source of this audio sample is difficult to pin down, but the date has to be October 1959, because the interviewer says Mead's *People and Places* (released that month) would be published the following day.

**190 However, no evidence indicates:** On Hubbard, see Dyck, *Psychedelic Psychiatry*, 106–17, who did careful work in the archives of the scientists Hubbard worked with most closely, Abram Hoffer and Humphry Osmond. His first correspondence with Hoffer, she notes, begins in May 1955. Betty Eisner fell especially hard for Hubbard's charm and storytelling abilities, and her belief that he was the first to use LSD on the West Coast (though she did not meet him until 1955 at the earliest) may have helped spread his mythos. "What a buccaneer Hubbard was," Eisner recalled. "Large, rambling, and with his own private plane and special island on Puget Sound (which some gossip said belonged to a mysterious sponsor; this was in no way ever confirmed). We all felt as though he traveled with pockets full of magic and gold." Eisner, *Remembrances*, 10.

**190 It was in this time:** Adams described his work on set and setting as inspired by "a principle known to many 'primitives' for centuries, namely, that the preparation and expectations of the subject and the conditions under which the drug were administered could modify the drug experience in many different directions, and that the resulting experiences could be very beautiful." Adams, *Secrets of the Trade*, 99. On the following page, Adams refers to trips in late 1958 with "laymen who had experimented informally for several years"—a likely reference to Al Hubbard.

**191 Ginsberg was not the only:** Frank Barron, too, participated in psychedelic experiments at the MRI in 1959. See "The Center" (n.d.), Folder 2, 3, Barron papers; Barron, "The Sacred Mushroom in Harvard Yard" (n.d.), Folder 1, 24, Barron papers.

**191 According to recent studies:** Abigail E. Calder and Gregor Hasler, "Towards an Understanding of Psychedelic-Induced Neuroplasticity," *Neuropsychopharmacology* 48, no. 1 (2023): 104–12.

**192 Years later, Theodore Schwartz:** All quotes from here to the end of the chapter are from an audiotape in the Theodore Schwartz papers at UC San Diego, "Notes for an autobiographical novel" (February 17, 1971), 18:35–20:44.

## Part III: THE NOISE

**195 "We began to have feelings":** John C. Lilly, "A Feeling of Weirdness," 1962, published in *Mind in the Waters: A Book to Celebrate the Consciousness of Whales and Dolphins*, ed. Joan McIntyre (New York: Scribner, 1974), 71.

## 19. THE FINE LINE

**197 "All the sadness and vanities":** Joe Hyams, "What Psychiatry Has Done for Cary Grant," *New York Herald Tribune*, April 20, 1959.

**197 "Psychiatrists called, complaining":** Bob Gains, "LSD: Hollywood's Status Symbol Drug," *Cosmopolitan*, November 1963, 79. For a far more comprehensive account of this era of media coverage, see Stephen Siff, *Acid Hype: American News Media and the Psychedelic Experience* (Champaign: University of Illinois Press, 2015), 99–105.

**197 "The changes in him":** Richard Gehman, "The Ageless Cary Grant," *Good Housekeeping*, September 1960, 66.

**198 "He, in a way, is the product":** Report of Cary Grant's seventy-second LSD trip, undated, with header "The following was dictated with a small portable recorder from time to time during a five-hour L.S.D. session," archives of Oscar Janiger (private collection). Based on the context, Grant's undated report could refer either to Alan Shepard's first suborbital flight (May 5, 1961), or John Glenn's first orbital flight (February 20, 1962). It is almost certainly referring to the latter, however, because Grant stated that the astronaut was "just being picked up at sea" when he arrived at his doctor's office to begin the psychedelic therapy session. Glenn's space capsule was recovered around noon Pacific time, whereas Alan Shepard splashed down at roughly 7:49 a.m.—a decidedly unlikely time for a patient to begin an LSD trip at his doctor's office.

**198 "For almost 58 years":** From Jay Michael Barrie's manuscript notes of Clare Luce Boothe's LSD session, March 11, 1959, Sylvia Jukes Morris papers (as cited in Sylvia Jukes Morris, *Price of Fame*, 686). The Nixon anecdote is from the same source.

**202 "out of their limited worlds":** "Dr. Adams to address Palo Alto Newcomers," *Peninsula Times Tribune*, June 9, 1959, 6.

**203 "transcendence and a new beginning":** Donald D. Jackson, "LSD and the New Beginning," *Journal of Nervous and Mental Diseases* 135, no. 5 (November, 1962): 435–39.

**204 In a speech:** See James Terrill, "The Nature of the LSD Experience," presented on January 16, 1960, and published in the *Journal of Nervous and Mental Disease* 135, no. 5 (November, 1962): 425–29; see also Charles Savage, "LSD, Alcoholism and Transcendence," in the same issue, pp. 429–35. Terrill's paper, which was funded by USPHS Grant MY 2621, describes studies involving "60 volunteers and 29 psychiatric patients."

**205 "speed up the process":** Eisner to Osmond, March 1, 1958, in Eisner, *Remembrances*, 72–73; see also Eisner's first appearance in print as a psychedelic therapist, "Researchers Tell of Using Drug in Psychotherapy," *Sacramento Bee*, June 26, 1958, 11.

**205 Two of the era's:** D. W. Woolley and E. N. Shaw, "Evidence for the the Participation of Serotonin the Mental Processes," *Annals of the New York Academy of Sciences* 66 (1957): 649. See also Hudson Hoagland, ed., *Hormones, Brain Function, and Behavior* (New York: Academic Press, 1957), where Woolley presents on "the lock and key analogy" for serotonin and its relationship to LSD and tranquilizers. A report on psychedelics in *Science* speculated in 1958 that "LSD could be introduced into the same lock as serotonin"; see Harold E. Himwich, "Pharmacologic Drugs," *Science* 127, no. 3289 (January 10, 1958): 63.

**206 "I think of it like":** Eisner to Osmond, March 1, 1958, as reprinted in Eisner, *Remembrances*, 72–73.

**206 In addition to Ginsberg:** Charlotte Olmsted Gill Kursh, "The Phoenix and the Roc, a Study of the Interconnection of Gambling, Bird Symbolism and Sex-Role Conflicts" (PhD thesis, Department of Anthropology, Stanford University, 1960), 1.

**206 Adams's trip was something different:** For instance, Jay Stevens's *Storming Heaven: LSD and the American Dream* (New York: Atlantic Monthly Press, 1987), a narrative history of LSD based on original interviews with many early psychedelic researchers, described it this way: "The Palo Alto Mental Research Institute had been studying the drug since 1958, and had been instrumental in introducing dozens of local psychiatrists and psychologists, as well as interested laymen like Allen Ginsberg, to the perplexities of the Other World. But the Institute's composure had been shaken by several terrifying incidents—colossal bad trips in which the subject returned from the Other World in questionable shape—and interest in LSD's therapeutic potential had diminished" (Stevens, *Storming*, 177).

**207 "immediately plunged into hellfire":** Oram, *Trials of Psychedelic Therapy*, 47, citing untitled manuscript, n.d., Folder 10, Box 7, Charles Savage papers, Purdue University Archives and Special Collections.

**207 "somewhat like an orgasm":** Adams, *Secrets of the Trade*, 4–5.

**208 "Frank had become passionate":** Leary, *Flashbacks*, 16.

**208 "William Blake revelations":** Leary, *Flashbacks*, 17.

## 20. ANTHROPOLOGISTS FROM THE TWENTY-FIRST CENTURY

**210 And he had welcomed:** Descriptions in this paragraph are derived from Leary, *High Priest*, 14–15; and Greenfield, *Leary*, 110. Dick Dettering may have been one of the sources for Leary's early interest in the writings of Gregory Bateson. In a 1953 article in the semantics journal *ETC*, entitled "Psychology as a Metalanguage," Dettering cited Bateson, alongside his collaborators Jurgen Ruesch and Norbert Wiener, as emphasizing "the transmissional processes involved in communication." Leary began developing what he called his "transactional" theory of doctor-patient communication soon afterward.

**211 "certain New England basements":** All quotes in this paragraph are from Leary, *Flashbacks*, 31–32.

**211 "narrow arrogance of scholars":** Leary, *Flashbacks*, 31.

**214 "He is a vampire":** Details of this event are derived from a personal communication with Michael Maccoby (October 25, 2022); and Maccoby as quoted in Ed Prideux, "Timothy Leary Turns 100: America's LSD Messiah, Remembered by Those Who Knew Him," *Vice*, October 23, 2020, www.vice.com/en/article/epdg3k/timothy-leary-lsd-acid-history. Note that "Dr. Beltran" is a pseudonym used by Maccoby to protect the anonymity of the psychoanalyst.

**215 "Tim had shared them":** Ram Dass, *Being Ram Dass* (Boulder, CO: Sounds True, 2021), 60.

**216 "The young kids":** Leary, *High Priest*, 44–45. Although Schwartz was not identified by name in Leary's memoirs, I was able to confirm that he and Lola were the anonymous "anthro and wife" at Cuernavaca via interviews with his surviving children and cross-referencing of the personal papers of Schwartz (at UC San Diego) and Leary (at the New York Public Library). Schwartz's obituary claimed, evidently confusing psilocybin with LSD, that "during his years in Mexico, he dropped acid with Timothy Leary" (www.anthropology-news.org/articles/theodore-schwartz/).

**216 "pleasant intellectual chap":** Leary, *High Priest*, 45.

**217 "We saw ourselves as":** Leary, *Flashbacks*, 190.

**217 He had simply stumbled:** On early psychedelic research in 1950s Mexico City, see Alexander S. Dawson, *The Peyote Effect: From the Inquisition to the War on Drugs* (Oakland: University of California Press, 2018), chapter 7; and Augusto Fernández Guardiola, *Las neurociencias en el exilio español en México* (Mexico City: Fondo de Cultura Económica, 1997), chapter 2. Guardiola, himself a distinguished Spanish Mexican neurologist who had also studied with Arturo Rosenblueth, wrote that he participated in an early experiment with psilocybin mushrooms alongside the Mexican scientists Dionisio Nieto, Carlos Guzmán, Humberto Mateos, Gunter Streker, and Carlos Beyer in late 1950s Mexico City. Lothar Knauth also became a psychedelic researcher around this time. In November 1960, Knauth wrote Leary from Mexico City to say that he was conducting experiments on the effects of psilocybin on creativity among a multicultural group of subjects who "switched from language to language without the least difficulty." See Knauth to Leary, November 29, 1960, Box 45, Leary papers, NYPL.

**217 "We inherit their secrets":** Humphry Osmond, "A Review of the Clinical Effects of Psychotomimetic Agents," *Annals of the New York Academy of Sciences* 66, no. 3 (1957): 418–34.

**217 The pharmacologist Joel Elkes:** Interview with Joel Elkes by David Healy, 1998, from *The Psychopharmacologists II* (London: Altman, 1998), 183–214.

**217 Even at the RAND:** See William H. McGlothlin, "Hallucinogenic Drugs: A Perspective with Special Reference to Peyote and Cannabis" (RAND Corporation white paper, July 1964), 10 and 19; and William H. McGlothlin, "Long-Lasting Effects of LSD on Certain Attitudes in Normals: An Experimental Proposal" (RAND Corporation report, May 1962), available at https://apps.dtic.mil/sti/citations/AD0604802.

**219 "Dave McClelland has told me":** Leary to Mead, March 22, 1961, Box 45, Leary papers, NYPL. A version of Mead's advice appears in her interview in "For God's Sake, Margaret: Conversation with Gregory Bateson and Margaret Mead," *CoEvolution Quarterly*, no. 10 (June 1976): 32–44. Leary cited the same remark in Timothy Leary, "How to Change Behavior," in Solomon, *LSD*, 104–5.

**220 "interesting research designs":** Timothy Leary, Memo to Dave McClelland, March 16, 1961, Box 45, Leary papers.

**220 "subtler warfare of control":** Margaret Mead, *The World Ahead: An Anthropologist Anticipates the Future* (The Study of Contemporary Western Cultures, vol. 6) (New York: Berghahn Books, 2005), 81.

**221 The *Oakland Tribune*:** "New Mind Control Pill Can Make Man Believe Anything," *Oakland Tribune*, January 29, 1961, 2.

**221 "a kind of 'therapeutic brainwashing'":** Emma Harrison, "Forecasts Made for Psychiatry," *NYT*, February 25, 1961.

**221 When it was pointed out:** *TCLW*, 229; and Michael Maccoby, personal communication, October 25, 2022.

**221 "In Israel I look":** Lutkehaus, *Mead*, 77.

**221 But she always kept:** *Atlanta Constitution*, November 10, 1962, 6.

**221 Leary had more success:** Gregory Bateson's name also came up, though it is unclear whether he was contacted. In January 1961, Alan Watts (who by this time had joined the Harvard psychedelic group) wrote to Leary with an update about their plans to give psilocybin to "opinion-making people"—people in a position to make possible a broad-based societal adoption of psychedelic therapy—whom "one could trust to handle it on their own." The first name on Watts's list of potential contacts was Gregory Bateson. "But then," Watts added, Bateson was "already turned on!" Watts to Leary, January 9, 1961, in *The Collected Letters of Alan Watts*, ed. Joan Watts and Anne Watts (Novato, CA: New World Library, 2017), 394.

**221 "Situation here worse":** William S. Burroughs to Brion Gysin, September 28, 1961, in *Rub Out the Words: The Letters of William S. Burroughs, 1959–1974*, ed. Bill Morgan (New York: Ecco, 2012).

**222 "money comes from Madame Luce":** William S. Burroughs to Allen Ginsberg in *Rub Out the Words*. Burroughs's claim about Clare Boothe Luce's anonymous funding of the project, though unsubstantiated, cannot be dismissed out of hand. Leary's archives indicate that he was in close contact with Gerald Heard during this period, and his correspondence with Ginsberg mentions at least one sizable anonymous donation from someone in the New York area.

**222 anonymous donor:** Alan Watts to "Mummy and Daddy," May 29, 1960, in *Collected Letters of Alan Watts*. For what is known about the early philanthropic donations to Leary's project, see Chris Elcock, "Psychedelic Philanthropy: The Nonprofit Sector and Timothy Leary's 1960s Psychedelic Movement," *Journal of the History of the Behavioral Sciences* 58, no. 1 (2022): 85–104.

## 21. PLANES OF FRACTURE

**224 sexism and racism:** Months before the thalidomide news broke, in a letter to Humphry Osmond, Eisner alluded to the "prejudice" that female scientists (and patients) faced in the male world of psychedelic therapy. Eisner to Osmond, April 24, 1961, in Eisner, *Remembrances.*

**225 "second daughter":** Author interview with Jean Houston, August 11, 2021.

**225 "Houston never joined":** Leary, *Flashbacks,* 158. Along with her husband, Houston would later establish an alternative center of psychedelic experimentation in New York City, focusing not on laboratory trials or psychological experiments but on exploring the edges of science itself.

**225 faced credible charges:** The team member was W. Madison Presnell; see Philip Smith, "Suit Raises Question of Doctor's Hiring at St. E's," *WP,* May 31, 1985, www.washingtonpost.com/archive/politics/1985/05/31/suit-raises-question-of-doctors-hiring-at-st-es/d7c83c0d-a73a-47b6-a914-81e87dc78b28.

**225 Moreover, despite their alliance:** The "sexual pathology" phrase comes from Richard Alpert, "Drugs and Sexual Behavior," *Journal of Sex Research* 5, no. 1 (1969): 50–56, which described what amounts to an attempted case of gay conversion therapy using LSD. Timothy Leary, likewise, argued in several different settings that psychedelics could cure homosexuality and other "sexual hang-ups." See Leary's 1967 interview with Ken Garrison, reprinted in Timothy Leary, *The Politics of Ecstasy* (New York: G. P. Putnam's Sons, 1968), 264.

**226 a groundbreaking documentary:** *The Rejected*, KQED (San Francisco Bay Area), September 11, 1961, available at https://diva.sfsu.edu/collections/sfbatv/bundles/225539.

**227 "imagination as an instrument":** "Cal Extension Series Studies Creative Man," *Concord Transcript*, October 11, 1962, 2.

**227 "revolution in women's consciousness":** *Oakland Tribune*, October 25, 1962, 17; and Barron in *Timothy Leary: Outside Looking In*, ed. Robert Forte (Rochester, VT: Park Street Press, 1999), 25–26.

**227 "the cortex can be cleared":** Leary, "How to Change Behavior," in Solomon, *LSD*.

**227 "make the spotlight of consciousness":** Alan Watts, "A Psychedelic Experience: Fact or Fantasy?" in Solomon, *LSD*. The 1962 Kefauver-Harris Drug Amendments meant that the baseline of scientific rigor that the U.S. Food and Drug Administration required of psychedelic experimenters dramatically increased during the period when Leary and his collaborators were moving away from seeing their work as scientific at all. On the changing FDA regulatory environment for psychedelics in the early 1960s, see Matthew Oram, "Efficacy and Enlightenment: LSD Psychotherapy and the Drug Amendments of 1962," *Journal of the History of Medicine and Allied Sciences* 69, no. 2 (2014): 221–50.

**228 "a phony swami":** As recounted in Alan Watts, "The Veil of Thoughts," tape recording from 1971, at 45:15–46:00, https://alanwatts.org/transcripts/the-veil-of-thoughts/.

**228 "curiously tasteless... blank white spaces":** Margaret Mead, "Towards More Vivid Utopias," *Science* 126, no. 3280 (November 1957): 958–59.

**229 "The group had sort of":** David Lipset interview with R. D. Laing, David Lipset papers, UCSC, 1; see also Lipset, *Bateson*, 238.

**229 "and go down to La Jolla":** Lipset, *Bateson*, 234

**230 names like "Valentino":** Octopus observation box, "Notes: Octopuses," envelopes containing Dictaphone disks dated October 29, 1963, GB papers.

**230 "This stuff is all":** Gregory Bateson and Mary Catherine Bateson, *Angels Fear: Towards an Epistemology of the Sacred* (New York: Macmillan, 1987), 70.

## 22. THE DEEP END OF THE POOL

**232 "His research involved":** Oral history interview with Dr. Virgil Carlson (interviewed by Dr. Ingrid Farreras), March 21, 2002, available at https://history.nih.gov/display/history/Carlson%2C+Virgil+2002.

**232 In May 1963:** Joseph S. Russin and Andrew T. Weil, "The Crimson Takes Leary, Alpert to Task," *Harvard Crimson*, May 28, 1963, www.thecrimson.com/article/1973/1/24/the-crimson-takes-leary-alpert-to/.

**233 "anyone wanted to purchase":** *Oakland Tribune*, June 4, 1964, 38. Al Hubbard may well have been the anonymous source who tipped off police, as his FBI file contains several references to contacts with narcotics officers in California during this period (for instance, see Department of Justice Character Reference dated October 7, 1966, on p. 56 of Hubbard's FBI file).

**233 "Since my illumination in August":** Leary, "The Religious Experience: Its Production and Interpretation," lecture at the 71st Annual Convention of the American Psychological Association, Philadelphia, August 30, 1963, reprinted in *Psychedelic Review* 1, no. 3 (1964): 324–46.

**234 Mexican authorities were tipped off:** The psychoanalyst in question was Dionisio Nieto, a refugee from Francoist Spain. See Cristina Sacristán, "En defensa de un paradigma científico: El doble exilio de Dionisio Nieto en México, 1940–1985," in *De la "Edad de Plata" al exilio: Construcción y reconstrucción de la psiquiatría española* (Madrid: Frenia SC, 2007). Nieto, who had done psychedelic research of his own in the 1950s, was later quoted in a *Saturday Evening Post* article on psychedelics as stating that "attempting to release creative mental activity with halluciogens is nonsense." See John Kobler, "The Dangerous Magic of LSD," *Saturday Evening Post*, November 2, 1963, 30.

**234 "an indissoluble family":** Lisa Bieberman, "The Psychedelic Experience," *New Republic*, August 5, 1967.

**235 Their criticisms carried weight:** As an example, see Henry K. Beecher, "Science, Drugs, Students," *Harvard Alumni Bulletin* 65, no. 8 (February 2, 1963).

**235 "administered the drug to themselves":** Roy R. Grinker, "Lysergic Acid Diethylamide," *Archives of General Psychiatry* 8, no. 5 (May 1963): 425, doi:10.1001/archpsyc.1963.01720110001001.

**235 "earnest desire to support":** Harold A. Abramson, "Extraordinary Behavior," *NYT*, January 3, 1965, 10 and 12.

**236 "our dislike of being":** From Margaret Mead's February 1967 speech at UC Davis, as quoted in June Peters, "Single Persons Are 'Lonely,'" *Sacramento Bee*, February 16, 1967, 28.

**236 "It was easy to forget":** Leary, *Flashbacks*, 157.

**236 "as high in the consciousness":** Greenfield, *Leary*, 213, citing an interview with Ralph Metzner.

**237 His professional résumé:** Lilly was a privileged and wealthy man who made extravagant demands of his trust fund, a lifelong legacy from his parents; in 1987, for instance, he proposed a personal annual budget of $245,000, approximately $550,000 in 2020 dollars. "In re Trust Created by Lilly," No. C6-97-299 (Minnesota Court of Appeals, September 16, 1997, available at https://casetext.com/case/in-re-trust-created-by-lilly).

**237 penchant for unusual headwear:** See Lilly's interview with Jeffrey Mishlove (1998), available on YouTube: "John Lilly: From Here to Alternity (excerpt)—A Thinking Allowed DVD w/ Jeffrey Mishlove," https://youtu.be/x6gLydCxJeA.

**237 Inside this soundproof:** Ted Nelson, "My Year at John Lilly's Dolphin Lab," YouTube, March 4, 2020, https://youtu.be/ONhnEmoSRfk.

**238 They have been retold:** D. Graham Burnett, *The Sounding of the Whale: Science and Cetaceans in the Twentieth Century* (Chicago: University of Chicago Press, 2012), chapter 6; Christopher Riley, director, *The Girl Who Talked to Dolphins*, BBC documentary, 2014; "Home Is Where Your Dolphin Is," *Radiolab*, August 21, 2014; "The Dolphin Who Learned to Speak," *Saturday Night Live*, November 11, 2017. On Lilly's cultural milieu, see David Kaiser and W. Patrick McCray, eds., *Groovy Science: Knowledge, Innovation, and American Counterculture* (Chicago: University of Chicago Press, 2016).

## 23. CARL SAGAN AT THE DOLPHIN LAB

**239 "Peter was insanely mad":** "Margaret Howe comments on Pamela's reaction to LSD, 1965 January 15," JCL papers (available at https://purl.stanford.edu/tx759bw1131).

**239 He summed it up:** Unreleased BBC interview with Eric Bateson, 4:18:36 (with thanks to Christopher Riley).

**240 "a fascinating person":** Personal communication, Jeff Bridges, December 15, 2022. "As a scientist he would experiment on himself à la Dr. Jekyll and Mr. Hyde," Bridges noted in a recorded response to questions for this book. "He would shoot LSD and get into that tank for *twenty-four hours*, man." Bridges added that in the 1970s he was "one of Lilly's guinea pigs" for sensory deprivation tank experiments. (With thanks to Becky Pedretti for arranging the interview.)

**242 "I hope from the dolphins":** Jack Oswald, "Mand and Dolphin Learn Together," *Miami News*, February 13, 1964, 21.

**242 One surviving clip:** John C. Lilly interview, Maui, "Hi Quality Japanese Footage, 8-26, 1993" [Betamax], Box 139, JCL papers. Lilly's remarkable archive has recently begun to be explored by scholars of science and technology studies and media studies; for instance, see Charlie Williams, "On 'Modified Human Agents': John Lilly and the Paranoid Style in American Neuroscience," *History of the Human Sciences* 32, no. 5 (2019): 84–107; and John Durham Peters, "'Memorable Equinox': John Lilly, Dolphin Vocals, and the Tape Medium," *Boundary: An International Journal of Literature and Culture* 47, no. 4 (2020): 1–24.

**242 Progress is slow:** Lilly interview, Box 139, JCL papers.

**244 Sagan's first encounter:** All quotes relating to Sagan's visit are from Carl Sagan, *The Cosmic Connection: An Extraterrestrial Perspective* (New York: Dell, 1973), 134–36.

**245 "*Oh, fuck! Fuck and damn!*":** Ted Nelson, "To Whatever Porpoise" (an excerpt from Nelson's unpublished autobiography, available at https://archives.obs-us.com/obs/english/papers/ted/tedbio12.htm).

**245 "brilliant and persuasive":** Ted Nelson, "My Year at John Lilly's Dolphin Lab," and personal correspondence.

**245 "Through the windows":** Ted Nelson, "Man and Dolphin: A Film Script," January 29, 1963, 10, available at https://archive.org/details/mananddolphinscript.

**246 "suddenly realized that":** John C. Lilly, *The Center of the Cyclone: An Autobiography of Inner Space* (New York: Julian Press, 1972), 10–11.

**246 "Did you ever do any":** Lilly to GB, April 28, 1964, Box 20, GB papers.

**248 By his own account:** This description has been pieced together from Lilly's *Center of the Cyclone*, 22–24, along with documentation from Lilly's archive and the detective work of D. Graham Burnett in *The Sounding of the Whale*, 615–16. There are numerous inconsistencies of fact and timing in Lilly's accounts, but the broad outlines can be corroborated. For instance, he describes his second trip as taking place on May 7, and then flying "across the continent" on the following day to give a talk in the evening. In fact, Lilly delivered a paper on the morning of May 6 (see "Program of the Sixty-Seventh Meeting of the Acoustical Society of America," May 6–9, 1964, Hotel New Yorker, New York, which describes Lilly's Wednesday morning, May 6, 1964, paper "Airborne Sonic Emissions of *Tursiops truncatus*").

**248 "cutting off the circulation":** Lilly, *Center of the Cyclone*, 25 and 35–36.

**248 From that point onward:** As Lilly writes (*Center of the Cyclone*, 43), "I proposed trying LSD on dolphins as an aid to my understanding of the substance and some of the physiological dangers of its use."

**248 "problems and potentialities":** "Hallucination Drug Topic for Discussion Saturday," *Spartan Daily* student newspaper, San Jose State College, May 7, 1964, 1.

**248 And watching from the audience:** John Markoff, *Whole Earth: The Many Lives of Stewart Brand* (New York: Penguin Press, 2022), 208.

**249 In May 1965:** See Lilly's comments in Harold A. Abramson, ed., *The Use of LSD in Psychotherapy and Alcoholism* (Indianapolis, IN: Bobbs-Merrill, 1967).

**249 "With man one is sometimes":** Charles C. Berdjis, ed., "Proceedings of a Contractors' Conference on Behavioral Sciences, 14 and 15 October 1965" (Edgewood Arsenal Special Publication, EASP 100-11, February 1967), 345–46. The proceedings were published for internal military use, and a note advised, "When this report is no longer needed, destroy it."

**250 The surviving tape recordings:** JCL papers, audiotape of dolphins Pam and Peter, January 15, 1965, roughly 5:00 to 7:45.

**250 Frances Kelsey, the FDA official:** "Memorandum of Conference," FDA, July 31, 1964, Frances Kelsey papers, LOC.

**251 One by one:** Katherine R. Bonson, "Regulation of Human Research with LSD in the United States (1949–1987)," *Psychopharmacology* 235, no. 2 (2018): 591–604. Matthew Oram provides considerable archival backing for the argument that the FDA pulled their support for psychedelic testing in Silicon Valley amid concerns about what one FDA representative called "the reliability of some of the people there"—perhaps a reference to Al Hubbard. See Oram, *Trials of Psychedelic Therapy*, 66–68.

**251 Bateson kept filling notebooks:** Undated page from Notebook 31 (summer 1964), headed "Problems + Questions," Box 72, GB papers.

**251 "To close it now":** Undated draft letter to Burt Boothe in Notebook 31, Box 72, GB papers.

**251 "a depressing moment":** Undated draft letter (circa August 1964) to Dr. Busse in Notebok 31, Box 72, GB papers.

**251 "Was it Patrick Henry":** Draft memo (undated, early 1965) to John C. Lilly, "Subject: Justification for Work at Oceanic Institute, Oahu," in a subsection titled "Personal aspects," Notebook 32, GB papers.

**251 "committed suicide by":** Lilly, *Center of the Cyclone*, 61.

**252 "I had no idea":** From an unpublished 2013 interview with Lois Bateson conducted by the research team of the BBC documentary *The Girl Who Talked to Dolphins* (2014). Many thanks to that film's producer, Christopher Riley, for sharing the transcript.

**252 "We did not get":** Author interview with Lois Bateson, April 15, 2022.

## 24. DIALECTICS OF LIBERATION

**253 "It looks groovy":** Box 119, Reel 7, "Bob Dylan, San Jose 1965 Dec 12," Ginsberg papers, Stanford. Portions of the Dylan tape are available on YouTube: "Bob Dylan—The Ginsberg Tapes (Live Recordings from 1965—San Francisco / San Jose," https://youtu.be/3rISCFDBh_s?t=75.

**253 Later that night:** Box 120, Reel 1, "1965 Dec 12. Side A: Conversation with Joe Adams (Saturday evening and acid test)," Ginsberg papers, Stanford (hereafter "Adams tape").

**254 "Hey, whatever happened":** Adams tape at approximately 4:27.

**254 "It was in January":** Adams tape, 39:00–39:50.

**254 "it was in *Menlo Park*":** All quotes from here until the end of the section are at Adams tape, 35:40 to 43:30 (the Cold War quote is at approximately 42 minutes).

**256 In Brazil, artist:** On psychedelic science and culture in 1960s Brazil, see Júlio Delmanto, *História social do LSD no Brasil: Os primeiros usos medicinais e o começo da repressão* (São Paulo: Editora Elefante, 2020).

**256 "a collective transformation":** Claudio Naranjo, *Ascenso y descenso de la montaña sagrada* (Providencia, Santiago: Penguin Random House Chile, 2019).

**256 And ancient psychedelic traditions:** For instance, see Beatiz Caiuby Labate, Brian Anderson, and Matthew Meyer, "1960s Media Coverage of Ayahuasca and the UDV," Erowid.org, December 3, 2009.

**256 "We are seeing accidents":** Sidney Cohen's speech at UCLA, March 2, 1966, available on YouTube: "Sidney Cohen Speaking at UCLA 3/2/1966," https://youtu.be/eeWTMUyPdXw.

**257 "Here is a drug":** All quotes from Robert F. Kennedy are from his preliminary remarks in "Organization and Coordination of Federal Drug Research and Regulatory Programs: LSD," May 24, 1966, Hearings Before the Subcommittee on Executive Reorganization, United States Senate (Washington, DC: U.S. Government Printing Office, 1966), 2–3.

**258 "I think in fairness":** Senator Dodd and Captain Trembley, "The Narcotic Rehabilitation Act of 1966," May 13, 1966, Hearings Before the Special Subcommittee on Narcotics, Committee on the Judiciary, United States Senate (Washington, DC: U.S. Government Printing Office, 1966), 218–19.

**258 "the so-called peril":** See Leary's remarks in "The Narcotic Rehabilitation Act," 239–41.

**258 "you definitely go out":** Leary in "The Narcotic Rehabilitation Act," 243–44.

**258 Leary's remarks were widely:** Leary was quoted as describing an "eerie power to release from the brain ancient energies" in Universal Press International, "Use of LSD Out of Hand," *Fort Lauderdale News*, May 13, 1966, 9.

**258 "I am trying to follow":** "The Narcotic Rehabilitation Act," 241.

**258 By November, even Allen Ginsberg:** "'Public Solitude': Allen Ginsberg Address Delivered at Arlington St. Church, Boston, Nov. 12, 1966," as printed in *International Times*, January 30–February 12, 1967, 4–5, available at https://content.wisconsinhistory.org/digital/collection/p15932coll8/id/43805.

**258 The *Saturday Evening Post*:** Bill Davidson, "The Hidden Evils of LSD," *Saturday Evening Post*, August 12, 1967; Nate Haseltine, "Worse Deformities Feared in LSD Than Thalidomide," *WP*, August 1, 1967.

**258 "the real goal of":** "Leary and Lettvin Clash on Drugs in M.I.T. Debate," *Harvard Crimson*, May 4, 1967, available at www.thecrimson.com/article/1967/5/4/leary-and-lettvin-clash-on-drugs/.

**258 Lettvin denounced efforts:** Carl Sagan was apparently an attendee of the debate, and a friend of Lettvin's. A decade later, he recalled discussing the possibility that LSD could cause brain damage with Lettvin at this time. See Sagan to Lester Grinspoon, August 28, 1978, Carl Sagan papers, LOC.

**258 "Faustian promise of a shortcut":** GB to Jerome Lettvin, March 26, 1968, GB papers.

**260 Allen Ginsberg appears next:** Allen Ginsberg to Gary Snyder, July 26, 1967, in *The Letters of Allen Ginsberg*.

**260 And then the camera:** Details in this and the preceding paragraph are based on an excerpt of the film *The Anatomy of Violence* (1967), directed by Peter Davies, a documentary about the conference, available at YouTube: "Dialectics of Liberation Preview," https://youtu.be/oN-Zksippj8?t=373.

**260 "enormous barn-like structure":** Angela Davis, *An Autobiography* (New York: International Publishers, 1988), 149.

**260 Nevertheless, many of those:** Barry Hugill, "We Are All the Children of '67," *Observer*, May 25, 1997, 18; "into a revolutionary center" quote is from David Cooper transcript, July 29, 1967, at https://villonfilms.ca/main/transcripts-dialectics-david-cooper-15-7-67.pdf.

**260 "Terror on a very high":** Undated transcript of outtakes from *An Anatomy of Violence* titled "Group Discussion: First Conversation," from Villon Films: https://villonfilms.ca/main/transcripts-dialectics-group-discussion-date-unknown-1967.pdf.

**261 "may well be the first":** Anthony Chaney, "As Deep a Question as Ever," *U.S. Intellectual History Blog*, July 22, 2017, https://s-usih.org/2017/07/as-deep-a-question-as-ever-guest-post-by-anthony-chaney/.

**261 "a sort of short cut":** "Challenge Film—Discussion—Gregory Bateson—19th July 1967," a thirty-six-page transcript of a symposium held on the fourth day of the Dialectics of Liberation conference, available at https://villonfilms.ca/main/transcripts-dialectics-gregory-bateson-19-7-67.pdf.

**261 "This week I've been":** Allen Ginsberg, "A Speech: Dialectics of Liberation," *The Digger Papers* (August 1968), p. 4, www.diggers.org/digpaps68/dialib.html.

**261 "whole fabric, top to bottom":** Ginsberg's words at the Dialectics of Liberation conference were recorded on camera as part of the Peter Davis documentary *Anatomy of Violence* (1967), at 20:35–21:10.

**262 "a visitor from another generation":** Quotes in this and the preceding paragraph are from *Hair: The American Tribal Love-Rock Musical*, book and lyrics by Gerome Ragni and James Rado, copyright 1967, 41–42.

**262 The B-plus grade:** Howard, *Mead*, 335; and Nicholas D. Kristof, "Ally of an Older Generation Amid the Tumult of the 60's," *NYT*, June 19, 2000.

**262 "on some project associated":** Charles B. Fahs diary, "Trip to the Far East," Jakarta, May 2, 1955, p. 41, Officer's Diaries, Rockefeller Archives.

**263 "I have put all my":** MM to Rhoda Métraux, February 22, 1967, *TCLW*, 188.

**264 The reader's letter:** Stephen V. Roberts, "Dr. Leary Fails to Sway Students," *NYT*, August 18, 1967.

**264 Though LSD's users could:** Margaret Mead, "*Margaret Mead* Answers: Should We Have Laws Banning the Use of *LSD*?" *Redbook* 130, no. 3 (January 1968): 30, 32.

**264 Mead penned a letter:** MM to Mysbergh, January 7, 1968, B14, MM papers.

**265 "two hours on my belly":** Mysbergh to MM, April 20, 1968, B14, MM papers.

## 25. "IF SOMEONE THROWS AWAY THE BOX, THEN WHAT?"

**267 "When you put a probe":** Gregory Bateson, as quoted in Richard H. Blum, *Society and Drugs: Social and Cultural Observations* (Hoboken, NJ: Jossey-Bass, 1969), 344.

**267 "He had started to care":** Lipset, *Bateson*, 262.

**268 "I think that Gregory":** Lipset, *Bateson*, 262.

**269 Bateson had maintained social ties:** For instance, Mysbergh visited Bateson in November 1966, writing happily to Mead that "he has changed very little" and enthusing about Bateson's "new underwater observation center for dolphins," and James Hamilton met with and corresponded with Bateson in 1967. But after this, judging from surviving archival records, he appears to have had little contact with either man. See Mysbergh to MM, November 11, 1966, B14, MM papers.

**269 "Much later I learned":** Steve J. Heims, *The Cybernetics Group* (Cambridge, MA: MIT Press, 1991), 168.

**269 The transcript begins:** "Interview with Margaret Mead," October 14, 1968, Heims papers, MIT.

**271 "It is my considered opinion":** "Margaret Mead Advocates Legalization of Marijuana," AP wire article, *Morning Herald*, October 28, 1969, 5.

**272 "All my life's hope for":** Richard A. Steudel to MM, October 5, 1969, K66:1, MM papers; other quotes are from letters in the same folder.

**272 "Psychologists ought to stop":** Stokely Carmichael transcript, Dialectics of Liberation conference, London, July 18, 1967, https://villonfilms.ca/main/transcripts-dialectics-stokely-carmichael-18-7-67.pdf.

**273 "human beings are reduced":** Margaret Mead, "Research with Human Beings: A Model Derived from Anthropological Field Practice," *Daedalus* 98, no. 2, Ethical Aspects of Experimentation with Human Subjects (Spring 1969): 367.

**274 "the scientist working":** Mead, "Research with Human Beings," 376.

**274 On the back of Blum's book:** Ralph Blum, *The Simultaneous Man* (New York: Little, Brown, 1970).

**275 It was part of a larger:** Matthew D. Tribbe, *No Requiem for the Space Age: The Apollo Moon Landings and American Culture* (New York: Oxford University Press, 2014).

**275 "immanent in the total":** Bateson, "Form, Substance, and Difference" (1970), in *Steps to an Ecology of Mind.*

**275 "a series of metaprograms":** Lilly, *Center of the Cyclone*, 109.

**276 "We lost control of the place":** Jeffrey J. Kripal, *Esalen: America and the Religion of No Religion* (Chicago: University of Chicago Press, 2008), 132.

**276 One of them was:** This account is adapted from Kripal, *Esalen*, 133, which offers the most carefully sourced and reliable description of Manson's visit to Esalen.

**276 "parents did not understand":** The letter from Kubie does not survive in Lilly's archive, so the date can only be guessed at. I have quoted here from Lilly's paraphrase of Kubie's letter, which he dates to the fall of 1969, in *Center of the Cyclone*, 117.

**276 "something out of a bad book":** Alan Watts to John C. Lilly, August 23, 1971, in *Collected Letters of Alan Watts.*

## 26. ESALEN SUNSET

**278 "Is it in us really":** GB to "Steve" [Elizabeth Stevenson Cubhold], O3:1, Papers of Colleagues: Gregory Bateson, MM papers.

**279 "I would not be surprised":** Lacey Fosburgh, "Leary Scored as 'Cop Informant' by His Son and 2 Close Friends," *NYT*, September 19, 1974.

**279 Amid the charged social climate:** While he was crashing a law enforcement conference about "dangerous drugs" in Las Vegas, for instance, the journalist Hunter S. Thompson's ears perked up at a question from the audience. A man asked the drug expert onstage if Margaret Mead's recent "strange behavior" might indicate she was addicted to marijuana. "I really don't know," Thompson recorded the man as saying, to roars of audience laughter. "But at her age, if she did smoke grass, she'd have one hell of a trip." The comment made its way into a series Thompson was writing for *Rolling Stone* magazine at the time, later to be immortalized as a book: *Fear and Loathing in Las Vegas*. See Hunter S. Thompson, "Fear and Loathing in Las Vegas," in *Fear and Loathing at Rolling Stone: The Essential Writings of Hunter S. Thompson* (New York: Simon & Schuster), 91.

**281 "You see," he said:** All quotes from Baldwin and Mead here are from *A Rap on Race* (Philadelphia: J. B. Lippincott, 1971), 237 and 252.

**281 "silence and ambiguity":** Naoko Wake, *Private Practices: Harry Stack Sullivan, the Science of Homosexuality, and American Liberalism* (New Brunswick, NJ: Rutgers University Press, 2011), 5–6.

**282 "there was some kind of connection":** Rick Gladstone, "Notes on a Hip History of Santa Cruz," December 30, 2015, https://hipsantacruz.org/stories/notes-on-a-hip-history.

**282 Stewart Brand published letters:** In one such letter, a commune resident asked Bateson how to create a mutually caring society. Bateson responded by saying that he no longer believed this was possible. All we could offer one another, he said, was "temporary protection from the cold winds of an insane civilization, some shared tears and laughter, and that's about it." It was a far cry from the long-ago dream of expanding the consciousness of the human species. Bateson letter, May 27, 1973, republished in *CoEvolution Quarterly*, no. 5 (Spring 1975): 137.

**283 Senator Frank Church:** On Church's role in uncovering MKULTRA, see James Risen, *The Last Honest Man: The CIA, the FBI, the Mafia, and the Kennedys—and One Senator's Fight to Save Democracy* (New York: Little, Brown, 2023).

**284 "It was a different time":** Testimony of Harris Isbell, "Biomedical and Behavioral Research, 1975," Joint Hearings Before the Subcommittee on Health of the Committee on Labor and Public Welfare and the Subcommittee on Administrative Practice and Procedure of the Committee on the Judiciary, United States Senate (Washington, DC: U.S. Government Printing Office, 1976), 255.

**284 "science of the human":** Margaret Mead, "Towards a Human Science," *Science* 191 (March 5, 1976): 903–9.

**285 It was a genuinely:** Nancy D. Campbell and Laura Stark, "Making Up 'Vulnerable' People: Human Subjects and the Subjective Experience of Medical Experiment," *Social History of Medicine* 28, no. 4 (November 1, 2015): 825–48.

**285 "Peyote, snakes, pocketknives":** "Prayer Breakfast Changes Sharply," *Sacramento Bee*, January 8, 1976, 3.

**286 Bateson later wrote:** Bateson, "Let Not Thy Left Hand Know," in Gregory Bateson and Mary Catherine Bateson, *Angels Fear: Toward an Epistemology of the Sacred* (New York: Macmillan, 1987), 79–80.

**287 "So the question is":** Allen Ginsberg, audiotape of a poetry reading on April 7, 1978, Allen Ginsberg papers; the poem is annotated on the tape as "LSD Prose (Minneapolis to L.A.) Journal"; a version was later published in Allen Ginsberg, *Poems All Over the Place, Mostly Seventies* (Cherry Valley, NY: Cherry Valley Editions, 1978), 51–52.

## 27. THE BLUE GLOW OF LIFE

**289 "And if you read":** Bateson, *Angels Fear*, 5.

**289 International media:** See, for instance, "We Have Ways of Making You Forget," *Observer* (UK), August 7, 1977, 21.

**289 In retirement, White had:** "Pot Smoking Suspects Pick the Wrong Spot," *Daily Independent Journal* (Marin County, CA), July 8, 1970, 28. The article describes White making an ad hoc drug bust of a group of hippies he had spotted smoking cannabis on his property.

**290 "served as a guinea pig":** White to Professor John Kaplan, June 12, 1970, GHW papers, Stanford.

**290 "the days of the Office of Strategic":** Bill Richards and John Jacobs, "CIA Conducted Mind-Control Tests Up to '72, New Data Show," *WP*, September 2, 1977.

**290 "MKULTRA Task Force":** Untitled CIA internal report for "Chief, MKULTRA Task Force," dated September 7, 1979, available at www.cia.gov/readingroom/docs/CIA-RDP86-00114R000100080003-9.pdf.

**291 As the new MKULTRA revelations:** Barbara Crossette, "A Film Tribute to Margaret Mead," *NYT*, September 16, 1977. A filmmaker spent the day after the festival making a documentary about Mead. In the raw footage, the camera pans to show her office, which has barely changed since the 1930s, still stuffed with books and artifacts, still dominated by an enormous oak desk overlooking the roofs of the American Museum of Natural History. But now it is filled with photographs. There is a Franz Boas, hanging next to a photo of her eight-year-old granddaughter, the child of Cathy, who is also pictured. Beside her is Ruth Benedict, white-haired and ethereal. Bateson is not visible.

**292 Brown, who recalled:** Author interview with Jerry Brown, May 11, 2020.

**292 "CIA-financed mind control":** Sigrid Bathen, "Drug Experiments...," *Modesto Bee*, October 13, 1977, 24.

**293 "The LSD movement was":** Walter Barney, "Grandfather of LSD Meets the Acid Children," *San Francisco Sunday Examiner and Chronicle*, October 16, 1977.

**293 In two letters:** GB to James Hamilton, January 30, 1968; and James Hamilton to GB, January 22, 1968, GB papers.

**293 Mead, meanwhile, was:** Howard, *Mead*, 408.

**294 "Mass on the body":** This account of Mead's illness is derived from Howard, *Mead*, 410–12.

**295 "I was in a laboratory":** MM to RB, March 14, 1926, *TCLW*, 139–40.

**296 He was invited:** GB to Skolowmoski, September 29, 1979, Box 9, GB papers.

**296 "I never traveled":** Bateson, "The Last Lecture," undated notes circa July 1979, Box 9, GB papers.

**297 "it is said secretly":** Bateson, *Naven*, 230–31.

**297 "endlessly turning upon":** Bateson, "Afterword," in John Brockman, ed., *About Bateson: Essays on Gregory Bateson* (New York: Dutton, 1977), 246.

**297 "It is so sad":** Author interview with Jean Houston, August 11, 2021 ("It was a birthday that they were having at Esalen. There were a lot of well-known people there, and he would listen to them. And then he started, it seemed, to cry. He said, 'It's such—' what did he say? 'It is so sad. It is *so, so sad*.' All these people [were] extolling him. And he did not feel the same way about himself").

**298 "Life is not perfectible":** Author interview with Ann Druyan, December 14, 2022.

# Index

NOTE: *Italic page numbers* indicate photographs

## About the Author

**Benjamin Breen** is the author of *The Age of Intoxication: Origins of the Global Drug Trade*, winner of the 2021 William H. Welch Medal from the American Association for the History of Medicine. He is an associate professor of history at the University of California, Santa Cruz and was previously a postdoctoral fellow at Columbia University. He lives in Santa Cruz, California.